# ExoFrontiers

Big questions in exoplanetary science

## About the program:

AAS-IOP Astronomy ebooks is the official book program of the American Astronomical Society (AAS), and aims to share in depth the most fascinating areas of astronomy, astrophysics, solar physics and planetary science. The program includes publications in the following topics:

Books in the program range in level from short introductory texts on fast-moving areas, graduate and upper-level undergraduate textbooks, research monographs and practical handbooks.

For a complete list of published and forthcoming titles, please visit iopscience.org/books/aas.

## About the American Astronomical Society

The American Astronomical Society (aas.org), established 1899, is the major organization of professional astronomers in North America. The membership (~7,000) also includes physicists, mathematicians, geologists, engineers and others whose research interests lie within the broad spectrum of subjects now comprising the contemporary astronomical sciences. The mission of the Society is to enhance and share humanity's scientific understanding of the universe.

# ExoFrontiers

Big questions in exoplanetary science

**Edited by**
**Nikku Madhusudhan**
*Institute of Astronomy, University of Cambridge, UK*

**IOP** Publishing, Bristol, UK

ISBN 978-0-7503-1472-5 (ebook)
ISBN 978-0-7503-1470-1 (print)
ISBN 978-0-7503-1889-1 (myPrint)
ISBN 978-0-7503-1471-8 (mobi)

DOI 10.1088/2514-3433/abfa8f

Version: 20211001

AAS–IOP Astronomy
ISSN 2514-3433 (online)
ISSN 2515-141X (print)

British Library Cataloguing-in-Publication Data: A catalogue record for this book is available from the British Library.

Published by IOP Publishing, wholly owned by The Institute of Physics, London

IOP Publishing, Temple Circus, Temple Way, Bristol, BS1 6HG, UK

US Office: IOP Publishing, Inc., 190 North Independence Mall West, Suite 601, Philadelphia, PA 19106, USA

Cover image: Illustration of an exoplanetary system. Artist: Amanda J Smith.

*To the human quest for worlds beyond...*

# Contents

## Part V New Horizons

# Preface

We live in one of the most exciting times in modern astronomy. For the first time in human history we are discovering and characterising worlds beyond the solar system and answering foundational questions on their demographics, diversity, origins, and potential for life. There is a realistic chance that we might even discover the first signs of life elsewhere within the lifetime of the present generation. The first exoplanets were discovered almost thirty years ago around a neutron star, soon followed by discoveries of giant exoplanets orbiting sun-like stars in close-in orbits, the so-called 'hot Jupiters', with no analogues in the solar system. Starting with such seemingly anomalous discoveries the field has seen a tremendous transformation in the past three decades over which thousands of exoplanets have been discovered with remarkable ubiquity and diversity. Present exoplanet demographics suggest that most stars likely host planets around them and that small planets, between Earth and Neptune in size, are the most common. The diversity in the observed properties of exoplanets, such as orbital parameters, masses, radii, temperatures, stellar hosts, atmospheric properties, and other aspects, far surpasses that witnessed in the solar system.

The field is rapidly undergoing a shift in focus from primarily exoplanet discovery and demographics toward more detailed characterisation of the panoply of planetary processes that underlie the vast diversity in observed macroscopic properties. These include atmospheric and interior processes, planet formation and evolution mechanisms, planetary habitability, and astrobiology. At least five space missions and numerous ground-based facilities in this decade promise unprecedented observations of exoplanets, paving the way for detailed insights into these various aspects. It is becoming increasingly clear that new interdisciplinary approaches are required to make holistic advances in the field, which has already motivated several large-scale initiatives worldwide.

Exoplanetary science is very exciting and burgeoning but what are the most important questions in the field currently? This is a question routinely asked by students, researchers and agencies on various occasions. While everyone witnesses the meteoric growth of the field there is a need for a coherent account of the important questions that we as a community are trying to answer in various areas of the field, and the challenges and opportunities therein. Such a resource has never been more important as we look forward to an extremely dynamic decade ahead with major efforts and investments in the field. This book attempts to provide such a resource.

The book is focused on serving as an easily accessible reference for the major questions, challenges, and opportunities in prominent areas of the field, both observational and theoretical. It is also expected to be a particularly useful resource for students and researchers to identify important problems across the field where they could potentially contribute depending on their interests and expertise as well as to identify the interlinks between different areas. We note, however, that the book is not a collection of comprehensive review papers on each of the areas; there are other great sources in the literature for such reviews. Instead, each chapter contains only a brief introduction to an important area followed by an exposition of the most important open questions, challenges, and opportunities in that particular area. The goal is for the reader to obtain a quick overview of an area before being exposed to the scene at the cutting edge.

We have aimed at an optimal balance between breadth and depth of topics, covering a combination of well-established areas and emerging areas with recognised growth potential. After a brief overview of the exoplanetary landscape, we begin with the frontiers of exoplanet detection and spectroscopy using different observational methods and facilities. We then explore the realms of exoplanet theory and characterisation, spanning areas of orbital dynamics, planet formation, circumstellar environments, exoplanetary interiors and atmospheres. This is followed by expositions on exoplanetary habitability and astrobiology. We conclude with a vision towards the future of spectroscopic searches for biosignatures in exoplanets. Each chapter presents the frontier landscape of an area as viewed through the lens of the authors who are experts in that area. Given the breadth of topics covered, the reader may witness some organic diversity in narrative and presentation across the book. I hope you enjoy the journey and views of the exciting landscape of exoplanetary science!

This book is a reflection of a vibrant field and would not have been possible without the contributions of many. I would like to thank the exoplanet research community for the pioneering advancements and tremendous progress the field has seen in the last three decades and for laying the foundation for an exciting future ahead. I thank all the authors for contributing their valuable time, effort, and vision in writing the chapters of this book as well as for engaging in discussions, attending meetings or providing valuable feedback wherever needed. I would also like to thank the members of the community who may have indirectly contributed to the chapters through discussions, comments, sharing figures or other means. The theme of this book was motivated by the Kavli ExoFrontiers Symposium series at Cambridge that I chaired between 2016–2019 with support from The Kavli Foundation. I would like to thank The Kavli Foundation, the Kavli Institute for Cosmology at Cambridge (KICC), and the Institute of Astronomy (IoA) at Cambridge, for their support. Special thanks to Christopher Martin, Director of Physical Sciences at The Kavli Foundation, for his enthusiasm and support for the initiative. I also thank all the participants, including many of the authors of the present book, volunteers and administrative staff at the IoA and KICC over the years for the lively symposia. I would like to thank Amanda Smith for the cover image. Finally, I would like to thank the AAS-IoP ebooks team for inviting the book project and for their support during the entire process. My special thanks to Leigh Jenkins, Sarah Armstrong, and Poppy Emerson at IoP Publishing and to Dawn Gelino on the AAS-IoP ebooks editorial advisory board.

Nikku Madhusudhan
Cambridge, UK

# Editor biography

## Nikku Madhusudhan

Nikku Madhusudhan is a Professor of Astrophysics and Exoplanetary Science at the Institute of Astronomy, University of Cambridge. His research interests span a wide range of areas in exoplanetary science, including exoplanetary atmospheres, interiors, formation and habitability. He was born in India and pursued his undergraduate studies at the Indian Institute of Technology, Banaras Hindu University, in Varanasi, India. He pursued his graduate studies at the Massachusetts Institute of Technology (MIT), USA, where he pioneered the techniques for atmospheric retrieval of exoplanets, for which he obtained his PhD in 2009. He pursued postdoctoral research at MIT (2009–2010), Princeton University (2010–2011), and Yale University (2012–2013) before joining the faculty of the Institute of Astronomy, University of Cambridge, in 2013 and has remained there since. He is widely known for his pioneering contributions to exoplanetary science, particularly for seminal advances in the understanding of exoplanetary atmospheres. His work has been recognised with various awards and honours, including the 2014 Vainu Bappu Gold Medal in Astrophysics by the Astronomical Society of India, the 2016 Young Scientist Medal by the International Union of Pure and Applied Physics (IUPAP) Commission on Astrophysics, the 2019 MERAC Prize in Theoretical Astrophysics by the European Astronomical Society, and the 2019 Pilkington Prize for excellence in teaching at the University of Cambridge.

# Contributor List

**Natalie Batalha**
University of California—Santa Cruz, USA

**Edward Bergin**
University of Michigan, USA

**Jayne Birkby**
University of Oxford, UK

**Matteo Brogi**
Warwick University, UK

**Jessie Christiansen**
California Institute of Technology, USA

**Magali Delieul**
Laboratoire d'Astrophysique de Marseille, France

**Drake Deming**
University of Maryland, USA

**Joanna Barstow Eberhardt**
The Open University, UK

**David Ehrenreich**
University of Geneva, Switzerland

**Daniel Fabrycky**
University of Chicago, USA

**Debra Fischer**
Yale University, USA

**Jonathan Fortney**
University of California—Santa Cruz, USA

**Luca Fossati**
Austrian Academy of Sciences, Austria

**B Scott Gaudi**
Ohio State University, USA

**Ravit Helled**
University of Zurich, Switzerland

**Christiane Helling**
University of St Andrews, UK

**Sasha Hinkley**
University of Exeter, UK

**Markus Kasper**
European Southern Observatory, Germany

**James Kasting**
Pennsylvania State University, USA

**Sylvestre Lacour**
Observatoire de Paris, France

**Nikku Madhusudhan**
University of Cambridge, UK

**Avi Mandell**
NASA Goddard Space Flight Center, USA

**Peter McCullough**
Johns Hopkins University, USA

**Victoria Meadows**
University of Washington, USA

**Michael Meyer**
University of Michigan, USA

**Alessandro Morbidelli**
CNRS, Observatoire de la Côte d'Azur, c, France

**Julianne Moses**
Space Science Institute, USA

**Nadine Nettleman**
German Aerospace Center (DLR), Germany

**John Papaloizou**
University of Cambridge, UK

**Ian Parry**
University of Cambridge, UK

**Don Pollaco**
Warwick University, UK

**Sascha Quanz**
ETH Zurich, Switzerland

**Heike Rauer**
German Aerospace Center (DLR), Germany

**George Ricker**
Massachusetts Institute of Technology, USA

**Aki Roberge**
NASA Goddard Space Flight Center, USA

**Sara Seager**
Massachusetts Institute of Technology, USA

**Adam Showman**
University of Arizona, USA

**Karl Stapelfeldt**
NASA Jet Propulsion Laboratory, USA

**Kevin Stevenson**
Johns Hopkins University, USA

**Jonathan Tennyson**
University College London, UK

**Amaury Triaud**
University of Birmingham, UK

**Diana Valencia**
University of Toronto, Canada

**Ewine van Dishoeck**
Leiden University, Netherlands

**Arthur Vigan**
Laboratoire d'Astrophysique de Marseille, France

**Mark Wyatt**
University of Cambridge, UK

**Sergei Yurchenko**
University College London, UK

# Part I

## Introduction

# Chapter 1

# The Exoplanetary Landscape

**Nikku Madhusudhan**

## 1.1 Introduction

From time to time in the history of the human species major scientific discoveries have revolutionized our understanding of the universe, from cosmic to the subatomic. We are witnessing such a revolution in modern astronomy. In just the past 30 years we went from not knowing of any planet outside the solar system to the overwhelming realization that nearly every star may have a planet around it. At the heart of exoplanetary science today is the human quest to answer three fundamental questions:

1. How diverse are planets?
2. How do planets form and evolve?
3. Is there life elsewhere?

The field started with pioneering discoveries of planets orbiting a neutron star (Wolszczan & Frail 1992) and 'hot Jupiters' around sun-like stars (Mayor & Queloz 1995). Today, thousands of exoplanetary systems are known whose diversity is unparalleled in the solar system—an experience both humbling and exhilarating at the same time. These exoplanets occupy a wide range in observable properties across and beyond that seen in the solar system. Recently, it has also become possible to observe high-fidelity atmospheric spectra of exoplanets using different methods, thereby allowing high-precision remote sensing of exoplanetary atmospheres. These revolutionary developments have been made possible by an array of extensive observational advances across the field, from radial velocity (RV), transit, direct imaging and microlensing surveys for exoplanet detection and characterisation to high-precision atmospheric spectroscopy from space and large ground-based facilities. At the same time, the combination of observational and theoretical efforts have led to important insights into planetary formation, evolution, interior, and atmospheric processes. Presently, the forefront of the field is also witnessing the detection and characterization of temperate low-mass planets orbiting nearby low-mass stars. The prospects of

doi:10.1088/2514-3433/abfa8fch1  

characterizing the atmospheres of habitable-zone rocky exoplanets orbiting M dwarfs seem increasingly within the reach of upcoming observational developments.

This vast exoplanetary landscape has opened an equally vast array of scientific questions on all aspects of planetary processes. The present book is an attempt to summarize the key questions in various areas of exoplanetary science. In this chapter we briefly discuss a broad view of the exoplanetary landscape focusing on some key aspects related to the overarching questions mentioned above. The subsequent chapters discuss various areas of the field, both observational and theoretical, in more detail.

## 1.2 Planetary Diversity

It is natural to wonder how unique is the solar system. Planetary diversity may be assessed across various macroscopic aspects such as bulk properties (masses, radii, temperatures), orbital parameters, and key atmospheric properties. A natural reference point in the search for exoplanets is the architecture of the solar system—an orderly arrangement of eight planets with specific properties. The planets fall in three broad categories based on their masses and radii (four terrestrials, two ice giants, and two gas giants) and their orbits are largely well aligned with the solar rotation and nearly circular. The terrestrial planets, Earth and smaller, are characterized by their rocky interiors with thin secondary atmospheres, which along with their closer-in orbits are consistent with their likely formation inside the $H_2O$ snow line in the solar protoplanetary disk. The outer giant planets, on the other hand, are characterized by their large gaseous envelopes and volatile-rich interiors, ostensibly consistent with their formation beyond the snow line, with low-temperature $H_2$-rich primary atmospheres. Even though enormous detail is known about all the eight planets and much more (minor bodies, satellites, interplanetary environment, etc.) in the solar system (de Pater & Lissauer 2015), the apparent simplicity of the above broad characteristics of the planets portrays a seemingly universal picture of planetary processes. For example, the well-defined planetary mass/size categories of solar system (SS) planets may imply a certain quantization in size as a natural outcome of planetary formation. Similarly, their orbital locations may imply that giant planets should be preferentially formed at larger orbital separations, among other such trends. However, even in just the macroscopic parameters, the demographics of the thousands of exoplanets known indicate that the solar system may arguably be the exception rather than the norm (see, e.g., Chapter 2).

At the outset, it is understandable that we are yet to find an exact exoplanetary analog of the solar system, as most detection methods are naturally biased against long-period planets orbiting main-sequence stars (see Chapters 2–6, and Fischer et al. 2014). However, a comparative assessment of exoplanets versus solar system planets can still be conducted for a wide range of factors, including bulk properties (e.g., masses, radii, insolation/temperature), orbital properties (e.g., eccentricities, stellar obliquity, multiplicity), and host stellar properties. Across all these factors, exoplanets have been found to span the entire range seen in the solar system and beyond. For example, in contrast to the three broad categories of planets mentioned above, exoplanets populate the continuum in both mass and radius, from Earth-like planets to super-Jupiters. This diversity in bulk properties maps on to the diversity in the internal structures, atmospheric processes, and formation mechanisms of exoplanets.

In particular, the planets occupying the gaps between the SS categories open particularly intriguing questions about their true nature, e.g., the sub-Neptunes (super-Earths/mini-Neptunes) that lie between Earth and Neptune in size, and the sub-Saturns that lie between the ice giants and gas giants. Even within the sub-Neptune regime, the recent discovery of a bimodal distribution in radii separated by a radius valley around 1.8 $R_\oplus$ has opened a new window into planetary formation in the low-mass regime unseen in the solar system (Fulton et al. 2017; see recent reviews in Mordasini 2020 and Bean et al. 2021). Similarly, the overwhelming population of known exoplanets in short periods, especially hot Jupiters, have challenged conventional theories of planetary formation, migration and evolution since the beginning of the field (Dawson & Johnson 2018). Other orbital parameters and dynamical architectures, including the eccentricities, stellar obliquities relative to the orbital plane, period-ratios in multi-planet systems, have all been found to be remarkably diverse compared to SS expectations (e.g. Winn & Fabrycky 2015) and provide important insights into the mechanisms of formation and dynamical evolution of planetary systems. Similarly, the bulk properties of exoplanets span the entire gamut of internal structures, from rocky terrestrial-like exoplanets to gaseous super-Jupiters (see, e.g., Chapter 16 and Baraffe et al. 2014). Moreover, the sub-Neptune class that encompasses super-Earths and mini-Neptunes allows for a vast range of internal structures between Earth and Neptune that are not encountered in the solar system, e.g., water worlds, magma ocean worlds, terrestrial planets with $H_2$-rich atmospheres, and more.

Extreme diversity is also becoming evident in the atmospheric properties of exoplanets. The extreme range of orbital conditions alone lead to a vast range in stellar irradiation with major implications for their atmospheric processes. For example, the equilibrium temperatures of observed exoplanetary atmospheres span ~300–4000 K, ranging from Earth-like temperatures to those of M and K dwarf stars. Furthermore, compositional differences from SS planets further accentuate the diversity in observable physico-chemical properties and processes in exoplanetary atmospheres. Recent years have seen major advances in atmospheric observations of exoplanets (see Chapters 5–8 and Birkby 2018; Kreidberg 2018; Madhusudhan 2019). Spectroscopic observations over a wide spectral range (ultraviolet to infrared) have allowed key insights into atmospheric chemical compositions, temperature structures, clouds/hazes, dynamical processes, exospheric/escape processes, and other disequilibrium processes. These insights have revealed an extreme diversity of exoplanetary atmospheres ranging from temperate habitable-zone sub-Neptunes to ultra hot Jupiters at temperatures comparable to low-mass stars. Such diversity is revealing both (a) fundamental differences in atmospheric processes operating under diverse planetary conditions beyond those experienced in the SS (see, e.g., Chapters 17–20), and (b) key departures in atmospheric compositions from SS expectations, which could be relics of their possibly diverse formation and evolutionary pathways (see Chapters 14 and 18).

## 1.3 Planetary Origins

Until exoplanets were discovered our understanding of planetary formation was naturally based entirely on the demographics and architecture of the solar system, e.g., planets of specific types at specific locations with specific orbital properties, along with their satellites and minor bodies. A canonical formation model of the solar system is guided by extensive information on various aspects, dynamical and chemical, of the planets, satellites, comets, asteroids, and more. Such information is obtained by numerous ground-based and space-borne observations, artificial satellites, *in situ* probes, and fly-by missions across the solar system. Such detail will likely never be obtained by humans for any exoplanetary system. However, the entirety of the solar system occupies only a small region in planetary parameter space, as discussed above. This is arguably one of the most profound realizations of exoplanetary science in the last two decades which has, in turn, revolutionized our understanding of planetary origins. Despite only broad macroscopic parameters known for exoplanets, their sheer number and diversity have exposed us to a vast space of possibilities for planetary formation, migration, and evolutionary mechanisms among which the solar system may just be one fortuitous occurrence. How planets form, migrate, and evolve are some of the most challenging questions in exoplanetary science. Even within the solar system there have been longstanding open questions regarding the formation mechanisms of giant planets, terrestrial planets, the architecture of the solar system, the mechanisms behind planetesimal formation, and other aspects (e.g., Raymond & Morbidelli 2020). The exoplanetary landscape has further unveiled even more open questions on various aspects of planetary origins, as discussed in Chapters 11–15 and 18. Here we briefly discuss some of the observational advancements that motivate this major frontier.

### 1.3.1 Orbital and Bulk Properties

The demographics and orbital architectures of exoplanetary systems necessitate considerations of a much broader range of planetary formation and migration mechanisms than those encountered in the solar system. We highlight a few specific examples here. The first decade of exoplanet discoveries around sun-like stars starting with 51 Peg b (Mayor & Queloz 1995) was dominated by "hot Jupiters"—giant exoplanets in very close-in orbits around their stars. Their orbital periods of few days and separations within ~0.05 au place them over a hundred times closer to their host stars than Jupiter is to the sun. The existence of such planets posed a challenge to conventional planetary formation theories, which require giant planets to form at large separations beyond the $H_2O$ snow line where adequate gas and ice-rich planetesimals facilitate their formation. As such, some form of inward migration is required to explain the present close-in orbits of hot Jupiters. Over the last two decades numerous facets of planetary formation and migration mechanisms have been investigated to explain the various dynamical properties of hot Jupiters, including the periods/separations, eccentricities, stellar obliquities, and presence of companions, opening vistas unseen in the solar system (Dawson & Johnson 2018). Beyond hot Jupiters, a range of other planetary architectures have also provided important constraints on formation theories and opened new avenues, including the presence of compact multi-planet systems, circumbinary planets, retrograde orbits, and more (Winn & Fabrycky 2015). At the same time, the discoveries of directly imaged planets at large orbital (tens of au) separations orbiting young stars have fuelled the debate of giant planet formation via core accretion versus gravitational instability (see e.g. Helled et al. 2014 and Chapter 12). In addition to the orbital properties, the bulk properties such as masses and radii, have also opened new dimensions in planetary formation theory. For example, as discussed above, the very existence of planets in the sub-Neptune regime, i.e., super-Earths and mini-Neptunes, pose a challenge with no such planet type in the solar system. Furthermore, explaining the presence of a radius valley in the sub-Neptune regime is one of the major open questions in exoplanetary science with various competing hypotheses (see e.g. recent reviews by Mordasini 2020; Bean et al. 2021).

### 1.3.2 Protoplanetary Environments

Alongside exoplanet detection, there has been a parallel revolution underway in our understanding of planetary formation environments and the building blocks of planets. In particular, the advent of the ALMA radio telescope has transformed observations of primordial circumstellar environments (see, e.g., recent review by Andrews 2020). For example, morphological features of gaps in protoplanetary disks possibly carved out by newly formed planets, as exemplified by the famous image of HL Tau, directly reveal planet formation in action. Such observations also allow direct constraints on the formation locations of planets, distributions of macroscopic disk properties such as mass, density, temperature, as well as more granular properties such as the size distribution of planet forming solids. These constraints, in turn, provide important insights into planetary formation and evolutionary processes, including aspects such as grain growth, mass transport, and accretion rate. Another important advancement in this direction has been the mapping of snowlines in protoplanetary disks, showing the phase boundaries of important chemical species between the gas and solids. These observations are revealing exquisite insights into planetary formation environments as well providing key constraints on planetary formation processes (see, e.g., Chapter 14). In addition to protoplanetary disks, there have also been similar advancements in the morphological and chemical characterization of debris disks, i.e., circumstellar disks of planetesimals, dust and gas that remain after planet formation (see, e.g., Chapter 15 and Hughes et al. 2018). Overall, one of the major goals in this area is to form a holistic understanding of planetary formation by combining all this information from various facets—of planet-forming environments, of planet-forming materials and their debris, and of the planets themselves.

### 1.3.3 Atmospheric Imprints

The chemical compositions of planetary atmospheres contain imprints of their formation and evolutionary histories. The native composition of a planet is governed by the compositions and relative amounts of solids and gas it accretes during formation which, in turn, depend on its formation location and migration history in the primordial disk. While the gas, like the nebula, is expected to be dominated by hydrogen (H) the solids are composed of condensates of heavier elements (e.g., O, C, N) in the form of prominent ices and rock. Thus the elemental abundance ratios of these elements relative to hydrogen and to each other (e.g., C/O, N/O ratios) can be important tracers of the formation history of a planet. This is particularly relevant for giant planets as their atmospheres are H-rich, largely as $H_2$, and the key heavy elements O, C, N are in molecules with strong spectroscopic signatures, e.g., $H_2O$, CO, $CH_4$, $NH_3$, depending on the temperature in the atmosphere. So, measuring the abundances of these molecules can provide a direct handle on the important elemental ratios in the atmosphere, which can then be used to trace their origins. It is more challenging for low-mass, rocky, exoplanets with secondary atmospheres that are H-poor and their primordial compositions may have been largely altered by surface-atmosphere interactions and atmospheric escape. Overall, using atmospheric chemical abundances to constrain the formation, migration, and evolutionary pathways of exoplanets is one of the major frontiers in exoplanetary science today (see Chapters 14 and 18 and recent reviews by Pudritz et al. 2018; Madhusudhan 2019).

## 1.4 Search for Life

Ultimately, the detection of life elsewhere is the holy grail of exoplanetary science. The search for extrasolar life involves two key steps: (a) the detection of a potentially habitable planet, and (b) the detection of a biosignature in such a planet. Both these aspects are confronted with uncertainties surrounding the notions of "habitability" and "biosignatures," which have been subjects of extensive theoretical investigations in recent years. A wide range of conditions, both stellar and planetary, can influence the habitability of an Earth-size planet (Meadows & Barnes 2018) even if it lies within the traditional habitable zone around its host star (Kasting et al. 1993; also see Chapter 22). Similarly, the prominent biosignature molecules on Earth, such as $O_2$, $CH_4$, and $N_2O$, can have potential false positives with geochemical or photochemical origins (e.g., Catling et al. 2018). Besides these challenges in robustly defining exoplanetary habitability and reliable biosignatures, there is the more basic problem of detectability in the first place. Finding an exact Earth-twin, i.e., with the same size and mass as Earth and orbiting at 1 au around a Sun-like star, is still a major observational challenge. Even if such planets are within the reach of long-term transit surveys, confirming them with radial velocity observations, for mass measurements, or with direct imaging are at the limit or beyond the reach of current instruments (see Chapters 4 and 5). Observing an atmospheric spectrum of such a planet is even more remote a possibility over the next decade.

Habitable-zone exoplanets orbiting low-mass stars (M dwarfs) have become the new hunting grounds in the search for life. The so-called "M dwarf Opportunity" favors the detection and characterization of low-mass habitable-zone planets around M dwarfs in many ways (see, e.g., Chapter 6). The low stellar temperature (~2000–3500 K) makes the habitable-zone much closer to the star, thereby increasing the detectability of such planets via both the transit and radial velocity methods. The shorter orbital separations increase the transit probability and frequency as well as the RV semi-amplitude. The smaller stellar radii increase the planet-star radius contrast and, hence, the observable transit depths. For reference, the size ratio of an Earth to a late M Dwarf is approximately the same as that of Jupiter to a Sun-like star. The fact that M dwarfs are the most common planets in the solar neighborhood makes them ideal targets in the search for habitable-zone planets. The discoveries of the TRAPPIST-1 transiting system with seven rocky planets (Gillon et al. 2017), including at least three that may be habitable, and Proxima Cen b (Anglada-Escudé et al. 2016) via the RV method serve as canonical examples of this emerging frontier. Detecting such planets has become the focus of major transit and RV surveys today. Dedicated efforts include several ground-based transit surveys as well the NASA TESS mission (Ricker et al. 2015) , which are already detecting temperate low-mass planets that are good targets for atmospheric characterization (see, e.g., Chapters 3 and 6). Similarly, a new generation of stable infrared spectrographs are being custom-built for radial velocity surveys of such targets (see, e.g., Chapter 6).

Such planets have also been demonstrated to be feasible targets for detecting atmospheric signatures with upcoming facilities. For example, terrestrial-size habitable-zone planets transiting nearby M dwarfs can be conducive for atmospheric characterization with JWST, though robustly detecting a potential biosignature such as $O_2$ or $O_3$ would require a substantial investment of JWST time (Kaltenegger & Traub 2009; Barstow & Irwin 2016; Lustig-Yaeger et al. 2019; Wunderlich et al. 2019). On the other hand, ground-based high resolution spectroscopy offers another promising avenue for atmospheric characterization of such planets. For example, the nearest habitable-zone exoplanets transiting M dwarfs would be favorable for an $O_2$ detection with the ELT (Snellen et al. 2013). Such observations of $O_2$ and other terrestrial biosignatures will need to be interpreted considering the possibility of potential false positives in exoplanetary conditions (Meadows 2017; Catling et al. 2018). At the same time, theoretical studies will continue to investigate whether life is actually possible under the environmental conditions in the proximity of M dwarfs. These factors are discussed in detail in Chapters 22 and 23. Similarly, extensive studies are also being dedicated to identifying a comprehensive list of potential biosignatures beyond the prominent ones on Earth (see e.g. Seager et al. 2016 and chapter 24). It is also possible that some sub-Neptune exoplanets with water-rich interiors and $H_2$-rich atmospheres may be habitable under the right conditions and may provide promising targets in the search for life elsewhere (Madhusudhan et al. 2020). Overall, it is conceivable that a potential biosignature in the atmosphere of a habitable-zone exoplanet orbiting an M dwarf may be detected within this decade.

## Epilogue

It is surreal to realise how far we have come in three decades. We have made tremendous progress in the pursuit of the three overarching questions on the diversity of exoplanets, their origins, and the search for life elsewhere. It is now firmly established that exoplanets are extremely common and diverse across the various orbital and bulk properties observed. The frontier in this area currently lies in filling important gaps in the orbital phase space of known exoplanets and probing the diversity of their atmospheres, interiors, and formation environments. Great progress has also been made in our understanding of planetary origins, motivating considerations of a wide range of formation pathways beyond those invoked to explain the solar system. Exoplanet discoveries across the vast phase space have also opened a plethora of fundamental questions on planetary origins, answering which is a forefront goal in the field. Finally, the prospects of detecting life on an exoplanet seem increasingly promising. There are now several habitable-zone low-mass exoplanets known orbiting nearby M dwarfs for which the prospect of detecting an atmospheric bio-signature appears feasible within this decade, paving the way for characterising an Earth-twin in the decades to come. With observations expected from at least five space missions (see e.g., Chapter 9) and various ground-based facilities, and extensive efforts in modelling and theory, this decade appears to be ushering in what is likely the golden era of exoplanet science. Only time will tell what wonders lie in store.

## References


Andrews, S. M. 2020, ARA&A, 58, 483

Anglada-Escudé, G., Amado, P. J., Barnes, J., et al. 2016, Natur, 536, 437

Baraffe, I., Chabrier, G., Fortney, J., & Sotin, C. 2014, Protostars & Planets VI, ed. H. Beuther, R. S. Klessen, C. P. Dullemond, & T. Henning (Tucson, AZ: Univ. Arizona Press), 763

Barstow, J. K., & Irwin, P. G. J. 2016, MNRAS, 461, L92

Bean, J. L., Raymond, S. N., & Owen, J. E. 2021, JGRE, 126, e2020JE006639

Birkby, J. L. 2018, Handbook of Exoplanets ed. H.J. Deeg, & J.A. Belmonte (Berlin: Springer), 16

Catling, D. C., Krissansen-Totton, J., Kiang, N. Y., et al. 2018, AsBio, 18, 709

Dawson, R. I., & Johnson, J. A. 2018, ARAA, 56, 175

de Pater, I., & Lissauer, J. J. 2015, Planetary Sciences (Cambridge: Cambridge University Press)

Fischer, D. A., Howard, A. W., Laughlin, G. P., et al. 2014, Protostars & Planets VI, ed. H. Beuther, R. S. Klessen, C. P. Dullemond, & T. Henning (Tucson, AZ: Univ. Arizona Press), 715

Fulton, B. J., Petigura, E. A., Howard, A. W., et al. 2017, AJ, 154, 109

Gillon, M., Triaud, A. H. M. J., Demory, B.-O., et al. 2017, Natur, 542, 456

Helled, R., Bodenheimer, P., Podolak, M., et al. 2014, Protostars & Planets VI, ed. H. Beuther, R. S. Klessen, C. P. Dullemond, & T. Henning (Tucson, AZ: Univ. Arizona Press), 643

Hughes, A. M., Duchêne, G., & Matthews, B. C. 2018, ARA&A, 56, 541

Kaltenegger, L., & Traub, W. A. 2009, ApJ, 698, 519

Kasting, J. F., Whitmire, D. P., & Reynolds, R. T. 1993, Icar, 101, 108

Kreidberg, L. 2018, Handbook of Exoplanets, ed. H. J. Deeg, & J. A. Belmonte (Berlin: Springer), 100

Lustig-Yaeger, J., Meadows, V. S., & Lincowski, A. P. 2019, AJ, 158, 27

Madhusudhan, N. 2019, ARA&A, 57, 617

Madhusudhan, N., Nixon, M. C., Welbanks, L., Piette, A. A. A., & Booth, R. A. 2020, ApJ, 891, L7

Mayor, M., & Queloz, D. 1995, Natur, 378, 355

Meadows, V. S. 2017, AsBio, 17, 1022

Meadows, V. S., & Barnes, R. K. 2018, in Handbook of Exoplanets, ed. H. J. Deeg, & J. A. Belmonte (Berlin: Springer), 57

Mordasini, C. 2020, A&A, 638, A52

Pudritz, R., Cridland, A. J., & Alessi, M. 2018, Handbook of Exoplanets, ed. H. J. Deeg, & J. A. Belmonte (Berlin: Springer), 144
Raymond, S. N., & Morbidelli, A. 2020, arXiv:2002.05756
Ricker, G. R., Winn, J. N., Vanderspek, R., et al. 2015, JATIS, 1, 014003
Seager, S., Bains, W., & Petkowski, J. J. 2016, AsBio, 16, 465
Snellen, I. A. G., de Kok, R. J., le Poole, R., Brogi, M., & Birkby, J. 2013, ApJ, 764, 182
Winn, J. N., & Fabrycky, D. C. 2015, ARA&A, 53, 409
Wunderlich, F., Godolt, M., Grenfell, J. L., et al. 2019, A&A, 624, A49
Wolszczan, A., & Frail, D. A. 1992, Natur, 355, 145

# Part II

## Exoplanet Detection and Spectroscopy

AAS | IOP Astronomy

ExoFrontiers
Big questions in exoplanetary science
**Nikku Madhusudhan**

# Chapter 2

## The Demographics of Exoplanets

**B. Scott Gaudi, Michael Meyer and Jessie Christiansen**

### 2.1 Introduction

The field of exoplanets can be thought of as resting on three broad pillars: demographics, characterization, and the search for habitable worlds and life.[1] Of course, these divisions are somewhat arbitrary, the boundaries between them are generally fuzzy, and all three are interrelated. Nevertheless, subdividing the field into these three categories of inquiry provides a useful way to frame the science goals and questions of the exoplanet field. In the broadest sense, the overall goal of exoplanet demographic surveys is to determine the frequency and distribution of planets as a function of as many of the physical parameters that may influence planet formation and evolution as possible, over as broad of a range of these parameters as possible. These parameters can include, but are not limited to: (1) properties of the planets, including the planet mass $m_p$, radius $r_p$, and orbital properties (e.g., semimajor axis $a$, period $P$, eccentricity $e$); (2) properties of the host stars, including mass $M_*$, radius $R_*$, effective temperature $T_{\rm eff}$, luminosity $L_*$, metallicity [Fe/H] and detailed elemental abundances, activity, and age; and (3) environmental properties of the system, including multiplicity and birth environment.

These distribution functions represent the ultimate empirical data set (e.g., the "ground truth") that all theories of planet formation and evolution must reproduce. To the extent that most planets form in a bottom-up scenario, *ab initio* theories of planet formation must explain how roughly micron-sized dust grains grow by ~12–14 orders of magnitude in radius and ~38–40 orders of magnitude in mass to the exoplanets we detect today. The physical mechanisms by which protoplanets accumulate mass and migrate as they cross through these many orders of magnitude

[1] Although the study of disks is obviously strongly linked to the study of exoplanets, particularly because the properties of protoplanetary disks provide the *ab initio* conditions for planet formation, we will restrict our discussion in this short essay to exoplanets in systems where the planets are fully formed and the protoplanetary disk is essentially gone (e.g., ages of $\gtrsim$10 Myr, see Yao et al. 2018).

doi:10.1088/2514-3433/abfa8fch2

in mass and radius vary dramatically, and may depend on the properties and environment of the host star, either directly or indirectly (see Raymond & Morbidelli 2002 for a review). The net result is that signatures of these physical mechanisms should be imprinted on these planet distribution functions.[2] Therefore, by comparing these planet distributions to the predictions of planet formation theories, we can begin to both test and refine these theories. An understanding of exoplanet demographics and the physical mechanisms that sculpted the planet distribution function is also essential for understanding planetary habitability. Of course, the first step is to determine the frequency of potentially habitable planets, often referred to as $\eta_{\oplus}$.[3] We discuss the most recent estimates of $\eta_{\oplus}$ and the opportunities for refining this estimate below. However, the habitability of any given planet can only be understood in the context of the entire planetary system, including its evolution. This is (in part) because classical theories predicted that planets that formed in the traditional habitable zone interior to the snow line of the disk were initially dry, thereby requiring giant planets to serve as a delivery mechanism for volatile species (e.g., carbon, nitrogen, and oxygen in the form of water) from beyond the snow line Morbidelli et al. (2016). On the other hand, more recent theories predict that water rich "pebbles" are likely to migrate interior to the snow line, which may result in the formation of many water worlds in planet population synthesis models (see Lichtenberg et al. 2019). Thus giant planets formed near the snow line could also act barriers to limit the inward migration of such volatile-rich material.

## 2.2 State of the Art

Although it was not the first exoplanet to be detected (Campbell et al. 1988; Latham et al. 1989; Wolszczan & Frail 1992); the discovery of the Jovian companion to 51 Pegasi by Mayor & Queloz (1995) ignited the field of exoplanets. There are now over 4300 confirmed exoplanets,[4] which were primarily detected with four techniques: radial velocities, transits, microlensing, and direct imaging. The distribution of the properties of known exoplanets in mass/period and radius/period space is shown in Figure 2.1.

Essentially all of the initial discoveries were made by the radial velocity (RV) method. For many years, the primary focus of the field was on discovering new exoplanets and improving the precision of RV surveys in order to expand the region of parameter space to which the method was sensitive. Eventually, the sample of known exoplanets detected in individual RV surveys was sufficiently large that a reasonably constraining statistical analysis could be performed. The first such

[2] The extent to which this is true may depend on the number of independent variables that contribute to the outcomes (e.g., the central limit theorem resulting in log-normal distributions), and which are scale free (e.g., resulting in power-law distributions).

[3] While there is no universally agreed-upon definition of $\eta_{\oplus}$, it generally refers to the frequency of terrestrial (rocky) planets in the traditional habitable zone of sunlike (FGK) stars. See, e.g., Kopparapu et al. (2013, 2014).

[4] As of 2020 November 23; https://exoplanetarchive.ipac.caltech.edu/.

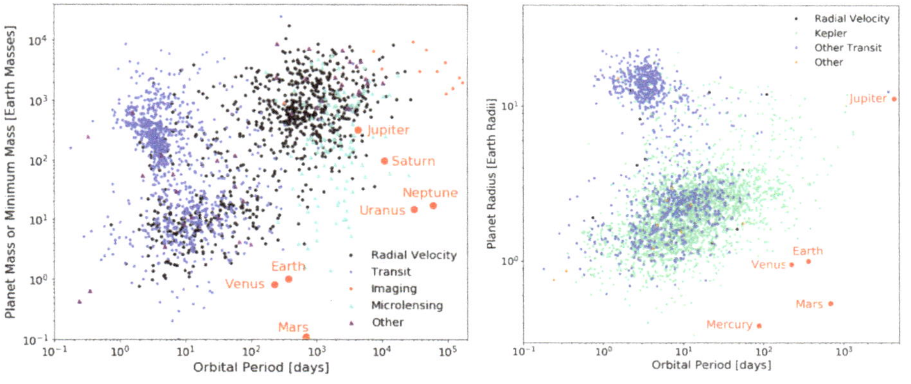


**Figure 2.1.** The distributions of the ~4300 confirmed low-mass companions to stars as a function of their mass and period (left) and radius and period (right) where semimajor axis is converted to period via Kepler's third law. The color coding denotes the method by which planets were detected. Some of the features in these diagrams are real, however many are due to the selection effects. For example, the large population of 1–4 $R_\oplus$ planets in the right panel is largely missing in the left panel due to the fact that these planets were primarily detected by Kepler, and thus typically have host stars that are too faint to enable a measurement of their mass via radial velocity. Similarly, the fact that there are nearly equal numbers of hot Jupiters as cold Jupiters in the left panel is due to the fact that ground-based transit surveys, which have larger sample sizes than radial velocity surveys, are primarily sensitive to hot Jupiters. Thus hot Jupiters are over-represented in the figures. Finally, the paucity of planets in the lower-right corner of each plot is purely a selection effect due to the fact that the radial velocity, transit, and direct imaging methods are not currently sensitive to planets in this region of parameter space. Note that ~25 directly-imaged planets are not shown in the left plot because they have periods that are greater than $\gtrsim 10^6$ days. This figure is based on data from the NASA Exoplanet Archive: https://exoplanetarchive.ipac.caltech.edu/.

statistical analysis was that of the Lick planet survey by Cumming et al. (1999), which contained 74 stars hosting 8 confirmed planetary companions, and several additional candidates. These results were superseded by the analysis of the Keck RV survey by Cumming et al. (2008), which included 585 stars hosting 48 planets. They fit a power-law mass–period distribution for planets with minimum mass $m_p \sin i > 0.3\ M_{\rm Jup}$ and $P < 2000$ days, finding

$$\frac{dN}{d\ln m_p \sin i\ \ d\ln P} \propto (m_p \sin i)^{-0.31\pm0.20}\ P^{0.26\pm0.10}, \tag{2.1}$$

with a normalization factor such that ~10% of solar-type stars host at least one planet with mass and period in the above range. Extrapolating, they estimated that ~20% of solar-type stars host a giant planet (>0.3 $M_{\rm Jup}$ within 20 au). In the interim, several authors made the first explorations of the dependence of giant planet frequency on host star mass and metallicity (e.g., Gaudi et al. 2002; Fischer & Valenti 2005; Jones et al. 2006; Johnson et al. 2007).

The precision of RV surveys has improved to the point that they are now sensitive to planets with mass as low as a few Earth masses with relatively short ($P \lesssim 50$ days) periods. Furthermore, the time baseline of some of the longest-running surveys is such that they are now becoming sensitive to Jovian planets with periods approaching 30 years. Some of the most recent statistical analyses of RV surveys can be found in Howard et al. (2010), Johnson et al. (2010), Mayor et al. (2011), Bonfils et al. (2013), Fernandes et al. (2019), and Fulton et al. (2021). The primary conclusions of these studies are broadly consistent and can be roughly summarized as: (1) for relatively short periods and solar-type hosts (setting aside hot Jupiters), the planet mass function rises dramatically toward low-mass planets, such that sub-Neptunes and super-Earths are at least an order of magnitude more common than giant planets in the same period range; (2) the frequency of giant planets orbiting solar-type stars increases with increasing period (in agreement with Cumming et al. 2008) up to ~5 years (~3 au), or roughly the location of the snow line in the solar protoplanetary disk (Morbidelli et al. 2016), and then appears to decline with increasing separation; (3) the occurrence rate for giant planets with semimajor axes of $\lesssim$2.5au increases with host star mass and metallicity (Fischer & Valenti 2005; Johnson et al. 2007).

Simultaneously, surveys using other exoplanet detection methods were being developed and conducted, including the transit, microlensing, and direct imaging methods. These surveys yielded the first detections of exoplanets by transits in 2003 (Udalski et al. 2002; Konacki et al. 2003), by microlensing in 2004 (Bond et al. 2004), and by direct imaging in 2008 (Marois et al. 2008).[5] Eventually, these methods also accrued sufficiently large samples of exoplanets to perform meaningful statistical analysis, see, e.g., Gaudi et al. (2002), Gould et al. (2010), Cassan et al. (2012), Gould et al. (2006), Mochejska et al. (2005), Burke et al. (2006), Hartman et al. (2009), Masciadri et al. (2005), and Metchev & Hillenbrand (2009).

To date, microlensing surveys have detected over 100 exoplanets. The most state-of-the art statistical analysis of the frequency of bound planets detected by microlensing was performed by Suzuki et al. (2016), based on the Microlensing Observations in Astrophysics II survey (MOA-II). Their sample consisted of 23 planet detections. Microlensing is generally sensitive to the planet/star mass ratio $q = m_p/M_*$ rather than the planet mass.[6] In addition, it is sensitive to the instantaneous projected separation of the planet in units of the angular Einstein ring radius $s$. Microlensing is generally sensitive to planets orbiting a range of hosts, including brown dwarfs, main-sequence stars, and stellar remnants. Although the majority of hosts are expected to be low-mass stars, a significant fraction will be due to solar-type stars, and indeed some solar-type exoplanet hosts have already been identified (e.g., Vandorou et al. 2020). Finally, microlensing surveys are most sensitive to planets with a projected separation from their parent star of roughly twice the snow line. See Gaudi (2012) for a review of microlensing searches for planets. The results from the MOA-II analysis provided strong evidence for a break in the mass ratio

[5] A directly-imaged planetary-mass companion was discovered around the brown dwarf primary, 2MASS 1207 (Chauvin et al. 2004), in 2005, but this system likely formed in manner more akin to a stellar binary, highlighting the importance of mass ratio in assessing the planetary nature of low-mass companions.

[6] Although, we note that it is possible to estimate the host star mass and thus the planet mass using a variety of techniques (see Bennett et al. 2007), and for the forthcoming survey by the Nancy Grace Roman Space telescope, such measurements should be routine (Bennett & Rhie 2002).

function (the distribution of mass ratios) at a mass ratio of $q_{\rm br} = 1.7 \times 10^{-4}$, roughly that of Neptune and the Sun (hereafter NSMR). Parameterizing the mass ratio $q$ and projected separation distribution, $s$, as a broken power law, they find:

$$\frac{dN}{d\log q\ \ d\log s} = 0.61^{+0.21}_{-0.16}\left[\left(\frac{q}{q_{\rm br}}\right)^{-0.93\pm0.13}\Theta(q - q_{\rm br}) + \left(\frac{q}{q_{\rm br}}\right)^{0.6^{+0.5}_{-0.4}}\Theta(q_{\rm br} - q)\right]s^{0.49^{+0.47}_{-0.49}}, \quad (2.2)$$

where $\Theta(x)$ is the Heaviside step function. This implies that the mass ratio function of cold exoplanets with mass ratios above that of the NSMR is steeper than by found by Cumming et al. (2008). In addition, Suzuki et al. (2016) found that the distribution of planets beyond the snow line is consistent with a log-uniform (Öpik 1924) distribution. Finally, the results of Suzuki et al. (2016) imply that NSMR planets are likely the most common planets beyond the snow line. We note that these results were subsequently refined by Udalski et al. (2018) and Jung et al. (2019).

The transit detection technique allows for a much larger number of stars to be surveyed at once, albeit with strict orientation requirements, and at greater orbital separations for low-mass planets, than the radial velocity technique (although still smaller orbital separations than are probed by microlensing). Some early transit demographics studies took advantage of this to examine nearby star clusters as laboratories for measuring occurrence rates as a function of controlled variables, e.g., metallicity and stellar density in 47 Tuc, $\omega$ Centauri, and other clusters (Mochejska et al. 2005; Burke et al. 2006; Gilliland et al. 2000; Weldrake et al. 2005, 2008). However, the first large-scale statistical studies with transiting exoplanets were performed with data from the NASA Kepler mission (Borucki et al. 2010). Prior to its launch in 2009 there were ~50 known transiting planets; to date an additional ~2400 planets have come from its prime mission data set, ushering in the statistical age of exoplanets.

Some important results from Kepler include the bi-modality of the size distribution of small planets (Fulton & Petigura 2018), the discovery of the hot Neptune desert (Mazeh et al. 2016), the clustering of planet sizes within planetary systems (Weiss et al. 2018; He et al. 2019), and the decreasing metallicity dependence of planet occurrence rate with decreasing planet size (Petigura et al. 2018). The primary goal of the mission was to measure $\eta_\oplus$, however a combination of the unexpectedly high noise properties of Sun-like stars and several hardware issues led to a final data set that is incomplete in the relevant parameter space. Nevertheless, there have been many increasingly sophisticated attempts to constrain $\eta_\oplus$. These are summarized in Figure 2.2 via a metric labeled $\Gamma_\oplus$, which is the value of the occurrence rate model (whether a functional fit in radius/period space or a grid of discrete bins over radius and period) evaluated around the values for Earth ($P = 365.25$ days, $R_p = 1.0R_\oplus$). The most recent investigations from 2019 and 2020 include robust corrections for completeness and reliability, and vary mostly in the functional form they fit to the underlying population, and the parameter space over which it is fit. From a spread of over three orders of magnitude in the first few years of analysis, the community now

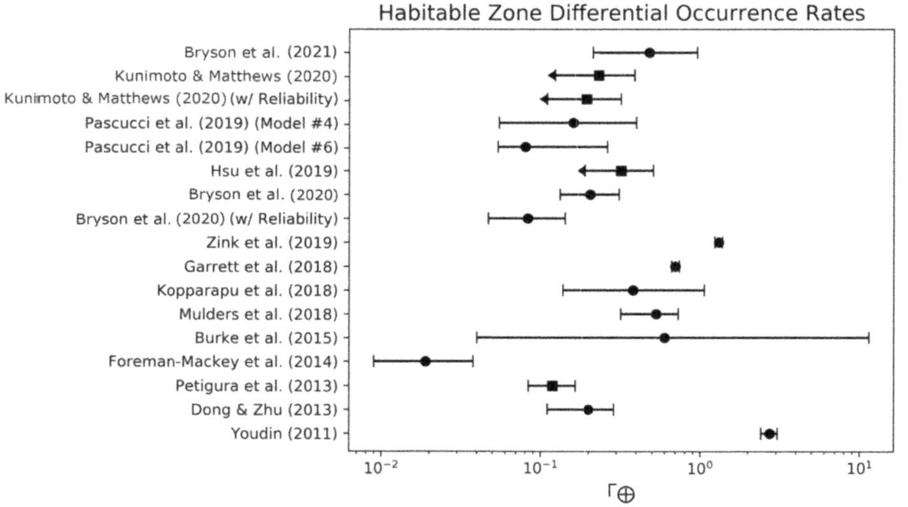


**Figure 2.2.** Adapted from Figure 14 of Kunimoto & Matthews (2020) (© 2020. The American Astronomical Society. All rights reserved), showing a collection of $\Gamma_\oplus$ values from the literature: Bryson et al. (2021), Kunimoto & Matthews (2020), Pascucci et al. (2019), Hsu et al. (2019), Bryson et al. (2020), Zink & Hansen (2019), Garrett et al. (2018), Kopparapu et al. (2018), Mulders et al. (2018), Burke et al. (2015), Foreman-Mackey et al. (2014), Petigura et al. (2013), Dong & Zhu (2013), and Youdin (2011). Squares indicate that grid-based occurrence rates were explored (Kunimoto & Matthews 2020; Hsu et al. 2019; Petigura et al. 2013), while circles indicate a functional form was fitted and integrated for the occurrence rate (all others).

appears to be converging on an $\eta_\oplus$ value between 5% and 50%, with the remaining uncertainly arising largely from the small number of detections in the relevant parameter space.

Direct-imaging surveys, enabled by state-of-the-art extreme (defined by actuator density on deformable mirrors) adaptive optics (AO) on 6–12m telescopes, are limited to gas giants >1 Jupiter mass beyond 10 au found around nearby young stars, as well as brown dwarf companions that could be a natural extension of the stellar companion mass ratio distribution (e.g., Reggiani et al. 2016). Early studies made predictions of expected yields based on extrapolations of the power-law models from Cumming et al. (2008) and fitted null results with an outer cutoff radius (Lafrenière et al. 2014; Nielsen & Close 2010; Heinze et al. 2010; Vigan et al. 2017). Such studies are done in the wavelength range from 1 to 4 microns as young planets (<300 Myr) are brighter and hotter than their older counterparts, and current AO systems perform well (Macintosh et al. 2015; Beuzit et al. 2019; Guyon 2018). It can be shown that the L-band (3.6–4.0 microns) is preferred for slightly older stars, which enable cohorts of closer stars to comprise target samples (Heinze et al. 2010). The youngest stars are rare in the solar neighborhood, therefore target stars are generally 15–150 pc away. Results to date can be summarized as: (a) gas giant planets beyond 10–50 au are rare; (b) wide-orbit gas giants from 1 to 15 Jupiter mass appear to be more common around higher mass stars, but are also consistent with a common mass ratio distribution (Vigan et al. 2007); (c) brown dwarf companions (20–76 Jupiter masses) are more common around M dwarfs compared to higher mass stars (Nielsen et al. 2019), though consistent with a common companion mass ratio distribution (Vigan et al. 2007); and (d) the eccentricity distribution may present a dichotomy where lower planetary-mass companions have lower eccentricities compared to brown dwarf companions (Bowler et al. 2020). Pioneering attempts are just being made to image planets that are in thermal equilibrium with the closest stars to the Sun (e.g., $\epsilon$ Eri b discovered by RV with imaging attempts by Mawet et al. 2019, and references therein). With temperatures of 100–500 K, set by stellar luminosity and orbital radius, observations are made from 5 to 13 microns around nearby stars with spectral type >K5. However, even detecting a super-Earth around Alpha Cen A at 1.3 pc would take many nights of observations (see Kasper et al. 2019).

## 2.3 Important Questions and Goals

As is clear from Figure 2.1, our census of exoplanets remains substantially incomplete, and as discussed below, practical obstacles have hindered attempts to combine statistical constraints from multiple surveys and detection methods. Many major questions about the demographics of exoplanets remain. We mention a few of the most notable here, but emphasize that this list is not meant to be complete.

- **How common are solar system analogs?** One of the most surprising results from Kepler is that the majority of stars appear to host relatively close-in, compact systems of super-Earths and/or sub-Neptunes (e.g., Howard et al. 2012; Youdin 2011; Dong & Zhu 2013; Swift et al. 2013). As our solar system does not host any analogs of such planets; this suggests that planetary architectures like our own (with small rocky planets in the temperate zone and gas giants beyond the ice line) may not be common. However, given that most detection methods are not yet sensitive to analogs of the majority of the planets in our solar system, it is still not clear how rare planetary systems like ours are. Even if gas giants beyond the ice line are realized in 10% of systems, and the frequency of habitable zone terrestrial planets is 10%, these properties could be correlated such that the joint probability could be >1%.
- **How does the shape of the exoplanet mass function depend on semimajor axis and host star mass?** The exoplanet mass function (or mass ratio function) has been measured over many orders of magnitude using RV and microlensing surveys. The radius distribution has been measured by Kepler, which can be converted to mass function by adopting a mass–radius relation (e.g., Chen & Kipping 2017). Generally, the mass function measured by these surveys over their full range is known to be inconsistent with a single power law (e.g., Suzuki et al. 2016; Pascucci et al. 2019). However, it is unclear if and how the normalization and detailed shape of the mass function (including any slopes and breaks of a piecewise power-law fit) depend on semimajor axis (particularly on either side of the snow line) and/or host star mass.

- **Which is the more fundamental parameter that determines the final architectures of planetary systems: mass or mass ratio?** The physical mechanisms that govern planet formation and evolution can depend more directly on either the planet mass or planet/star mass ratio. For example, the critical mass for nucleated runaway gas accretion is more directly dependent on the core mass (Mizuno 1980; Pollack et al. 1996; Rafikov 2006), and thus (along with the disk lifetime) may influence the frequency of "failed Jupiters". On the other hand, the final mass of a giant planet formed by core accretion likely depends on many factors, including the Hill radius of an exoplanet, as well as the properties and lifetime of the disk (Rosenthal et al. 2020). In the traditional model of solid accretion, the "isolation mass" depends on the mass ratio (via the Hill radius). However, when accounting for migration and/or pebble accretion, the importance of the isolation mass is less clear (see Rosenthal & Murray-Clay 2020, and references therein). Thus determining whether mass or mass ratio is the primary physical parameter that governs planet demographics in various regimes may help elucidate the physical of planet formation at various stages. We note that there is preliminary evidence that the mass ratio function may be more universal (Pascucci et al. 2019).
- **How do the detailed architectures of planetary systems depend on the properties and environment of the host star?** There is substantial evidence that the frequency of relatively short-period giant planets depends on the mass and metallicity (Fe/H) of the host star (e.g., Gonzalez 1997; Fischer & Valenti 2005; Johnson et al. 2010). However, the precise dependence of the frequency of less-massive short-period planets on these quantities is less clear and likely more complex (e.g., Howard et al. 2012; Dressing & Charbonneau 2013; Fressin et al. 2013; Petigura et al. 2018; Zhu et al. 2016; Lee & Chiang 2015). For wide-separation giant planets detected by direct imaging, there is some evidence for a dependence of the occurrence rate on spectral type, where higher mass stars are more likely to have gas giant companions (Vigan et al. 2007). In the most general terms, setting aside zones of dynamical instability, the bulk frequency of planets does not appear to depend strongly on stellar multiplicity, although significant work is needed to determine the effects of stellar companions (mass, semimajor axis, and eccentricity) on the detailed architectures of planetary systems (see the discussion in Bonavita & Desidera 2020 and references therein). The demographics of planetary systems as a function of stellar age has barely been explored (see Berger et al. 2020 and references therein), primarily because of the notorious difficultly of determining precise ages of isolated field stars, particularly in the pre-Gaia era. Finally, there have been relatively few explorations of the effect of detailed abundances (e.g., [$\alpha$/Fe]) on exoplanet demographics, primarily because the host stars for the largest statistical sample of planets available, i.e., Kepler, are generally too faint for detailed abundances determinations. One well-posed question that has not yet been unambiguously answered is whether there is a minimum metallicity below which planet formation is inefficient or impossible (Gilliland et al. 2000; Sozzetti et al. 2009; Faria et al. 2016).

- **Are there multiple formation channels for giant exoplanets?** The two most common models for the giant planet formation[7] are the bottom-up core accretion or nucleated instability model (Pollack et al. 1996), and the top-down gravitational instability model (Boss 1997). Several theoretical studies have argued that the latter mechanism is only likely to operate in the outer parts of massive disks, and typically forms fairly massive giant planets (Rafikov 2005; Kratter et al. 2010). A related question is whether or not the wide-separation giant planets that have been detected by direct-imaging surveys are a distinct population of objects, or are simply the long-period, high-mass tail of planets indirectly detected by RV surveys (Bowler 2016), or the low-mass tail of the stellar companion mass ratio distribution (see Vigan et al. 2017, 2007). There is suggestive evidence based on segregation in the mass functions, metallicity, and orbital eccentricity distributions of giant planets that at least two formation mechanisms may operate (Schlaufman 2018; Bowler et al. 2020).
- **How many planets become unbound from their host stars, and what is the mass function of these "free floating" planets?** The detection of the first Hot Jupiters provided evidence of large-scale migration of giant planets (Lin et al. 1996; Rasio & Ford 1996), with several models to explain this migration invoking dynamical instabilities. These would not only form hot Jupiters, but would also lead to the ejection of protoplanets or planets. Classical models of planet formation include competition and interaction of multiple planetary embryos (5–15 $M_{\oplus}$), some of which get ejected (e.g., Barclay et al. 2017). Indeed, it has since been proposed that our own solar system may have experienced a much less quiescent formation history than has previously been assumed; such a formation history would also lead to the ejection of a significant number of planetary-mass bodies (see Raymond & Morbidelli 2020 for a review). In addition, other mechanisms including star cluster dynamics have been proposed that can unbind planets from their host stars (Adams et al. 2006; Veras et al. 2011). Therefore, in principle, the abundance and mass function of free-floating planetary-mass objects can provide a constraint on models of the formation and evolution of planetary systems (as well as the low mass end of the initial mass function of brown dwarfs).

## 2.4 Challenges

There are many challenges to determining the demographics of exoplanets over a broad region of parameter space. These challenges have been identified by several authors, and methods to mitigate these challenges have also been identified. Here we will simply provide a list of these challenges, and highlight a few studies that have attempted to address them.

[7] We will assume that low-mass planets, and in particular rocky planets, must primarily be formed in a bottom-up scenario.

- **Intrinsic biases and sensitivities of different exoplanet detection methods.** The primary exoplanet detection methods at our disposal (RV, transits, microlensing, direct imaging, and astrometry) are all endowed with their unique biases and planet detection sensitivities. Fortunately, these methods are largely complementary. As a result, by properly combining the results from diverse surveys, it is possible to derive constraints on nearly the entire region of parameter space. However, combining the results of different surveys using the same detection method, let alone combining the results of surveys based on different detection methods, is not trivial. In particular, it is crucial that both the **completeness** and **reliability** of individual surveys be carefully quantified. Furthermore, given that different methods are sensitive to different planetary properties, there are often auxiliary constraints or priors that must be adopted and/or marginalized over to combine results from different methods. As a specific example, combining the results from transit surveys and radial velocity surveys requires adopting a mass–radius relation (Chen & Kipping 2017; Rogers 2015). An additional complication is that mass–radius relationships for exoplanets are not completely deterministic, and thus the dispersion in these relations must also be considered (Wolfgang et al. 2016). Likely due to these challenges, there have been few attempts to combine demographic constraints from multiple detection methods, with some notable exceptions (Howard et al. 2012; Clanton & Gaudi 2014, 2016; Meyer et al. 2018).
- **Comparing constraints from microlensing surveys to those from other techniques.** Microlensing is relatively unique among exoplanet detection methods in that (in general) the host star mass and distance is not constrained by the routine observables (specifically the microlensing light curve). Generally, the only direct observables are the planet/host mass ratio and the projected separation of the planet in units of the angular Einstein ring of the host. However, for planetary microlensing events, it is typically possible to estimate the angular Einstein ring radius. Furthermore, with high angular resolution observations with large telescopes with adaptive optics (e.g., Keck or Very Large Telescope), it is possible to measure the flux from the host star lens (e.g., Bhattacharya et al. 2018). Combining this with the measurement of the angular Einstein ring radius, it is possible to estimate the mass and distance to the lens, and the mass of the planet and projected separation of the planet in physical units. With the Nancy Grace Roman Space Telescope (née WFIRST) Galactic Exoplanet survey (Spergel et al. 2015), it will be possible to estimate the mass and distance to the primary lens, and thus the mass of the planet, for the majority of the detected planets (Bennett et al. 2007). This will allow the direct comparison of microlensing demographics constraints to those from other techniques, and in particular the dependence of exoplanet demographics on host star mass.
- **Determining the demographics of exoplanets as a function of the properties and environmental conditions of the host star.** It is observationally expensive to precisely constrain the parameters (e.g., mass, radius, abundance, age) and

environment (e.g., multiplicity), of exoplanet host stars, particularly for relatively faint hosts. In some cases the multiplicity itself has an impact on the quality of the data one is able to obtain (e.g., RV and direct imaging). The largest statistical survey undertaken, i.e., Kepler, targeted a relatively narrow and relatively faint population of main-sequence FGK stars, whose parameters were initially derived using only broad and narrow band photometry (Batalha et al. 2010). These photometrically-derived parameters turned out to be significantly biased, particularly for the cooler stars (Mann et al. 2012). Radial velocity surveys typically target an even narrower range of quiet FGK main-sequence stars that have been screened for low stellar activity and relatively slow rotation rates. Larger surveys like K2 and TESS will greatly expand the types of stars that are targeted for exoplanet searches, but the problem of accurately (and uniformly) deriving their stellar parameters remains. Several large-scale spectroscopic surveys (e.g., LAMOST: Luo et al. 2015; APOGEE: Alam et al. 2015; GALAH: De Silva et al. 2015) are publicly releasing hundreds of thousands to millions of stellar spectra. In combination with the precise distances from the ESA Gaia mission, careful treatment of these spectra, including calibration across surveys to remove systematic offsets, and cross-matching with target lists of K2 and TESS, will enable us to constrain exoplanet demographics much more robustly as a function of stellar parameters in the near future.

- **Determining the effect of exoplanet multiplicity on their demographics.** All of the methods that have been used to detect exoplanets (pulsar timing, radial velocities, transits, microlensing, and direct imaging) have detected multi-planet systems, and there is little doubt that astrometry will also detect multi-planet systems as well; multi-planet systems therefore appear to be common. Indeed, it is not always widely appreciated that one of the most fundamental questions one can ask about exoplanet demographics, namely "How common are exoplanets?", is not well-posed. This question could be rephrased as either "What fraction of stars host (at least one) exoplanet?", or "What is the average number of exoplanet per star?". These are two different questions with two distinct answers. For instance, there are a number of lines of evidence from Kepler that multi-planet systems may be distinct from systems with only one detected planet (Lissauer et al. 2011; Johansen 2012; Ballard & Johnson 2016; Zink & Hansen 2019). In addition, there is evidence that Kepler's multi-planet systems tend to be "peas in a pod", e.g., that the planets in such systems tend to have similar radii and are regularly-spaced (Weiss et al. 2018); or to be clustered in size and spacing (He et al. 2019; although see Murchikova & Tremaine 2020; Zhu 2020; Lauren et al. 2020). Accounting for the effect of planetary system multiplicity may change $\eta_{\oplus}$ by an amount that is larger than the currently stated uncertainties on that value (see, e.g., Zink & Hansen 2019).
- **Calibrating the mass–age–luminosity relationship of young exoplanets.** Observations of young, self-luminous, directly-detected planets can provide estimates of their luminosity, and temperature, derived from spectral energy

distributions as well as spectroscopy, the combination of which provides a coarse radius estimate. While some spectral features are known to be sensitive to surface gravity, inferring the mass of such directly-imaged planets requires an estimate of the age of the system, as such planets gradually cool as they radiate residual energy from formation (Burrows et al. 1997). With an age and luminosity, it is possible to use models to infer the mass of the planet, modulo an assumption of the specific entropy of the forming planet. In principle, there is a unique temperature associated with a luminosity and age as a function of specific entropy. However, these models are only now being calibrated with direct imaging observations of objects with dynamical mass estimates and known age (e.g., Snellen & Brown 2018; see Section 3 of Bowler 2016). Legacy radial velocity surveys as well as Gaia provide opportunities to rectify the situation.

- **Direct Detection of Earth Analogs.** Mature planets that are in thermal equilibrium with their host star have two primary components to their spectrum: the reflected light and the thermal emission from reprocessed stellar radiation. The amount of reflected light depends on the product of the radius and the albedo, modified by the (wavelength-dependent) scattering phase function. The thermal emission spectrum of such planets depends on the temperature of the planet (which in turn depends on the scattering albedo). Thus, by measuring both the reflected light and thermal emission of planets in equilibrium, it is possible to infer the radius, albedo, and temperature of mature directly-imaged planets. Unfortunately, directly detecting mature planets in either reflected light or thermal emission from the ground will require future Extremely Large Telescopes (ELTs) with adaptive optics, and even then will only be possible for a relatively small sample of the most nearby targets. Furthermore, the best targets for reflected light observations are planets orbiting low-mass stars, whereas the best targets for thermal emission are planets orbiting FGK stars. Nevertheless, it may be possible to detect both signals for a handful of planets with radius $>2\ R_{\oplus}$ orbiting inside the habitable zones of the most nearby stars (should the planets exist; Wang et al. 2019). True Earth analogs orbiting in the habitable zones of FGK stars may be out of reach of ground-based facilities. Detection of significant samples of such planets will require specially-designed space-based missions operating in the UV/optical/near-infrared such as the Habitable Exoplanet Observatory (HabEx; Gaudi et al. 2020), or the Large UV/Optical/InfraRed Mission (LUVOIR; The LUVOIR Team 2019), or space-based missions operating in the mid-infrared ($\sim 10 \mu$m) such as the Large Interferometer for Exoplanets (LIFE; Quanz 2019; thermal emission). We also note that direct detection of true Earth analogs (around Sun-like stars) via secondary eclipse is well beyond the capabilities of JWST.
- **Measuring the masses of potentially habitable directly-imaged planets orbiting nearby solar-type stars.** There are several proposed mission concepts that have as one of their primary goals to directly detect and take the spectrum of terrestrial planets in the habitable zones of nearby sunlike stars (e.g., HabEx,

LUVOIR, and LIFE). The interpretation of the spectra of such planets obtained by these missions would be greatly simplified if the masses of these planets were known. Unfortunately, the RV signal of an Earth–Sun analog is only ~9 cm $s^{-1}$, which is a factor of 4–5 smaller than the current state of the art for ground-based RV surveys. As a result, substantial effort and investments have been put into improving the precision and accuracy of RV surveys to ~10 cm $s^{-1}$ and below (Fischer et al. 2016; Plavchan et al. 2015). The astrometric signal of an Earth–Sun analog is only ~0.3 $\mu$as, which is almost certainly not achievable from the ground. Several mission concepts studies that may enable astrometric precisions well below these levels have been proposed (e.g., Malbet et al. 2016).

## 2.5 Opportunities

In this section, we briefly summarize some of the future opportunities for refining and expanding the demographics of exoplanets. In particular, we will focus on the prospects for exploring the region of planet parameter space (planet mass, radius, semimajor axis, and period) for which exoplanet demographics can be determined, as well as the region of host star parameter space (mass, temperature, abundance, multiplicity, and age). As we are focusing on demographics, we will **not** discuss the many exciting future prospects for detailed atmospheric characterization of exoplanets using, e.g., current and future ground-based facilities (including ELTs), current and future space-based facilities such as the Hubble Space Telescope (HST), the CHaracterising ExOPlanet Satellite (CHEOPS), the James Webb Space Telescope (JWST), Twinkle, the Nancy Grace Roman Space Telescope (Roman), and the Atmospheric Remote-sensing Infrared Exoplanet Large-survey (ARIEL), or proposed missions such as Origins.

- **The Transiting Exoplanet Survey Satellite (TESS):** The NASA TESS mission (Ricker et al. 2015) will observe almost the entire sky at 30-minute cadence with observing baselines of 27–351 days; transiting exoplanet detections are expected down to $T_{\rm mag} = 14$. While the range in planet radius and period covered by TESS is generally narrower than that covered by Kepler, the target list is several orders of magnitude larger. Thus, although the primary goal of TESS is to provide a sample of small ($<4R_{\oplus}$) transiting planets orbiting bright stars, it will nevertheless contribute significantly to our understanding of exoplanet demographics. In particular, the TESS target sample includes stars covering a much larger range of stellar properties. This fact, combined with the fact that the host stars will be relatively bright and thus amenable to detailed characterization, will enable more in-depth investigations of planet demographics as a function of, e.g., stellar age, mass, temperature, and metallicity. This information will provide important constraints on the dominant physical processes sculpting the planet distributions we see. In addition, the large sample of planets with well-constrained masses and radii amenable to atmospheric characterization will allow us to expand our demographic studies to include atmospheric properties.

- **Gaia:** ESA's Gaia mission possesses revolutionary capabilities to discover planets down to Neptune masses around nearby low-mass stars at relatively large orbital radii via astrometry. Previous estimates that extrapolate radial velocity distributions (Cumming et al. 2008) predict Gaia will discover thousands of planets below a few Jupiter masses (e.g., Perryman et al. 2014). However, if the orbital radius distributions peak at $<10$ au these estimates are likely too optimistic. Nonetheless, Gaia will provide dozens to hundreds of planets with dynamical masses and full Keplerian orbital elements (up to a two-fold degeneracy). These systems can be used to explore the architectures and (in particular) the coplanarity of planetary systems. Furthermore, many of these systems will be amenable to characterization by direct imaging, offering unprecedented opportunities to compare the properties of wide-orbit planets with close-in planets discovered via transit.
- **JWST:** In the background limit, JWST's ability to directly detect exoplanets will outperform ground-based telescopes by orders of magnitude in the infrared, enabling the discovery of planets $>10$ Earth masses around nearby (10–50 pc), young (30–300 Myr), late-type stars (>M0). However, in the contrast-limit, it is likely that ground-based telescopes with extreme AO will outperform JWST within $\sim 10\ \lambda/D$. Spectral retrieval studies will benefit tremendously from the higher signal-to-noise ratio and wavelength coverage of JWST relative to ground-based telescopes, providing constraints on volatile composition to confront planet formation theory.
- **Extremely Large Telescopes (ELTs):** ELTs will provide the angular resolution and sensitivity to detect planets below 1 Jupiter mass around nearby stars, many of which will have dynamical masses from radial velocity measurements or Gaia. This will be the first opportunity to characterize a large sample of mature wide-orbit planets to compare with gas giants found around young stars. ELTs will enable imaging at angular scales around stars close enough that imaging and radial velocity (and astrometric) samples will have large overlap in orbital radius, on both sides of the local maxima between 1 and 10 au. In addition, ELTs will be able to directly detect planets down to $<2$ Earth radii around the very nearest stars, including possibly in the habitable zone. Thermal infrared imaging will prefer stars earlier than K5 while reflected light studies will be best around M dwarfs, particularly as AO performance pushes into the visible (e.g., Males et al. 2019).
- **The Nancy Grace Roman Space Telescope (Roman):** Roman will perform the Roman Galactic Exoplanet Survey (RGES) using the microlensing detection method. Initial estimates from the RGES survey predict that it will detect $\sim 1500$ cold planets (planets beyond the snow line), with masses down to the mass of Ganymede (Penny et al. 2019). When combined with Kepler, Roman will complete the statistical census of planets with masses and radii greater than roughly that of the Earth, for semimajor axes from zero to infinity; see Figure 2.3. Indeed, Roman will be sensitive to free-floating planets (planets that are not bound to any star) with masses down to the masses of Mars (Johnson et al. 2020). Together, Kepler and Roman will provide the empirical

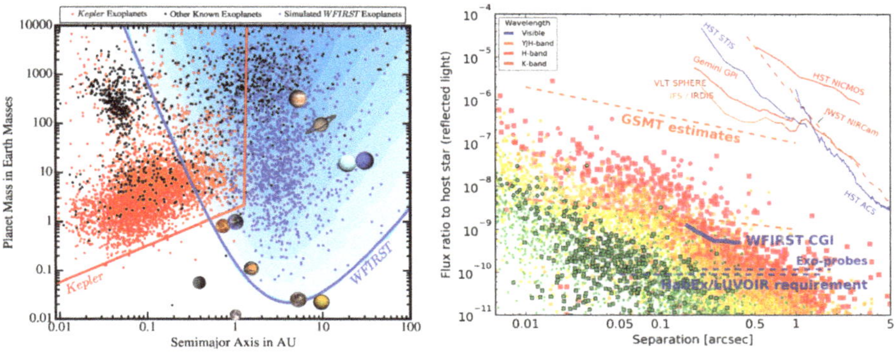


**Figure 2.3.** (Left) The region of sensitivity of the Kepler prime mission (red solid lines) versus the predicted region of sensitivity of the *Roman* (née WFIRST) Galactic Exoplanet Survey (blue solid lines). Courtesy of Penny et al. (2019). AAS. Reproduced with permission. (Right) Reflected light flux ratio versus angular separation for current and future direct-imaging surveys. The markers denote simulated planet populations within 27 pc using occurrence rates from Kopparapu & Hébrard (2018). Various cuts have been applied to more realistically portray the detectability of these populations. The marker size is proportional to planet size and colors denote specific bins: giant planets (red); Neptunes (orange), mini-Neptunes (yellow); super-Earths (dark green); temperate Earths and super-Earths (light green). Round points indicate planets orbiting cool stars ($T_{\rm eff}$ < 4000 K), while square points denote planets around warmer stars. Republished with permission from Figure 4.3 in National Academies of Sciences (2018), see that reference for additional details. Permission conveyed through Copyright Clearance Center, Inc. Courtesy of D. Mawet, B. Macintosh, T. Meshkat, V. Bailey, and D. Savransky.

ground truth that any *ab initio* planet formation theories must match. Roman will also be able to constrain the frequency of potentially habitable planets orbiting sunlike stars ($\eta_\oplus$), complementing and bolstering the estimates from Kepler.

- **The PLAnetary Transits and Oscillations of stars (PLATO) mission:** PLATO (Rauer et al. 2014) is a selected medium-class ESA mission whose objective is to discover and characterize a large number of transiting exoplanets, as well as perform asteroseismology on a large number of stars, including the host stars of a significant fraction of the detected transiting planets. PLATO can be thought of as a next-generation Kepler-like survey: in particular, it will improve upon the primary Kepler survey by observing a larger area of the sky. Therefore, PLATO's target stars will typically be brighter than that of Kepler, thereby facilitating follow-up of the planetary systems, including the ability to measure masses of detected transiting planets with radial velocities. Notably, PLATO will enable a detailed exploration of planet demographics as a function of the stellar age, accessible with asteroseismology, and may also provide an improved estimate of $\eta_\oplus$.
- **Space-based direct-imaging surveys:** Future proposed direct-imaging surveys, such as HabEx (Gaudi et al. 2020), LUVOIR (The LUVOIR Team 2019), and LIFE (Quanz 2019), will enable a direct census of nearby ($d \leqslant 10$ pc) planetary systems over nearly the entire range of semimajor axes spanned by the planets in our solar system (see Figure 2.3(b) for the parameter space to

which HabEx and LUVOIR are expected to be sensitive). In particular, these missions would enable the detection, as well as orbital and atmospheric characterization, of analogs of our own Venus, Earth, and Jupiter. Depending on the precise architecture and starlight suppression technique, these missions may also be able to detect, and potentially characterize, analogs to Mars and Saturn. Depending on the contrast floor and outer working angle, they may also be able to detect and possibly characterize gas giants at separations corresponding to the ice giants in our solar system (see, e.g., Gaudi et al. 2020). In other words, these missions have the potential to detect **and** characterize analogs to nearly all of the planets in our solar system, *in individual systems*. As a result, these missions will enable us to ask such fundamental questions as: "What is the conditional probability that, given the existence of a terrestrial world in the habitable zone of a nearby star, that the system hosts a giant planet beyond the snow line?".

## Epilogue

The future of exoplanet demographics looks bright. Of the three pillars of the study of exoplanets outlined in the introduction of this chapter, exoplanet demographics is likely the first to reach fruition. This is as it should be. The other two pillars of exoplanet studies: detailed characterization of exoplanets, and the search for habitable worlds and life, fundamentally rely on our basic knowledge of the architectures of planetary systems, and how these architectures depend on the host star properties.

## Acknowledgments

We would like to thank Radek Poleski for assistance with improving Figure 2.1. Partial support for B.S.G. was provided by the Thomas Jefferson Chair for Space Exploration endowment at the Ohio State University. This research has made use of the NASA Exoplanet Archive, which is operated by the California Institute of Technology, under contract with the National Aeronautics and Space Administration under the Exoplanet Exploration Program.

## References

Adams, F. C., Proszkow, E. M., Fatuzzo, M., & Myers, P. C. 2006, ApJ, 641, 504
Alam, S., Albareti, F. D., Allende Prieto, C., et al. 2015, ApJS, 219, 12
Ballard, S, & Johnson, J. A. 2016, ApJ, 816, 66
Barclay, T., Quintana, E. V., Raymond, S. N., & Penny, M. T. 2017, ApJ, 841, 86
Batalha, N. M., Borucki, W. J., Koch, D. G., et al. 2010, ApJL, 713, L109
Bennett, D. P., Anderson, J., & Gaudi, B. S. 2007, ApJ, 660, 781
Bennett, D. P., & Rhie, S. H. 2002, ApJ, 574, 985
Berger, T. A., Huber, D., Gaidos, E., van Saders, J. L., & Weiss, L. M. 2020, AJ, 160, 108
Beuzit, J. L., Vigan, A., Mouillet, D., et al. 2019, A&A, 631, A155
Bhattacharya, A., Beaulieu, J. P., Bennett, D. P., et al. 2018, AJ, 156, 289
Bonavita, M., Desidera, S., et al. 2020, Galax, 8, 16

Bond, I. A., Udalski, A., Jaroszyński, M., et al. 2004, ApJL, 606, L155
Bonfils, X., Delfosse, X., Udry, S., et al. 2013, A&A, 549, A109
Borucki, W. J., Koch, D., Basri, G., et al. 2010, Sci, 327, 977
Boss, A. P. 1997, Sci, 276, 1836
Bowler, B. P. 2016, PASP, 128, 102001
Bowler, B. P., Blunt, S. C., & Nielsen, E. L. 2020, AJ, 159, 63
Bryson, S., Coughlin, J., Batalha, N. M., et al. 2020, AJ, 159, 279
Bryson, S., Kunimoto, M., Kopparapu, R. K., et al. 2021, AJ, 161, 36
Burke, C. J., Christiansen, J. L., Mullally, F., et al. 2015, ApJ, 809, 8
Burke, C. J., Scott Gaudi, B., DePoy, D. L., & Pogge, R. W. 2006, AJ, 132, 210
Burrows, A., Marley, M., Hubbard, W. B., et al. 1997, ApJ, 491, 856
Campbell, B., Walker, G. A. H., & Yang, S. 1988, ApJ, 331, 902
Cassan, A., Kubas, D., Beaulieu, J. P., et al. 2012, Natur, 481, 167
Chauvin, G., Lagrange, A. M., Dumas, C, et al. 2004, A&A, 425, L29
Chen, J., & Kipping, D. 2017, ApJ, 834, 17
Clanton, C., & Gaudi, B. S. 2014, ApJ, 791, 91
Clanton, C., & Gaudi, B. S. 2016, ApJ, 819, 125
Cumming, A., Paul Butler, R., Marcy, G. W., et al. 2008, PASP, 120, 531
Cumming, A., Marcy, G. W., & Paul Butler, R. 1999, ApJ, 526, 890
De Silva, G. M., Freeman, K. C., Bland-Hawthorn, J., et al. 2015, MNRAS, 449, 2604
Dong, S., & Zhu, Z. 2013, ApJ, 778, 53
Dressing, C. D., & Charbonneau, D. 2013, ApJ, 767, 95
Faria, J. P., Santos, N. C., Figueira, P., et al. 2016, A&A, 589, A25
Fernandes, R. B., Mulders, G. D., Pascucci, I., Mordasini, C., & Emsenhuber, A. 2019, ApJ, 874, 81
Fischer, D. A., Anglada-Escude, G., Arriagada, P., et al. 2016, PASP, 128, 066001
Fischer, D. A., & Valenti, J. 2005, ApJ, 622, 1102
Foreman-Mackey, D., Hogg, D. W., & Morton, T. D. 2014, ApJ, 795, 64
Fressin, F., Torres, G., Charbonneau, D., et al. 2013, ApJ, 766, 81
Fulton, B. J., Rosenthal, L. J., Hirsch, L. A., et al. 2021, ApJS, 255, 14
Fulton, B. J., & Petigura, E. A. 2018, AJ, 156, 264
Garrett, D., Savransky, D., & Belikov, R. 2018, PASP, 130, 114403
Gaudi, B. S., Albrow, M. D., An, J., et al. 2002, ApJ, 566, 463
Gaudi, B. S. 2012, ARA&A, 50, 411
Gaudi, B. S., Seager, S., Mennesson, B., et al. 2020, arXiv:2001.06683
Gilliland, R. L., Brown, T. M., Guhathakurta, P., et al. 2000, ApJL, 545, L47
Gonzalez, G. 1997, MNRAS, 285, 403
Gould, A., Dong, S., Gaudi, B. S., et al. 2010, ApJ, 720, 1073
Gould, A., Dorsher, S., Gaudi, B. S., & Udalski, A. 2006, AcA, 56, 1
Guyon, O. 2018, ARA&A, 56, 315
Hartman, J. D., Gaudi, B. S., Holman, M. J., et al. 2009, ApJ, 695, 336
He, M. Y., Ford, E. B., & Ragozzine, D. 2019, MNRAS, 490, 4575
Heinze, A. N., Hinz, P. M., Kenworthy, M., et al. 2010, ApJ, 714, 1570
Howard, A. W., Marcy, G. W., Bryson, S. T., et al. 2012, ApJS, 201, 15
Howard, A. W., Marcy, G. W., Johnson, J.-A., et al. 2010, Sci, 330, 653
Hsu, D. C., Ford, E. B., Ragozzine, D., & Ashby, K. 2019, AJ, 158, 109

Johansen, A., Davies, M. B., Church, R. P., & Holmelin, V. 2012, ApJ, 758, 39
Johnson, J.-A., Aller, K. A., Howard, A. W., & Crepp, J. R. 2010, PASP, 122, 905
Johnson, J. A., Butler, R. P., Marcy, G. W., et al. 2007, ApJ, 670, 833
Johnson, S. A., Penny, M., Gaudi, B. S., et al. 2020, AJ, 160, 123
Jones, H. R. A., Butler, R. P., Tinney, C. G., et al. 2006, MNRAS, 369, 249
Jung, Y. K., Gould, A., Zang, W., et al. 2019, AJ, 157, 72
Kasper, M., Arsenault, R., Käufl, U., et al. 2019, Msngr, 178, 5
Konacki, M., Torres, G., Jha, S., & Sasselov, D. D. 2003, Natur, 421, 507
Kopparapu, R. K., Hébrard, E., Belikov, R., et al. 2018, ApJ, 856, 122
Kopparapu, R. K., Ramirez, R., Kasting, J. F., et al. 2013, ApJ, 765, 131
Kopparapu, R. K., Ramirez, R. M., & SchottelKotte, J 2014, ApJL, 787, L29
Kratter, K. M., Murray-Clay, R. A., & Youdin, A. N. 2010, ApJ, 710, 1375
Kunimoto, M., & Matthews, J. M. 2020, AJ, 159, 248
Lafrenière, D., Jayawardhana, R., van Kerkwijk, M. H., Brandeker, A., & Janson, M. 2014, ApJ, 785, 47
Latham, D. W., Mazeh, T., Stefanik, R. P., Mayor, M., & Burki, G. 1989, Natur, 339, 38
Lee, E. J., & Chiang, E. 2015, ApJ, 811, 41
Lichtenberg, T., Golabek, G. J., Burn, R., et al. 2019, NatAs, 3, 307
Lin, D. N. C., Bodenheimer, P., & Richardson, D. C. 1996, Natur, 380, 606
Lissauer, J. J., Ragozzine, D., Fabrycky, D. C., et al. 2011, ApJS, 197, 8
Luo, A. L., Zhao, Y.-H., Zhao, G., et al. 2015, RAA, 15, 1095
Macintosh, B., Graham, J. R., Barman, T., et al. 2015, Sci, 350, 64
Malbet, F., Léger, A., Anglada Escudé, G., et al. 2016, Proc. SPIE, 9904, 99042F
Males, J., Close, L. M., Guyon, O., et al. 2019, BAAS, 51, 236
Mann, A. W., Gaidos, E., Lépine, S., & Hilton, E. J. 2012, ApJ, 753, 90
Marois, C., Macintosh, B., Barman, T., et al. 2008, Sci, 322, 1348
Masciadri, E., Mundt, R., Henning, T., Alvarez, C., & Navascués, D. B. Y. 2005, ApJ, 625, 1004
Mawet, D., Hirsch, L., Lee, E. J., et al. 2019, AJ, 157, 33
Mayor, M., Marmier, M., Lovis, C., et al. 2011, arXiv:1109.2497
Mayor, M., & Queloz, D. 1995, Natur, 378, 355
Mazeh, T., Holczer, T., & Faigler, S. 2016, A&A, 589, A75
Metchev, S. A., & Hillenbrand, L. A. 2009, ApJS, 181, 62
Meyer, M. R., Amara, A., Reggiani, M., & Quanz, S. P. 2018, A&A, 612, L3
Mizuno, H. 1980, PThPh, 64, 544
Mochejska, B. J., Stanek, K. Z., Sasselov, D. D., et al. 2005, AJ, 129, 2856
Morbidelli, A., Bitsch, B., Crida, A., et al. 2016, Icar, 267, 368
Mulders, G. D., Pascucci, I., Apai, D., & Ciesla, F. J. 2018, AJ, 156, 24
Murchikova, L., & Tremaine, S. 2020, AJ, 160, 160
National Academies of Sciences, 2018, Exoplanet Science Strategy (Washington, DC: National Academies Press)
Nielsen, E. L., & Close, L. M. 2010, ApJ, 717, 878
Nielsen, E. L., De Rosa, R. J., Macintosh, B., et al. 2019, AJ, 158, 13
Öpik, E. 1924, PTarO, 25, 1
Pascucci, I., Mulders, G. D., & Lopez, E. 2019, ApJL, 883, L15
Penny, M. T., Gaudi, B. S., Kerins, E., et al. 2019, ApJS, 241, 3
Perryman, M., Hartman, J., Bakos, G, & Lindegren, L. 2014, ApJ, 797, 14

Petigura, E. A., Howard, A. W., & Marcy, G. W. 2013, PNAS, 110, 19273
Petigura, E. A., Marcy, G. W., Winn, J. N., et al. 2018, AJ, 155, 89
Plavchan, P., Latham, D., Gaudi, S., et al. 2015, arXiv:1503.01770
Pollack, J. B., Hubickyj, O., Bodenheimer, P., et al. 1996, Icar, 124, 62
Quanz, S. 2019, EPSC-DPS Joint Meeting 2019, Vol. 2019, EPSC-DPS2019-327
Rafikov, R. R. 2005, ApJL, 621, L69
Rafikov, R. R. 2006, ApJ, 648, 666
Rasio, F. A., & Ford, E. B. 1996, Sci, 274, 954
Rauer, H., Catala, C., Aerts, C., et al. 2014, ExA, 38, 249
Raymond, S. N., & Morbidelli, A. 2020, arXiv:2002.05756
Reggiani, M., Meyer, M. R., Chauvin, G., et al. 2016, A&A, 586, A147
Ricker, G. R., Winn, J. N., Vanderspek, R., et al. 2015, JATIS, 1, 014003
Rogers, L. A. 2015, ApJ, 801, 41
Rosenthal, M. M., Chiang, E. I., Ginzburg, S., & Murray-Clay, R. A. 2020, MNRAS, 498, 2054
Rosenthal, M. M., & Murray-Clay, R. A. 2020, ApJ, 898, 108
Schlaufman, K. C. 2018, ApJ, 853, 37
Snellen, I. A. G., & Brown, A. G. A. 2018, NatAs, 2, 883
Sozzetti, A., Torres, G., Latham, D. W., et al. 2009, ApJ, 697, 544
Spergel, D., Gehrels, N., Baltay, C., et al. 2015, arXiv:1503.03757
Suzuki, D., Bennett, D. P., Sumi, T., et al. 2016, ApJ, 833, 145
Swift, J. J., Johnson, J. A., Morton, T. D., et al. 2013, ApJ, 764, 105
The LUVOIR Team, 2019, arXiv:1912.06219
Udalski, A., Ryu, Y. H., Sajadian, S., et al. 2018, AcA, 68, 1
Udalski, A., Zebrun, K., Szymanski, M., et al. 2002, AcA, 52, 115
Vandorou, A., Bennett, D. P., & Beaulieu, J.-P. 2020, AJ, 160, 121
Veras, D., Wyatt, M. C., Mustill, A. J., Bonsor, A., & Eldridge, J. J. 2011, MNRAS, 417, 2104
Vigan, A., Bonavita, M., Biller, B., et al. 2017, A&A, 603, A3
Vigan, A., Fontanive, C., Meyer, M., et al. 2021, A&A, 651, A72
Wang, J., Meyer, M., Boss, A., et al. 2019, BAAS, 51, 200
Weiss, L. M., Marcy, G. W., Petigura, E. A., et al. 2018, AJ, 155, 48
Weiss, L. M., & Petigura, E. A. 2020, ApJL, 893, L1
Weldrake, D. T. F., Sackett, P. D., & Bridges, T. J. 2008, ApJ, 674, 1117
Weldrake, D. T. F., Sackett, P. D., Bridges, T. J., & Freeman, K. C. 2005, ApJ, 620, 1043
Wolfgang, A., Rogers, L. A., & Ford, E. B. 2016, ApJ, 825, 19
Wolszczan, A., & Frail, D. A. 1992, Natur, 355, 145
Yao, Y., Meyer, M. R., Covey, K. R., Tan, J. C., & Da Rio, N. 2018, ApJ, 869, 72
Youdin, A. N. 2011, ApJ, 742, 38
Zhu, W. 2020, AJ, 159, 188
Zhu, W., Wang, J., & Huang, C. 2016, ApJ, 832, 196
Zink, J. K., & Hansen, B. M. S. 2019, MNRAS, 487, 246

# Chapter 3

## Transit Surveys

**Don Pollacco, Natalie Batalha, Magali Deleuil and George Ricker**

### 3.1 Introduction

Transits of giant planets have long been proposed to explain dips in brightness in variable stars (e.g., Piggot 1785) but it was Struve's seminal paper proposing exoplanet discovery through radial velocity variations (Struve 1952) that first connected the two techniques. While many other discovery techniques have been used, transits have remained the only way to determine accurate planetary radii in a relatively model independent way. Transit models can also be used to establish the orbital inclination and when combined with radial velocity information, Kepler's equations can be solved and the planetary bulk density determined. This can then be compared to models of planetary population formation and evolution. In principle, the planetary parameters can be determined with great accuracy but only when the corresponding host star parameters are also accurately known. Transits have become a favored technique over the last 25 years and planet discovery through transits have dominated the area since the first specialist space missions (CoRoT and Kepler), see Figure 3.1.

Transits also gave us the first insights into the atmosphere properties of the close-in population. As with other aspects of Exoplanet research the overriding goals of transit research are:

- The ultimate goal of exoplanet research is often seen as the detection of extra-solar life. While it is probable that other techniques will be required to achieve that goal, there is a small but non-zero chance that evidence of a habitable environment could be detected via transmission spectroscopy sometime in this decade.
- Detection and characterization of planets of all types within the inner parts of their solar systems (~1–2 au) and especially rocky planets in the habitable zones (HZ) of their host stars (of all spectral types). Transit surveys allow us to study the demographics of exoplanets.

doi:10.1088/2514-3433/abfa8fch3

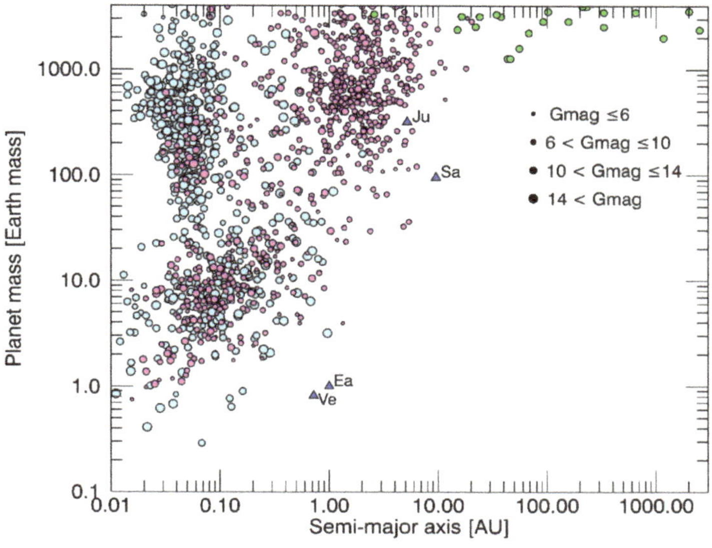


**Figure 3.1.** Current state of exoplanet discovery with known or estimated mass. The color of the symbol indicates the discovery method: transit (blue), radial velocity (pink), and direct imaging (green) and the size of the symbol scales with the GAIA magnitude. Four planets of the solar system are shown for comparison. (Data downloaded on 2020 August from the exoplanets.org database).

- Understanding the architectures of planetary systems through the detection of multi-planet systems.
- Understanding the interactions between host stars and their planets.
- Detection of exoplanetary moons and rings to depict the planets environment.
- Identification of nearby targets for atmospheric studies and especially HZ rocky planets.

## 3.2 State of the Art

The Kepler mission has profoundly changed our view and understanding of planetary systems. The survey demonstrated that (1) on average, every Sun-like star has at least one planet, (2) that one in five will have a temperate, terrestrial-size planet, and (3) the diversity of planets in the Galaxy exceeds the diversity of planets in the solar system (Batalha 2014). The data showed large numbers of multi-transiting systems: multi-planet systems with low mutual inclinations and with relatively short periods. The Kepler population also revealed structures in the distribution of bulk planetary properties (Figure 3.2). The size distribution, for example, is not a simple power law as it is for solar system objects. For example, a deficit of planets at $\sim 1.8 R_e$ reveals physical processes (photo evaporation or core-powered mass loss) that separate the super-Earth and sub-Neptune populations—the so-called radius valley (Fulton et al. 2017).

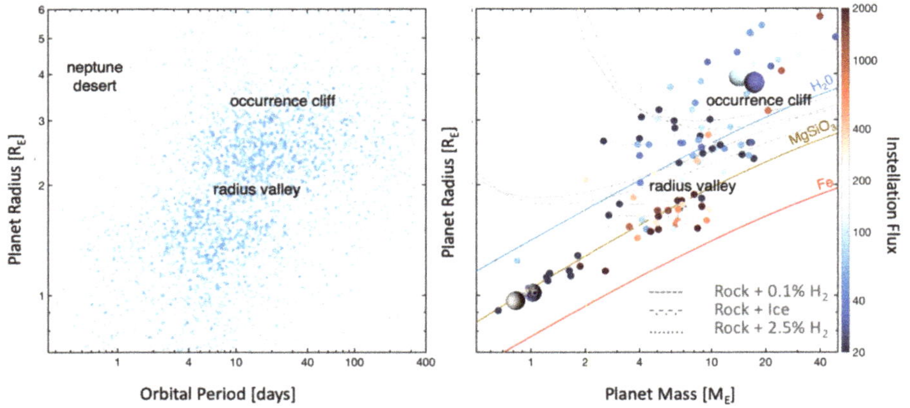


**Figure 3.2.** Distributions of the bulk properties of transiting exoplanets. The period–radius diagram (left) is shown for the high-reliability sub-sample of the homogeneous Kepler candidate catalog with updated radius values based on Gaia parallaxes. The mass–radius diagram (right) is shown for all confirmed transiting planets that have a high confidence mass measurement.

One of the lessons from the CoRoT and Kepler space surveys is that really bright stars are needed for a complete characterization of the planet's properties, including their atmosphere's chemical and physical properties. Consequently, the space agencies have selected a new generation of space missions which aim at overcoming this limitation. The TESS (NASA) (Ricker et al. 2015) and CHEOPS (ESA) (Benz et al. 2020) missions successfully launched in 2018 and 2019 respectively, and target close-in Neptune or Super-Earth sized planets. On the longer term, PLATO (ESA) (Rauer et al. 2014) will explore the diversity of planets at longer orbital periods.

The observation strategy is different from one mission to the others: TESS, now in its first extended mission, will survey the whole sky, giving us a complete census of the close-in population. In particular, it explores the HZ of M dwarf hosts. As a result of the TESS prime and first extended mission, TESS will provide light curves for more than 10 million bright (T<13.5 mag) stars with spectral types ranging from FV to M5 spectral types. CHEOPS is not a transit survey but has the flexibility to be pointed at targets of interest for an extended period. This way, it will observe a few hundred bright stars, with known or suspected planets. Taking frequent measurements, even at less than 1 minute cadence, CHEOPS can accurately measure radii and even identify whether a planet has a significant atmosphere, as illustrated by its observation of the occultation of the hot-Jupiter WASP-189 b (Lendl et al. 2020). Within the next five years, we thus expect to achieve a complete census of the short-period planet populations, some of them with parameters precise enough to provide us with insights into their compositions, and thereby constrain their formation history.

With the advent of these space missions, many of the original ground based surveys have refocused on other science areas or have been decommissioned. However, their databases of observations are still proving useful in constraining ephemerides of newly discovered objects. Beyond these, more specialized surveys

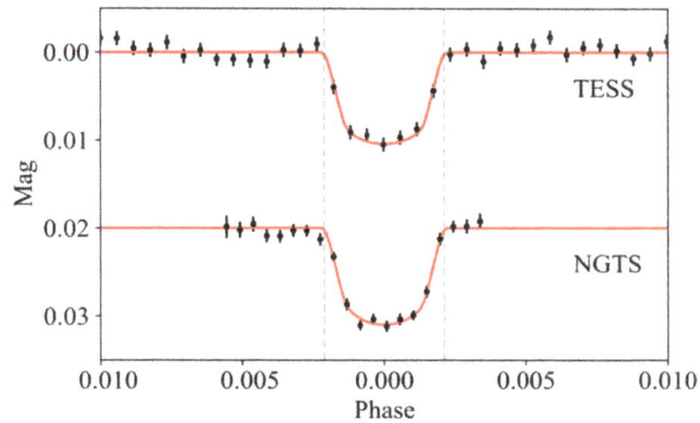


**Figure 3.3.** NGTS-11b is one of the first recovered single transiting TESS planets, TIC-54002556 b (adapted from Gill et al. 2020).

have emerged. For example, the SPECULOOS survey (Gillon 2018; Delrez et al. 2018), which is optimized for late-type red dwarfs and builds upon the Trappist survey and the important TRAPPIST-1 system discovery. These stars, while local, are faint at optical wavelengths and not suited to observations by the current generation of space missions. SPECULOOS, like its predecessors TRAPPIST and MEarth, is a pointed survey observing specific single targets throughout the night.

NGTS is another specialist survey (Wheatley et al. 2018). This facility is composed of multiple small telescopes each on their own mount and each tuned to give the best wide field photometric performance. NGTS has conducted its own survey and found several Neptune sized planets, e.g., NGTS-4b (West et al. 2019). For specific objects, data from multiple NGTS telescopes can be combined to produce a final data set with an accuracy that can rival that of TESS. NGTS's leading contribution is likely to be the detection of additional transits from TESS single-transit candidates thereby increasing the known population of temperate planets. Figure 3.3 shows the recovery of a 34.45 day Saturn sized planet (Gill et al. 2020) originally discovered as a single transit with TESS observations. It's also likely that NGTS could be used to detect long period transiting planets predicted from Gaia observations (Voss et al. 2013).

## 3.3 Important Questions/Goals

Built upon the initial discoveries from the Kepler prime mission, the challenges for the next decade are:

- Determination of the frequency of different classes of planets around stars of all types. This should enable identification of any dependency of planet formation on host star properties, system architecture, and even Galactic location. The first step is to go beyond simple detection by measuring precise and accurate planetary masses and radii, especially for small planets such as Earth–Sun analogs. This requires combining the best precision in terms of photometry and radial velocity, over a wide range of host star spectral types. In addition, as the planetary parameters are always measured relative to those

of their host, precise fundamental stellar parameters not only for the host but also for the parent population in the survey are needed (including stellar ages). Longer period planets are likely to come from GAIA and the microlensing surveys such as WFIRST or EUCLID. Even still our knowledge of planet populations is likely to be limited by small samples of objects.
- Determination of the various formation and evolution pathways that lead to planet and planetary system diversity. Besides accurate bulk properties, this requires the determination of the complete architecture of planetary systems over a large range of orbital periods, and thus needs a combination of data from all the various detection techniques. This will allow us to link the occurrence of small close-in planets to those of cold giants, and to perform a complete census of the planetary systems and thus place our solar system planets in a broader context. For that to be successful, we need an understanding of the biases that affect each method. This effort will be challenging.
- Exploration of the HZ around stars of all spectral types. The question of how to define the habitable zone and how the HZ evolves with the temporal evolution of the star is essential to know which exoplanets could have biospheres (and when). Understanding the life question will be one of the major topics over the next several decades. Currently, there are claims of *potentially habitable* planets that are based almost solely on the flux they receive from their host star. More work is needed both in the laboratory and at the telescope. The overarching question of planetary habitability will require advances in many fields. In astrophysics, pertinent questions include understanding the planetary internal composition as this greatly effects the planetary surface geology and atmospheric chemistry.
- Identification of a number of extremely bright targets with HZ planets for atmospheric studies with JWST and ELT spectroscopy. These targets are likely to be M dwarfs initially, and eventually solar-type stars.
- Ability to fully include the effects of stellar activity into the discovery and characterization techniques. When investigating the small-size planet population properties, stellar activity is certainly a major limitation. It affects both the transit profiles and the planet radial velocity signatures and hinders the characterization of the planet. In the presence of activity, it is thus mandatory to properly correct the planet's signatures from the activity induced effects so that the planet parameters can be measured with the best possible precision, hence its real composition.

## 3.4 Opportunities

In the near future, opportunities not only arise from new facilities but also from refined analysis of existing data sets:

- The Trappist-1 system (Gillon et al. 2016) has been at the center of much research effort over the last few years (and will be with JWST/ELTs, etc.). SPECULOOS is continuing to look for similar systems and while rare, we stand to learn much from these nearby objects. SPECULOOS is also currently undergoing a northern extension with several facilities now in use.

- For archived data sets (e.g., Kepler, K2, and TESS), opportunities exist to develop new algorithms/techniques and apply additional computing resources to detect additional planets, understand survey biases, and continue to explore demographics. The latest techniques needed to minimize the effects of stellar activity using sophisticated (and computationally expensive) techniques that will become more tractable as computing costs decline.
- CHEOPS is now starting its main mission. With a typical visit of a few days, over the mission duration it will observe a few hundred nearby stars, already known to host planets. CHEOPS will yield accurate radii measurements of planets with already known masses, thus constraining their composition and possible internal structure.
- TESS was launched into an inclined 2:1 resonance orbit with the Moon. This orbit is expected to remain stable for decades so minimal spacecraft consumables are required to extend the mission. The spacecraft is well into its first 2 year extended mission and will likely continue beyond that timeframe. Given the restrictive coverage for most of the TESS observations (<30 days) many single-transit candidates have and will be identified. CHEOPS will play a significant role establishing the orbital periods of such transiting planets.

  During its first extended mission that is currently in progress, TESS will observe most of the ecliptic plane for the first time, providing coverage for 15 of the 20 K2 campaign fields. These observations will refine the ephemerides of dozens of bright host K2 exoplanet systems to enable spectroscopic follow-up by JWST and the upcoming ELTs. Moreover, the recent addition of a 20 s observing cadence for TESS should enable a better understanding of stellar activity associated with M dwarf exoplanet hosts.

  It has also been suggested that relatively inexpensive clones of the TESS spacecraft could be relatively cheaply produced and launched into similar orbits (Ricker 2019). This approach would greatly extend the time on sky finding longer period and possibly smaller planets.
- In the Gaia DR3 we expect the first planetary candidates to be released. This could be a rich source of long period transiting planets. While rare, these objects could be followed up by facilities like NGTS to establish accurate bulk properties.
- PLATO is due to launch in 2026. It is designed to detect and characterize Earth–Sun analogs and accurately characterize the host stars, including their age, through asteroseismology. While stars in the input catalog have already been characterized through GAIA data, the prime targets will all be brighter than 11th Mag and will be good radial velocity targets for ESO's ESPRESSO spectrograph at the VLT. This capability should be optimal for making progress in many of the areas highlighted earlier.

## 3.5 Challenges

Looking for long period small planets through the transit technique is a difficult task. However, despite the low probability of the transit geometry for any given system, transits have a proven track record in HZ planet detection and when combined with radial velocity observations, remain the only techniques able to determine accurate planetary parameters. While the low statistical occurrence will remain a problem there are a number of other areas that are challenging:

- Our knowledge of the host star physical parameters are often limiting the accuracy of the derived planetary parameters including the planetary age. For solar-type stars this is minimized if asteroseismic data is available and this is one of the aims for PLATO in the second half of this decade. This remains a problem for late-type stars where the asteroseismic signals are small or non-existent
- When extending transit survey searches to smaller planets, the effects of stellar activity become an issue. In the absence of corrections of the actual observations for activity, photometric contamination can be offset through longer observation sequences (i.e., more transits). However, there is good evidence that activity is lower for later type red dwarfs (Robertson et al. 2013) and given the easier detectability of HZ rocky planets around M dwarfs this will certainly be an area of great interest during this decade.
- Transit observations can be used to determine the bulk internal compositions of rocky planets but the transits must enable radii with a precision better than 3%–5% (Noack et al. 2014) to be derived. This will require multiple transits (especially if some of the individual transits are affected by activity) and for HZ planets, space based observations spread over several years.
- The effects of stellar activity on radial velocity observations is perhaps a greater problem. This arises, in part, because activity levels for any particular star are constantly changing, with varying effects on individual absorption lines. Hence, the normal cross-correlation techniques for radial velocity determination can be unreliable. This will be exacerbated for planets on orbital periods that are comparable to the hosts star activity cycle. However, there is much research being devoted to activity and its proxies. It is likely that this research will lead to significant progress over the next few years.

Transit surveys have evolved to become focused on specific targets. The current and near futures surveys are examining the red dwarf population where the HZ is at short periods and the relative planet/star diameter is favorable for detection. Discovering and characterizing (including planetary age) relatively nearby Earth–Sun analogs is already possible but challenging in part because of the effects of stellar activity. However, future missions will be optimized for this experiment.

## References

Batalha, N. M. 2014, PNAS, 111, 12647

Benz, W., Broeg, C., Fortier, A., et al. 2020, ExA, 51, 109

Delrez, L., Gillon, M., Queloz, D., et al. 2018, Proc. SPIE, 10700, 107001I
Fulton, B. J., Petigura, E. A., Howard, A. W., et al. 2017, AJ, 154, 109
Gill, S., Wheatley, P. J., Cooke, B. F., et al. 2020, ApJL, 898, L11
Gillon, M. 2018, NatAs, 2, 344
Gillon, M, Jehin, E., Lederer, S. M., et al. 2016, Natur, 533, 221
Lendl, M., Csizmadia, Sz., Deline, A., et al. 2020, A&A, 643, A94
Noack, L., Godolt, M., von Paris, P., et al. 2014, P&SS, 98, 14
Piggot, E. 1785, RSPT, 75, 127
Rauer, H., Catala, C., Aerts, C., et al. 2014, ExA, 38, 249
Ricker, G. 2019, TESS Science Conference I, July 2019, https://tsc.mit.edu/talks/George%20Ricker.pdf
Ricker, G. R., Winn, J. N., Vanderspek, R., et al. 2015, JATIS, 1, 014003
Robertson, P., Endl, M., Cochran, W. D., & Dodson-Robinson, S. E. 2013, ApJ, 764, 3
Struve, O. 1952, Obs, 72, 199
Voss, H., et al. 2013, in Highlights of Spanish Astrophysics VII: Proc. of the X Scientific Meeting of the Spanish Astronomical Society (SEA), ed. J. C. Guirado, L. M. Lara, V. Quilis, & J. Gorgas (Sociedad Española de Astronomía)
West, R. G., Gillen, E., Bayliss, D., et al. 2019, MNRAS, 486, 5094
Wheatley, P. J., West, R. G., Goad, M. R., et al. 2018, MNRAS, 475, 4476

AAS | IOP Astronomy

ExoFrontiers
Big questions in exoplanetary science
Nikku Madhusudhan

# Chapter 4

## Radial Velocity Surveys

**Debra A. Fischer**

### 4.1 Introduction

Radial velocity (RV) measurements opened the floodgates for the field of exoplanets, beginning with the discovery of 51 Peg b (Mayor & Queloz 1995). Between 1995 and 2010, RV surveys were the source of most detected exoplanets as ground-based photometric programs began to ramp up, first working collaboratively with RV teams (Bakos et al. 2004; Henry 1999) and then making independent discoveries (Bakos et al. 2007; Charbonneau et al. 2009). After 2010, the detection of transiting exoplanets rapidly outpaced RV-detected exoplanets, fueled by extensive data sets from space-based missions: CoRoT (Carpano et al. 2009), Kepler (Borucki et al. 2010) and TESS (Ricker et al. 2015) and RV measurements began to play a subordinate role, providing mass measurements to calculate the bulk density of transiting planets.

More than 4000 exoplanets have now been discovered by transit, RV, and other methods; however, our understanding of the diversity of exoplanets and planetary architectures is still woefully incomplete. RV surveys with new, extreme precision radial velocity (EPRV) spectrographs and more intense observing cadence have a critical and unique role to play in answering some of the big science questions:

- How do exoplanets form and evolve? What mechanisms stall the fatal migration of meter-size planetesimals? What are the timescales for forming rocky and gas giant planets?
- What is the diversity in planet interior compositions, atmospheres, and environments?
- How do planetary architectures compare? Are they dynamically full?
- What is the role that host stars play (i.e., the influence of metallicity, stellar mass, stellar evolution)?
- How common are habitable conditions? Are there biosignatures on other planets?

doi:10.1088/2514-3433/abfa8fch4  

## 4.2 State of the Art

The ground-breaking discovery of 51 Peg b and other "hot Jupiter" exoplanets was met with both excitement and skepticism. Jupiter-like planets orbiting their host stars in just a few days (Mayor & Queloz 1995; Butler et al. 1997) or exoplanets in highly eccentric orbits (Marcy & Butler 1996; Cochran et al. 1997) were not easily reconciled with the solar nebula model, which aimed to explain planet formation based on the architecture of the solar system. There was speculation that astrophysical phenomena such as pulsations or spectral line profile variations from starspots (Gray 1997) were being misinterpreted as planetary signals, or that the measured RVs might be attenuated by the inclination of nearly face-on binary star orbits (Black 1997). In 1999, two observations clinched the planetary nature of RV detections: a multi-planet system was detected around $\upsilon$ Andromedae where dynamical arguments constrained the masses of the three planets to the substellar regime (Butler et al. 1999), and photometry was used to detect the transit of a hot Jupiter, HD 209458 b (Charbonneau et al. 2000; Henry et al. 2000), eliminating the $\sin i$ ambiguity in the Doppler mass measurement and proving that the mass and radius were consistent with a gas giant planet. As this example shows, the combination of information from different techniques is more than the sum of the parts.

The distribution of known exoplanets for RV and transit surveys is shown in Figure 4.1 (left and right) with dashed lines to indicate the location of Earth, Neptune, and Jupiter in the mass and radius parameter space respectively. It is immediately apparent that there is a higher density of data points in the parameter space of transiting exoplanets. There are ~10 times more transiting exoplanets than RV-detected planets (Figure 4.1, right); however, this is because space-based transit surveys have monitored ~100 times more stars than ground-based RV surveys. While radial velocity observations are labor intensive (stars must be observed one at a time), ultimately the potential planet yield per star is roughly 10 times greater with RV data due to the stringent inclination requirement for transiting planets.

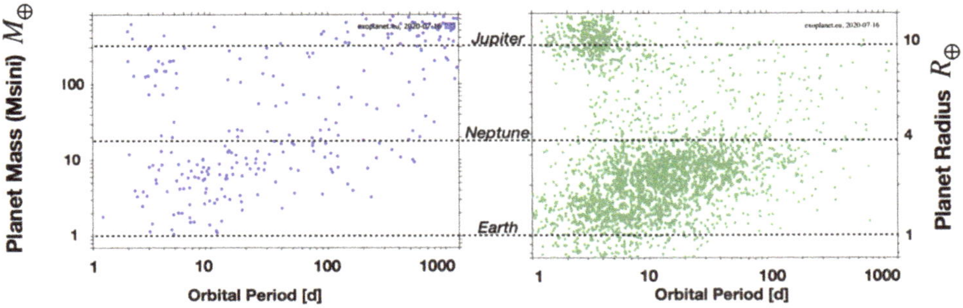


**Figure 4.1.** (Left) The distribution of planet mass as a function of orbital period for exoplanets detected with RV measurements. (Right) The ensemble of transit detections with planet radius as a function of orbital period. Dashed lines indicate the approximate location in the relevant parameter space of mass or radius versus orbital period for planets similar to Earth, Neptune, and Jupiter (*data source:* exoplanet.eu; Schneider et al. 2011).

It is also clear from Figure 4.1 that the transit method is more sensitive to smaller planets. There are eight times more planets with radii smaller than Neptune compared to planets larger than Neptune $N(R < 4R_{\oplus})/N(R > 4R_{\oplus}) \sim 8$. Many sub-Neptune-mass planets have been missed by RV surveys because of inadequate RV precision. This is about to change with a new generation of recently commissioned spectrographs that are reaching ~20 cm $s^{-1}$ single measurement precision on bright stars (Pepe et al. 2013; Suárez Mascareño et al. 2020; Jurgenson et al. 2016; Blackman et al. 2020; Petersburg et al. 2020).

The dependence of the RV amplitude ($K$) on planet mass ($m_{pl}$), orbital inclination ($i$), stellar mass ($M_{\star}$), orbital period ($P$), and orbital eccentricity ($e$) is described by the following expression:

$$K = \frac{8.95 \text{ cm s}^{-1}}{\sqrt{1-e^2}} \frac{m_{pl} \sin i}{m_{\oplus}} \left(\frac{M_{\star}}{M_{\odot}}\right)^{-2/3} \left(\frac{P}{1yr}\right)^{-1/3}. \tag{4.1}$$

And the radial velocity equation is given by:

$$RV = K[\cos(\omega + f) + e \cos \omega] \tag{4.2}$$

where $\omega$ is the argument of periastron and $f$ is the true anomaly (the angle between the direction to periastron and the position of the orbiting planet; Lovis & Fischer 2010). RV detections require well-sampled phases over at least one full orbital period with a detection signal-to-noise ratio ($D_{SNR}$) > 5.8 (Traub et al. 2010), where the $D_{SNR}$ is defined as:

$$D_{SNR} = K\sqrt{N_{obs}}/\sigma. \tag{4.3}$$

Here, $N_{obs}$ is the number of observations and $\sigma$ is the single measurement uncertainty. At face value, Equations (4.1) and (4.2) imply that an Earth analog with an RV amplitude of 8.95 cm $s^{-1}$ could be detected with 1 m $s^{-1}$ RV uncertainties if 4200 measurements are taken ($N_{obs} = (5.8 \times 1.0/0.0895)^2$). With an RV precision of 20 cm $s^{-1}$, the required number of observations plummets to 168. However, this calculation misses two important factors: the need for well-sampled orbital phases will drive up the number of required observations, and velocities originating in the stellar photosphere and lower chromosphere act as an additional noise source ("stellar jitter") that is not captured by $\sigma$, the single measurement precision.

Figure 4.2 summarizes the progress of RV detections. In the early years, the best RV precision was between 3 and 15 m $s^{-1}$, sufficient to detect hot Jupiters with $K > 50$ m $s^{-1}$. The diagonal arrow in this figure suggests smooth, steady progress toward higher RV precision and lower mass planets. The major technical advance occurred in 2003 when the High Accuracy Radial velocity Planet Search (HARPS) spectrograph was commissioned (Mayor et al. 2003), ushering in the era of 1 m $s^{-1}$ RV precision. HARPS set a new bar for other instruments and the state of the art precision has been 1 m $s^{-1}$ for the past decade (Fischer et al. 2016). The steady improvement seen in Figure 4.2 also occurred because RV programs began surveying lower mass stars (which yield a larger RV amplitude for a given planet)

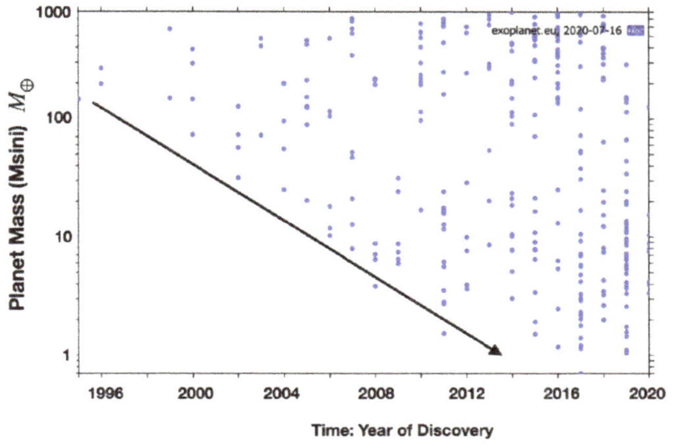


**Figure 4.2.** The mass of RV planet detections over time. The diagonal arrow shows that from 1996 to 2012, there was a steady improvement in the ability to detect low-mass planets, driven both by improved measurement precision (especially for HARPS) and improved observing strategies (*data source:* exoplanet. eu; Schneider et al. 2011).

and increasing the number of observations (augmenting the $D_{SNR}$). Another factor actor that contributed to the lower envelope of detections was the announcement of planets at progressively lower S/N as confidence in the reality of exoplanets was secured. The detection significance of lower bound planets in 2010 was lower than 51 Peg b.

Figure 4.2 also shows that progress seems to slow around 2012. Why didn't RV surveys further reduce measurement uncertainty and detect even lower mass planets? The RV error budget is roughly the quadrature sum of several terms that can be classified under the umbrellas of: instrumental, analytical (including barycentric corrections and modeling of telluric line), wavelength calibration, and stellar jitter. When one term in an error budget dominates, it is easier to identify and to attack that error source. As the tall poles in the RV error budget were addressed between 1995 and 2012, the magnitude of individual terms in the error budget began to converge. It became more difficult to make rapid progress since improving a single term in the RV error budget did not significantly impact the overall measurement uncertainty. This made it difficult to see cause and effect, and difficult to disentangle contributions from individual components of the error budget. Astronomers working on RV surveys knew this day was coming, yet they continued to push the frontiers, almost with a blind faith. They improved the stability and resolution of spectrograph designs; they tackled wavelength calibration with laser frequency combs; and they refined their analysis codes. It was no longer sufficient to design a spectrograph to measure Doppler shifts. New spectrographs would need to allow astronomers to interrogate the multiple sources of stellar jitter.

The sensitivity of RV surveys to lower mass planets is about to be reinvigorated again with the recent commissioning of two next-generation spectrographs, ESPRESSO (Pepe et al. 2013; Damasso et al. 2020; Suárez Mascareño et al. 2020) and EXPRES (Jurgenson et al. 2016; Blackman et al. 2020; Petersburg et al. 2020; Brewer et al. 2020). The instrumental precision of these spectrographs is <10 cm $s^{-1}$ and the measurement precision on stars is 10–30 cm $s^{-1}$. The goal of these spectrographs is to isolate stellar jitter as the last tall pole in the error budget. While stellar jitter dominates EPRV errors, it is now clear that the magnitude of the stellar jitter term has been overestimated, at least in the case of chromospherically quiet stars (Brewer et al. 2020).

## 4.3 Important Questions/Goals

Can we detect Earth analogs? The 1-year orbit of the Earth induces a reflex velocity of about 9 cm $s^{-1}$ in the Sun. This is close to the single measurement precision of EPRV spectrographs, so it is time to revisit the best targets from long-standing surveys to search for low-amplitude signals. We know from Equation (4.3) that a winning strategy will couple EPRV precision with a large number of observations (high cadence). The search for Earth-class exoplanets is an important focus of ESPRESSO (Hojjatpanah et al. 2019) and EXPRES (Brewer et al. 2020) and a number of promising statistical methods are now being employed to disentangle the photospheric velocity jitter from the Keplerian signals (see below). The community should not pause and wait for progress on statistical methods for two reasons: (1) dense, high-cadence and high-fidelity data sets will help to train statistical methods and (2) it will be possible to apply statistical modeling retroactively to previously acquired spectra. Success here seems imminent. If stellar jitter can indeed be mitigated, it is likely that EPRV surveys offer the most efficient way to fast-track the detection of Earth analogs.

Can we disentangle the effect of stellar jitter on radial velocity measurements when modeling Keplerian signals? In reality, stellar jitter is more than a constant term in the error budget; it is a *signal* originating from various physical processes within the star. For example, a star's rotation moves cool spots and hot faculae into and out of view. The effect on RVs is a "quasiperiodic" pattern with timescales and amplitudes similar to those of Keplerian signals. Shapes of spectral lines, as well as simultaneous photometry, can be used to probe and approximately remove this component (Rajpaul et al. 2015; Davis et al. 2017). Another powerful approach is to assume that stellar jitter is random, but correlated over short timescales (Tuomi et al. 2013; Haywood et al. 2014; Rajpaul et al. 2015) related to the rotation period and lifetimes of active regions. This technique has proved extremely successful through dozens of recent RV detections of low-mass planets (stellar jitter is often modeled with a Gaussian Process). These methods are promising in the search for Earth-like exoplanets, but their ability to separate stellar jitter from Keplerians is closely tied to the precision of the data. Applying new models of stellar jitter to data from forthcoming spectrographs might be the only way forward (Langellier et al. 2020).

Can we reconcile the statistics from space-based transit missions such as Kepler with the exoplanet distribution from RV surveys (shown in Figure 4.1)? Using the NASA Kepler mission, astronomers learned that the most common type of planet (at least, with orbital periods $P < 100$ days) have radii between 2 and 2.8 $R_\oplus$ (Petigura et al. 2013). As noted above, the fraction of these planets detected in RV surveys is smaller by a factor of eight or more. There is also a suggestion that the mass of planets and stars is positively correlated. Low-mass rocky planets appear to be more common around later type M dwarfs (Batalha 2014; Howard et al. 2010), while gas giants are preferentially found around the more massive late F and G type stars. Since planets with radii of 2 and 2.8 $R_\oplus$ will have masses between 3 and 7 $M_\oplus$ (Weiss et al. 2016), the reflex stellar velocities will range between 0.5 and 1.5 m $s^{-1}$. While many of these planets would have been missed with past RV surveys because of lower RV measurement precision and lower cadence observations, the detection of ~1 m $s^{-1}$ RV signals is low hanging fruit the new EPRV spectrographs.

How do planets respond to evolution of the host star from the main sequence to the giant branch? Radial velocity surveys for exoplanets around more massive (A and early F type) host stars are difficult because these stars have few absorption lines (high stellar $T_{\rm eff}$) and rotational line broadening reduces the Doppler content of the spectra. However, these searches can be carried out with RV surveys of subgiants, whose progenitors were A and F type stars on the main sequence (Johnson et al. 2007). Here the combination of improved EPRV precision and effective modeling of stellar jitter could make substantial progress toward understanding the scaling of planet-star mass and orbital evolution.

What mechanism is responsible for the inflated size of some hot Jupiters? Long before the detection of the first transiting hot Jupiter, physical models suggested that the radius of Jupiter was an upper limit for gas giant planets. Intermediate mass brown dwarfs can be smaller than Jupiter (Carmichael et al. 2020) and the lowest mass M dwarfs have radii that are about the same as Jupiter (Parsons et al. 2018). Therefore, it was a surprise when transit surveys revealed that some (but not all) short-period, gas giant planets are puffed up, sometimes to more than twice the radius of Jupiter. Several mechanisms for radius inflation have been proposed: hot Jupiters born *in situ* may be assembled with inflated radii, or the radii of migrating gas giants may be re-inflated, either by tidal dissipation, or the evolution of host stars (Guillot 2005; Hartman et al. 2016). Careful determinations of mass, radius, age, chemical composition, and orbital parameters—especially eccentricity and obliquity—should help to differentiate the origin of radius inflation. RV measurements with precision of 1 or 2 m $s^{-1}$ are sufficient to determine the mass of a gas giant planet; however, EPRV measurements are needed to search for smaller, more distant planets that could pump up orbital eccentricity. EPRV data could also make more precise measurements of orbital eccentricity.

Why is the size and spacing so similar for planets in Kepler multi-planet systems? The planet radius is a good proxy for the internal composition of an exoplanet and the structure of exoplanet architectures provides key information about planet formation and evolution. The empirical "peas in a pod" phenomenon (i.e., planets have similar size and spacing) is remarkable (Ciardi et al. 2013; Weiss et al. 2018; Weiss & Petigura 2020) and must have a physical explanation. One possibility is that compact and nearly co-planar planetary systems may commonly experience resonance-driven spin–orbit coupling (Millholland & Laughlin 2019) that produces the characteristic planet size and spacing through a process called "obliquity sculpting." Most knowledge about the similarity of these exoplanets comes from transit data. The use of EPRVs could derive exoplanet masses, useful for understanding the nature of these planets and for advancing dynamical modeling. Kepler target stars are typically faint, so follow-up will require large aperture telescopes for these EPRV measurements.

Why is there a gap in the radius-period distribution for Kepler exoplanets? The radius-period distribution of transiting exoplanets with orbital periods shorter than 100 days and well-determined radii exhibits a striking deficit of planets with 1.5–2 $R_\oplus$ (Fulton et al. 2017). In adjacent radius bins, the number of planets with radii from 2 to 3 $R_\oplus$ is nearly the same as the number of planets with radii from 1 to 2 $R_\oplus$. This "hot super-Earth desert" was produced in simulations of atmospheric stripping by photo-evaporation when the incident flux exceeded 650 $F_\oplus$ (Lundkvist et al. 2016); however, in this case, the radius gap was slightly shifted to 2.2–3.8 $R\oplus$. Since the gravitational potential of the exoplanet should be a factor in the ability of a planet to maintain an atmosphere, EPRV measurements with large aperture telescopes could contribute to understanding this phenomenon.

## 4.4 Challenges

There are still a number of challenges for EPRV programs. Some of the challenges are technological. Astronomers do not represent a large market share so collaboration with industry partners is largely driven by the serendipitous interest of the commercial vendor in scientific research. Instrument builders worry about discontinued industrial production of echelle gratings, charge-coupled devices, and laser frequency combs.

Some of the challenges are financial. Even when is possible to fund instruments through federal grant programs (NSF, NASA, DOD, DOE), it can be extremely difficult to fund operations after the instrument is built. This is a real challenge for EPRV programs where high-cadence observing is critical, both for detecting planets and for modeling stellar jitter. Spectrographs are needed on telescopes with a range of apertures and our community has not been good at coordinating efforts to optimize observations. There is a danger that multiple telescopes will focus on the same targets. While some redundancy is good, EPRV surveys make extended investments in targets and science might progress more efficiently with some coordination.

Some of the challenges are political or environmental. Our planet is facing accelerated climate change. Warming trends degrade seeing at some astronomical sites and fires have threatened ground-based observatories in Australia and California. We can hope that the discovery of habitable worlds and detections of biosignatures may inspire global efforts to save our own planet. The consortium of Astronomers for Planet Earth (White et al. 2021) is raising awareness and providing information and materials for teaching about this topic in introductory astronomy courses.

## 4.5 Opportunities

After more than a decade of 1 m $s^{-1}$ precision, radial velocity surveys have taken the next leap forward (Suárez Mascareño et al. 2020; Brewer et al. 2020) toward ~10 cm $s^{-1}$. This order of magnitude advance in measurement precision offers exciting scientific opportunities. Perhaps the most exciting goal is the detection of exoplanets in potentially habitable environments that are similar to the Earth. However, EPRV surveys will also begin to unravel the structure of planetary architectures, laying the foundation for the next flagship missions like LUVOIR or HabEx to search for biosignatures (Wang et al. 2018). Observations over the coming few years should reconcile current differences in the distribution of transiting exoplanets and RV-detected exoplanets. We will also obtain more clues about planet formation and orbital evolution by detecting more components in multi-planet systems and improving the accuracy of orbital parameters such as eccentricity. This in turn will reveal more about how planet formation and orbital evolution is affected by environmental parameters such as the protoplanetary disk mass, and the chemical composition, spectral type, and evolution of the host star.

## Acknowledgments

Thanks to Sam Cabot and Greg Laughlin for reviewing this essay and for providing helpful comments.

## References

Bakos, G., Noyes, R., Kovács, G., et al. 2004, PASP, 116, 266
Bakos, G. Á., Noyes, R. W., Kovács, G., et al. 2007, ApJ, 656, 552
Batalha, N. M. 2014, PNAS, 111, 12647
Black, D. C. 1997, ApJL, 490, L171
Blackman, R. T., Fischer, D. A., Jurgenson, C. A., et al. 2020, AJ, 159, 238
Borucki, W. J., Koch, D., Basri, G., et al. 2010, Sci., 327, 977
Brewer, J. M., Fischer, D. A., Blackman, R. T., et al. 2020, AJ, 160, 67
Butler, R., Marcy, G. W., Fischer, D. A., et al. 1999, ApJ, 526, 916
Butler, R. P., Marcy, G. W., Williams, E., Hauser, H., & Shirts, P. 1997, ApJL, 474, L115
Carmichael, T. W., Quinn, S. N., Mustill, A. J., et al. 2020, AJ, 160, 53
Carpano, S., Cabrera, J., Alonso, R., et al. 2009, A&A, 506, 491
Charbonneau, D., Berta, Z. K., Irwin, J., et al. 2009, Natur, 462, 891
Charbonneau, D., Brown, T. M., Latham, D. W., & Mayor, M. 2000, ApJL, 529, L45
Ciardi, D. R., Fabrycky, D. C., Ford, E. B., et al. 2013, ApJ, 763, 41
Cochran, W. D., Hatzes, A. P., Butler, R., Marcy, G. W., et al. 1997, ApJ, 483, 457
Damasso, M., Sozzetti, A., Lovis, C., et al. 2020, arXiv:2007.06410
Davis, A. B., Cisewski, J., Dumusque, X., Fischer, D. A., & Ford, E. B. 2017, ApJ, 846, 59
Fischer, D. A., Anglada-Escude, G., Arriagada, P., et al. 2016, PASP, 128, 066001
Fulton, B. J., Petigura, E. A., Howard, A. W., et al. 2017, AJ, 154, 109
Gray, D. F. 1997, Natur, 385, 795
Guillot, T. 2005, AREPS, 33, 493
Hartman, J., Bakos, G., Bhatti, W., et al. 2016, AJ, 152, 182

Haywood, R., Collier Cameron, A., Queloz, D., et al. 2014, MNRAS, 443, 2517
Henry, G. W. 1999, PASP, 111, 845
Henry, G. W., Marcy, G. W., Butler, R. P., & Vogt, S. S. 2000, ApJL, 529, L41
Hojjatpanah, S., Figueira, P., Santos, N., et al. 2019, A&A, 629, A80
Howard, A. W., Marcy, G. W., Johnson, J. A., et al. 2010, Sci., 330, 653
Johnson, J. A., Fischer, D. A., Marcy, G. W., et al. 2007, ApJ, 665, 785
Jurgenson, C., Fischer, D., McCracken, T., et al. 2016, Proc. SPIE, 9908, 99086T
Langellier, N., Milbourne, T., Phillips, D., et al. 2020, arXiv:2008.05970
Lovis, C., & Fischer, D. 2010, in Exoplanets, ed. S. Seager (Tucson, AZ: Univ. Arizona Press), 27
Lundkvist, M., Kjeldsen, H., Albrecht, S., et al. 2016, NatCo, 7, 11201
Marcy, G. W., & Butler, R. P. 1996, ApJL, 464, L147
Mayor, M., Pepe, F., Queloz, D., et al. 2003, Msngr, 114, 20
Mayor, M., & Queloz, D. 1995, Natur, 378, 355
Millholland, S., & Laughlin, G. 2019, NatAs, 3, 424
Parsons, S., Gänsicke, B., Marsh, T., et al. 2018, MNRAS, 481, 1083
Pepe, F., Cristiani, S., Rebolo, R., et al. 2013, Msngr, 153, 6
Petersburg, R. R., Joel Ong, J., Zhao, L. L., et al. 2020, AJ, 159, 187
Petigura, E. A., Howard, A. W., & Marcy, G. W. 2013, PNAS, 110, 19273
Rajpaul, V., Aigrain, S., Osborne, M., Reece, S., & Roberts, S. 2015, MNRAS, 452, 2269
Ricker, G. R., Winn, J. N., Vanderspek, R., et al. 2015, JATIS, 1, 014003
Schneider, J., Dedieu, C., Le Sidaner, P., Savalle, R., & Zolotukhin, I. 2011, A&A, 532, A79
Suárez Mascareño, A., Faria, J., Figueira, P., et al. 2020, A&A, 639, A77
Traub, W., Beichman, C., Boden, A., et al. 2010, in EAS Publications Series, Vol. 42, ed. K. Goździewski, A. Niedzielski, & J. Schneider, 191
Tuomi, M., Jones, H., Jenkins, J., et al. 2013, A&A, 551, A79
Wang, J., Mawet, D., Hu, R., et al. 2018, JATIS, 4, 035001
Weiss, L. M., Marcy, G. W., Petigura, E. A., et al. 2018, AJ, 155, 48
Weiss, L. M., & Petigura, E. A. 2020, ApJL, 893, L1
Weiss, L. M., Rogers, L. A., Isaacson, H. T., et al. 2016, ApJ, 819, 83
White, J. A., Dalgleish, H., Burtscher, L., et al. 2021, BAAS, 53, 0202

# Chapter 5

# Direct Imaging and Spectroscopy

**Sasha Hinkley, Arthur Vigan, Markus Kasper, Sascha P. Quanz and Sylvestre Lacour**

## 5.1 Introduction

As the indirect transit and radial velocity (RV) exoplanet detection methods are inherently much less sensitive to wide-separation planets with long orbital periods, the technique of exoplanet direct imaging will be the leading method to fully define the outer extents of planetary systems (Figure 5.1), and provide a more complete understanding of the true frequency of planetary mass companions to nearby stars (Bowler 2016; Nielsen et al. 2019; Vigan et al. 2021). Perhaps most importantly, by spatially separating the light of the host star and the extremely faint planet, this technique is naturally suited to *direct* spectroscopy of planets themselves, making the directly imaged planets among the most valuable exoplanet population for atmospheric characterization (Macintosh et al. 2015; Chauvin et al. 2017), and studying physical processes such as disequilibrium atmospheric chemistry, vertical atmospheric mixing, and cloud physics. Further, upcoming exoplanet characterization work in the next 2–5 years will be an essential step on the "roadmap" to characterizing *terrestrial* exoplanets, and the accompanying search for biosignatures in a small number of very nearby planets 7–10 years from now. Consequently, the direct imaging technique has emerged as one of the leading routes for exoplanet characterization, and is a key science driver for numerous ground-based instruments as well as upcoming flagship space missions.

For a large handful of planetary mass companions, the direct imaging method has allowed powerful characterization through *direct* spectroscopy, returning not only effective temperatures and surface gravities (Hinkley et al. 2013; Macintosh et al. 2015), but also unambiguous detections of numerous molecular species, which in turn have provided basic estimates of the atmospheric compositions (e.g., Konopacky et al. 2013). However, most of this characterization has taken place using ground-based observatories over the relatively narrow ~1–2 μm wavelength range. There has been some limited coverage between 3 and 5 μm, but the

doi:10.1088/2514-3433/abfa8fch5

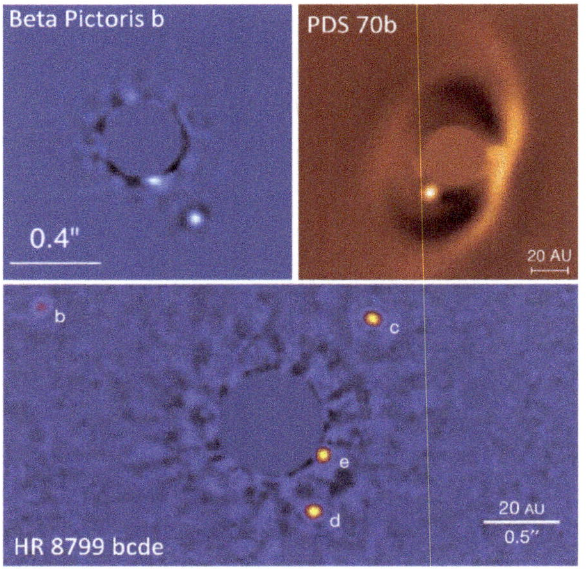


**Figure 5.1.** An image gallery showing examples of directly imaged exoplanets including: the ~10 $M_{Jup}$ planet beta Pictoris b (top left, Lagrange et al. 2010), the 5–9 $M_{Jup}$ companion PDS 70b (top right, Keppler et al. 2018), which is embedded in a circumstellar disk; and the multi-planet HR 8799 system hosting four directly imaged planets with masses 5–7 $M_{Jup}$ and separations ranging from 15 to 68 au (bottom, Marois et al. 2010). For clarity, the inner residual scattered starlight close to the star has been masked out.

exceedingly high telluric background at wavelengths longer than 5 μm, have made observations from the ground impossible. Further, when targeting very young systems, the direct imaging method has so far only demonstrated sensitivity to objects more massive than ~2 $M_{Jup}$. Given this state of the field, the following three goals begin to summarize the primary science drivers for exoplanet direct imaging in the coming decade.

- **Completing the demographics:** The upcoming James Webb Space Telescope (JWST) is predicted to have sensitivity to much lower-mass planets (Saturn/Neptune analogs) than have been probed to date by direct imaging, at the widest orbital separations (tens to hundreds of au). This is the last remaining portion of the mass/orbital separation parameter space that has so far remained completely out of reach. Further, the 30–40m telescopes achieving "first-light" in the next decade will offer a transformational change in the achievable angular resolution of direct imaging efforts, enabling detections of Jovian planets at a few au (Quanz et al. 2015) orbiting nearby stars, and finally probing the population of planets identified by RV surveys.
- **Broader characterization:** To date, due to the very high telluric background in the mid-infrared, most of the full luminous range of massive planets has remained completely out of reach to ground-based observatories. Accessing multiple molecular features with wavelength coverage out to ~28 μm, the characterization abilities of JWST will provide precise abundances of the

dominant molecules in exoplanet atmospheres (e.g., $CH_4$, CO, $CO_2$, $H_2O$, and $NH_3$), reveal the presence of disequilibrium atmospheric chemistry, as well as possibly the existence of silicate clouds. At the same time, most of the spectroscopic characterization of massive planets has been limited to the spectral resolution ($R \sim 50$–200) of dedicated ground-based exoplanet imaging spectrographs (Hinkley et al. 2011, 2013; Macintosh et al. 2015; Chauvin et al. 2017), although a few important exceptions have obtained much higher resolution (e.g., Konopacky et al. 2013). While JWST will provide spectral resolution of a few thousand, upcoming efforts at the Keck and VLT observatories to couple high resolution spectrographs to high contrast imaging platforms that already use Adaptive Optics (AO) and sensitive infrared cameras will allow direct characterization of exoplanets at resolutions up to 20,000–100,000 (Vigan et al. 2018; Mawet et al. 2018).

- **Direct Detection of habitable zone Earth-like planets:** All of the discoveries via the direct imaging technique have so far been Jovian-mass planets with masses greater than $\sim 2\ M_{\rm Jup}$, but upcoming, planned instrumentation such as METIS (Brandl et al. 2018) on the 39 m Extremely Large Telescope observing into the thermal infrared ($\sim 10\ \mu$m), is predicted to provide sensitivity to much lower mass, perhaps terrestrial, exoplanets in the habitable zones of the nearest solar-type stars. At the same time, the design phase is under way for the Planetary Camera and Spectrograph (PCS, Kasper et al. 2010, 2013) operating at visible wavelengths, that will have the combination of achievable high contrast and angular resolution to be sensitive to temperate exoplanets orbiting the nearest M-type stars, where the habitable zones are significantly closer to their host stars.

In this chapter, we present the state-of-the-art of exoplanet direct imaging, some important goals for the field, some key upcoming opportunities, as well as several fundamental challenges that will need to be understood and conquered going forward.

## 5.2 State of the Art

The last decade has ushered in the era of exoplanet imaging. This has primarily been driven by the commissioning of customized instrumentation specifically dedicated for the detection and characterization of massive exoplanets on 8m-class telescopes such as the Very Large Telescope (VLT), Gemini, and Subaru as described below. However, much of the discoveries that took place prior to the commissioning of these dedicated instruments, such as the Gemini Planet Imager (GPI, Macintosh et al. 2014) and the Spectro-Polarimetric High contrast imager for Exoplanets REsearch (SPHERE, Beuzit et al. 2019) at the VLT, were made possible by *multi-purpose* instrumentation with some form of particular observing configuration which enabled the necessary planet-to-star contrast of $\sim 10^4$–$10^6$ at separations within $\sim 1$ arcsecond to be reached, probing masses as low as a $\sim$few Jupiter masses for systems with ages of $\sim 10$–100 Myr. Further optimization of key technologies such as

AO and coronagraphy that have shaped the field of ground-based exoplanet direct imaging over the last 20 years has largely been dictated by the pace of the advancements in deformable mirrors, wave front sensors, and high performance coronagraphic masks.

Initial work on the direct imaging concept began to be taken seriously in the mid-1990s (e.g., Angel 1994; Labeyrie 1995), and first efforts using adaptive optics coupled with rudimentary coronagraphs began to come online resulting in initial detections of substellar objects (e.g., Nakajima et al. 1995; Oppenheimer et al. 2001). In tandem with these initial surveys, optimization of AO systems, and the identification of the necessary observing configurations, soon became a major focus (e.g., Racine et al. 1999; Steinbring et al. 2002; Bloemhof et al. 2001; Marois et al. 2006; Hinkley et al. 2007), with the goal of achieving the high planet-to-star contrast needed to detect planetary mass companions by suppressing and calibrating the dominant instrumental systematics: the residual, uncorrected light from the host star. At about the same time, early surveys using the Hubble Space Telescope (e.g., Lowrance et al. 2005; Luhman et al. 2005) targeted a few dozen nearby stars, resulting in early detections of brown dwarf companions. Shortly after this, other developments in "high-contrast" observing techniques were brought forth by a series of surveys for substellar companions to nearby stars using the multi-purpose NaCO instrument at the VLT (e.g., Masciadri et al. 2005; Kasper et al. 2007; Chauvin et al. 2010). The latter survey initially identified perhaps the first direct image of a truly planetary mass companion to the brown dwarf 2M1207-3932 (Chauvin et al. 2005), a $5 \pm 2\ M_{\rm Jup}$ object with remarkably red colors, giving a first glimpse at the effects of clouds on the colors of substellar objects.

Other efforts at Keck, VLT, MMT, Gemini (Biller et al. 2007; Nielsen et al. 2008; Vigan et al. 2012) and Palomar (Carson et al. 2009; Metchev & Hillenbrand 2009) were contemporaneous with these. These early surveys primarily also used existing, multi-purpose instrumentation, resulting in several breakthrough detections of planetary mass companions including $\beta$ Pic b (Lagrange et al. 2010), a $\sim$12 $M_{\rm Jup}$ companion to a nearby, $\sim$23 Myr star, as well as HR 8799bcde, a $\sim$40 Myr system hosting four planetary mass companions (Marois et al. 2008). Despite their use of multi-purpose instrumentation with a particular observational set up allowing the high contrast to be achieved, these efforts in the mid/late 2000s made clear the importance of using a careful statistical analysis to provide the first estimates of the frequency of massive, planetary mass companions at large orbital separations to nearby stars. Following these initial discoveries, a subsequent generation of *dedicated* instrumentation was commissioned in the mid-2000s with the explicit goal of achieving high contrast and basic characterization of any detected companions. These efforts resulted in their own surveys at observatories such as Gemini (e.g., "NICI", Nielsen et al. 2008; Biller et al. 2013), Palomar (Metchev & Hillenbrand 2009; Hinkley et al. 2011), and Subaru (the "SEEDS" survey, Brandt et al. 2014; Tamura 2016).

This initial generation of instrumentation and surveys made clear the benefits of commissioning dedicated suites of instrumentation equipped with high-order adaptive optics and state-of-the-art coronagraphs for the purpose of directly imaging wide-separation Jovian-mass planets. SPHERE at the VLT (Beuzit et al. 2019) and GPI at Gemini are two of the leading instruments for the direct detection of planetary mass companions orbiting nearby stars, along with SCExAO at Subaru (Jovanovic et al. 2015). The GPI Exoplanet Survey resulted in the detection of 51 Eridani b, a very low mass (~2 $M_{Jup}$) methane-rich planet orbiting at 13 au (Macintosh et al. 2015), while the SPHERE GTO survey presented HIP 65426b,a 6–12 $M_{Jup}$ planet orbiting at 92 au (Chauvin et al. 2017). The initial science return from these dedicated surveys has only recently become public and Figure 5.2 shows the "completeness" of a subset of stars from the SpHere INfrared survEy (SHINE) program from Vigan et al. (2021), demonstrating much greater sensitivity to lower-mass planets, as well as sensitivity at smaller orbital separations, than the pre-cursor NaCO Large Program survey at the VLT (Vigan et al. 2017). The initial results from SHINE also indicate that the frequency of all companions in the mass range 1–75 $M_{Jup}$ and semimajor axis in the range 5–300 au is $23.0^{+13.5}_{-9.7}\%$ for stars with B and A spectral types, but M stars are approximately half as likely to host planets, and FGK stars are nearly four times less likely to be planet hosts (Vigan et al. 2021). Similarly, the results from the corresponding survey from GPI similarly finds a statistically significant occurrence between stellar mass and giant planets (Nielsen et al. 2019).

Most recently, the GRAVITY instrument at the VLT interferometer has shown that it is capable of similarly high contrast observations enabling the direct detection of massive planets, but also providing spectral resolution of $R$~500 in the $K$-band (Gravity Collaboration et al. 2019; Mollière et al. 2020). However, Figure 5.3 shows

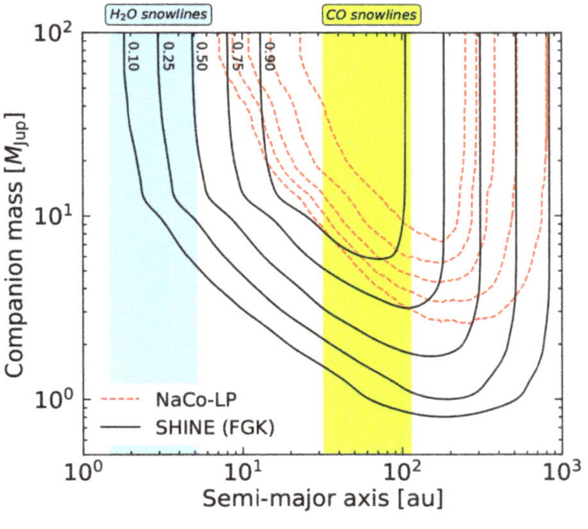


**Figure 5.2.** A plot using data from Vigan et al. (2021) showing the survey "completeness" for the first-generation NaCO Large Program (red dashed lines, Vigan et al. 2017), as well as a subset of FGK stars from the SPHERE SHINE survey that is in progress (dark lines, Vigan et al. 2021). The figure shows a factor of ~3–10 improvement in both sensitivity to both lower-mass planets and smaller orbital separations when SPHERE, a dedicated instrument customized for exoplanet direct detection, is used instead of a multi-purpose instrument such as NaCO.

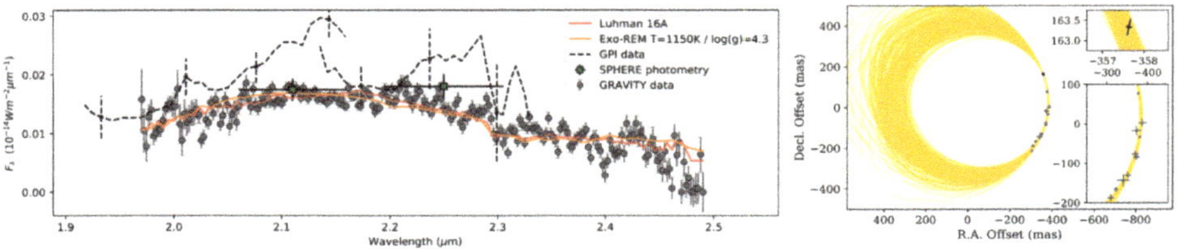


**Figure 5.3.** An example taken from Gravity Collaboration et al. (2019) of the K-band spectrum of the innermost planet in the HR8799 system afforded by the GRAVITY instrument (left panel). Also shown is the exquisite astrometric precision on the orbit (right panel), allowing much more stringent constraints on the family of permissible orbits. The GRAVITY measurement is shown by the black cross, while previous measurements are shown by the gray crosses. Reproduced from Gravity Collaboration et al. (2019) 

an example of the astonishing ~100 microarcsecond astrometric precision delivered by GRAVITY demonstrated on the innermost planet in the HR 8799 system, allowing much more stringent constraints on the orbital geometry of this planet. Such information on the eccentricities of the orbits of substellar companions is proving to be especially powerful, giving hints at possible formation mechanisms. Specifically, a recent statistical study (Bowler et al. 2020) of the eccentricity distributions of a population of both directly imaged planets and brown dwarfs shows that the eccentricities of giant planets are drawn from a different distribution of orbital eccentricities (i.e., they are more circular) than the eccentricities of brown dwarf systems, suggesting that giant planets may form in massive, extended protoplanetary disks, rather than being the low-mass byproducts of the star formation process.

## 5.3 Important Remaining Questions for This Discipline

Below we summarize some of the most important questions and goals going forward for the direct imaging of extrasolar planets.

1. *What is the frequency of Saturn/Neptune analogs at the widest orbital separations (50–1000 au)?*

   The exoplanet discovery space spanned by mass and orbital separation (see Bowler 2016 for an example) shows that the RV and transit methods have achieved very good completeness at orbital separations within ~10au, but are inherently less sensitive beyond. At the same time, direct imaging surveys using ground-based telescopes have not yet detected any companions less massive than ~2 $M_{Jup}$. Thus, although valuable initial studies have been carried out with Spitzer (e.g., Janson et al. 2015; Durkan et al. 2016) the region of this parameter space beyond ~10au, and below ~2 $M_{Jup}$ remains largely unexplored. Figure 5.4 shows the predicted sensitivity of JWST, verifying that it will place the strongest constraints in this region of parameter space. Specifically, these simulations (Carter et al. 2021) incorporate the latest estimates of JWST contrast performance (Perrin et al. 2018), as well as the latest theoretical evolutionary models of the brightness of planetary mass companions (Phillips et al. 2020) when operating at 4.4 and

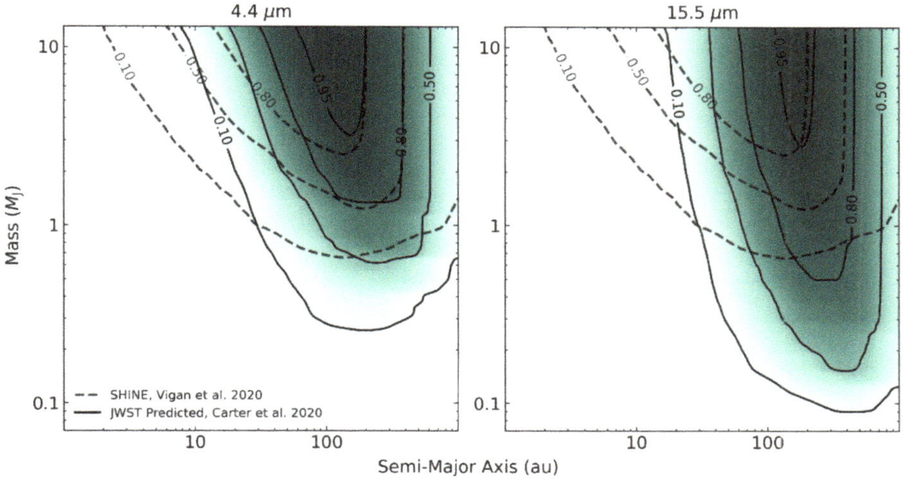


**Figure 5.4.** Demonstrations of the sensitivity of JWST to planetary mass companions when operating at 4.4\μm (left panel) and 15.5 μm (right panel) around a set of 94 stars that are members of the $\beta$ Pictoris and TW Hya young moving groups. These calculations incorporate the latest estimates of JWST contrast performance (Perrin et al. 2018), and theoretical models of exoplanet luminosities (Phillips et al. 2020). JWST will be sensitive to bona fide planets, and at 15.5 μm JWST will be capable of reaching Saturn or Neptune analogs at wide orbital separations (figure courtesy of Aarynn Carter, and adapted from Carter et al. 2021).

15 μm around a set of 94 actual stars that are members of the ~23 Myr $\beta$ Pictoris and ~10 Myr TW Hya young moving groups. Regardless of observing filter, coronagraphic observations with JWST will be sensitive to bona fide planets, but at 15 μm JWST is predicted to be capable of reaching Saturn or Neptune analogs (Carter et al. 2021). Thus, JWST will be well poised to place the first statistical constraints on the frequency of these objects at such wide separations, and the JWST Early Release Science Program (Hinkley et al. 2018) will provide the first estimates of the in-orbit performance of the observatory, such that a broad, community-driven effort can make such a measurement.

2. *How can the direct imaging technique probe into the snow lines of the youngest stars to detect planets shortly after the era of planet formation has ceased?*

   Studying the characteristics of exoplanets shortly after their formation will be a particularly important goal for the direct imaging technique, and provide the best observational constraints on planet formation models. Luminosity measurements of these nascent objects will give an essential assessment of their initial entropy, constraining their earliest thermal histories, still an area of substantial uncertainty (Fortney et al. 2008; Spiegel & Burrows 2012). At the same time, the initial orbital placement of these objects will not have been confused by any subsequent dynamical processes, and therefore may encode an imprint of their formation. Recent theoretical works (e.g., Fernandes et al. 2019; Frelikh et al. 2019), as well as recent observational results incorporating decades of observations from RV surveys (Fulton et al. 2021) have independently shown that the peak of the orbital distribution of massive planetary companions to solar-type stars is expected to be near the expected location of the water ice line at ~3 au, and decline sharply at orbital separations beyond this. Further, the preliminary results of the recent high contrast imaging surveys are consistent with a peak in the occurrence of giant planets between ~1 and 10 au (Vigan et al. 2021). Besides simply showing the completeness of the SHINE survey, Figure 5.2 also shows the improvement in sensitivity to lower-mass planets, and smaller orbital separations, when an instrument *customized* for exoplanet direct imaging (i.e., SPHERE) is used instead of a *multi-purpose* infrared instrument (NaCO). However, even such dedicated instrumentation on 8 m telescopes only begins to probe into snow-line separations (a few au) around the youngest stars. In Vigan et al. (2021), the authors show that for the 150 systems observed with SPHERE, only 15–20 of these have the combination of youth and close proximity to provide sensitivity to ~10$M_{Jup}$ planets at water ice line separations of ~3 au, and less than five of these 150 stars allow sensitivity to ~3 $M_{Jup}$ planets at 3 au. The fundamental limiting factor to the improved sensitivity at small separations is the 8 m telescope aperture: larger telescope apertures are needed to provide resolution sufficient to more completely sample ice line separations. As an added benefit, all error sources that limit contrast on large telescopes that employ "extreme" AO systems scale with $D^{-2}$ where $D$ indicates the telescope aperture. So, for the same set of AO system parameters (e.g., wave front sensor type, deformable mirror actuator spacing, etc.), larger telescopes also deliver improved contrast. However, despite the extremely large apertures (~30–40m) of upcoming Giant Segmented Mirror Telescopes (GSMTs), the resolutions afforded by these behemoths operating in the mid-infrared (where the planet/star contrast is most favorable) is still not sufficient to directly probe within orbital separations of ~3 au for stars in the nearest star-forming regions at typical distances of ~140 pc. Nonetheless, with interferometric techniques such as Sparse Aperture Masking (Hinkley et al. 2011, 2015) operating at 3.8 μm on the GSMTs, "molecule mapping" (Snellen et al. 2010, 2014; Hoeijmakers et al. 2018), or long baseline interferometry (Nowak et al. 2020), accessing planets at orbital separations <10 au may be within reach.

3. *What new observational techniques and theoretical atmosphere models are needed to assess more precisely the atmospheric compositions of the directly imaged planets?*

   The few examples to date of exoplanet atmospheric characterization using ground-based observations at wavelengths >2 μm have provided a much-needed wavelength "lever" for more precisely assessing atmospheric compositions, but have also proved to be essential for discriminating between models incorporating equilibrium and disequilibrium atmospheric chemistry (e.g., Skemer et al. 2012). Upcoming JWST observations of known planetary mass companions will serve as some of the first observations of extrasolar planets over their full luminous range, using a much longer wavelength baseline and sampling multiple molecular features (Hinkley et al. 2018). These first-ever observations will provide clues not only to the presence of equilibrium/disequilibrium chemistry, but also will likely detect the presence of silicate clouds in the atmospheres of planetary mass companions. However, transformative data sets like these can only reach their maximum scientific potential if the best theoretical models are available to interpret the data. Previous modeling work applied to directly imaged planets has largely used an existing grid of models originally intended for brown dwarfs (Barman et al. 2011; Zahnle & Marley 2014; Moses et al. 2016), but perhaps with a reduced gravity to reflect the youth of directly imaged planets, and possibly an ad hoc prescription of clouds that is not fully connected to the chemical networks (Charnay et al. 2018). At the same time, it has been well established (e.g., Barman et al. 2011) that the degree of vertical mixing in an exoplanet atmosphere is a key driver of the chemical abundances at a given atmospheric height, which, in turn dictates the features imprinted on a spectrum. Therefore, the exoplanet direct imaging community should work to develop a set of theoretical models that are specifically customized for the directly imaged planets, using a treatment of clouds that is physically self-consistent with the underlying chemical networks, as well as uses a physically motivated treatment of atmospheric mixing.

At the same time, the KPIC effort at the W. M. Keck Observatory (Mawet et al. 2016) as well as the HiRISE effort at VLT (Vigan et al. 2018; Otten et al. 2021) aim to couple the high contrast capabilities of the AO systems with high resolution spectrographs such as NIRSPEC (Keck) or CRIRES (VLT), boosting the spectral resolution to 30,000–100,000. Commonly obtaining spectroscopy at these resolutions for directly imaged planets will potentially be transformative for this discipline. Therefore, theoretical atmospheric models that incorporate up-to-date lists of molecular lines (Tennyson & Yurchenko 2012), as well as a treatment of disequilibrium chemistry that is self-consistent with the treatment of condensate formation, will be especially important for the correct interpretation of spectra at this resolution. Similarly, improved theoretical models of planet formation that take into account the underlying disk chemistry will also be especially crucial to achieve the goal of testing the formation locations of giant planets.

4. *What techniques need to be developed in the next 8–10 years to ensure that an Earth-like planet can be directly imaged on one of the Giant Segmented Mirror Telescopes?*

One of the singular achievements of exoplanetary science in the last several years is the determination that small ("super-Earth", or "mini-Neptune") planets are common (Dressing & Charbonneau 2013), and the identification of numerous more such objects is expected to proceed in the coming years. Figure 5.5 shows that the contrast relative to a solar-type host star between a ∼300 K planet in the optical wavelengths is expected to be $10^8$ or more. However, in the thermal infrared (∼10 μm) the declining flux of the host star in the Rayleigh–Jeans tail of its Spectral Energy Distribution

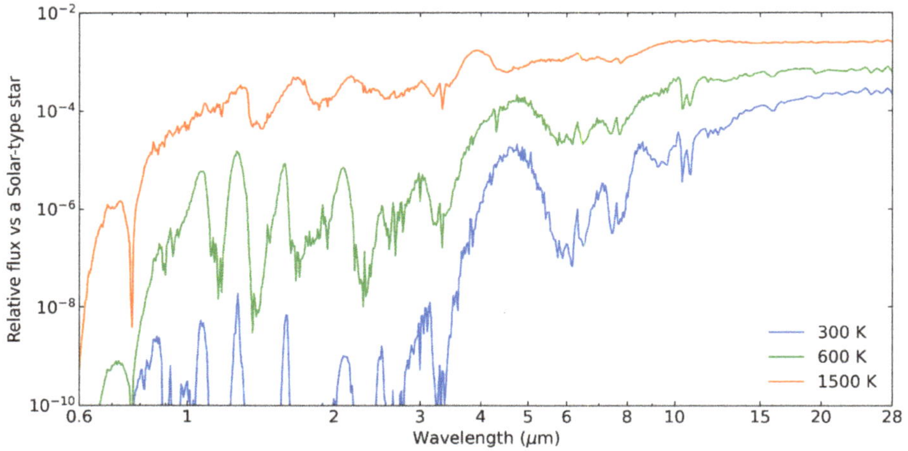


**Figure 5.5.** The predicted contrast ratio relative to a solar-type star for planets of varying effective temperatures computed using the exoplanet atmosphere models of (Phillips et al. 2020) and the PHOENIX stellar models of (Husser et al. 2013). At visible and near-infrared wavelengths, the required contrast to image a temperate (~300 K) exoplanet can be prohibitively large ($10^8$ or greater). However, in the thermal infrared (~10 μm), the contrast ratio is much more favorable, and within reach of the theoretical predicted performance of instruments such as METIS on the upcoming 39 m Extremely Large Telescope (Brandl et al. 2018). Figure courtesy of Aarynn Carter (Exeter University).

(SED), combined with the increasing intrinsic thermal emission from the planet itself, results in a predicted relative contrast that should be achievable with upcoming thermal infrared instruments such as METIS on the European Extremely Large Telescope (ELT, Brandl et al. 2018). Indeed, the left panel of Figure 5.6 shows the relative 10 μm contrasts for a collection of known, and hypothesized terrestrial planets orbiting nearby stars, as well as the expected discovery space for METIS, which encompasses several of these. Furthermore, as Figure 5.5 shows, instruments that are able to reach relative contrasts of $10^5$–$10^6$ in the thermal infrared will have sensitivity to a broad range of planetary mass companions ranging from ~300 K objects to ~1500 K super-Jovian planets analogous to the directly imaged planets currently being detected. At the same time, Figure 5.6 demonstrates that PCS will be designed to have the capability (Kasper et al. 2010, 2013) to achieve the very high contrast ($>10^{-7}$) needed in the optical and near-infrared to detect "exo-Earth" planets orbiting the nearest M-type stars, where the habitable zones are significantly closer to their host stars (Figure 5.6, right panel).

## 5.4 Opportunities

This chapter has so far touched on several potential opportunities for the direct imaging of extrasolar planets. In this section we summarize those opportunities that

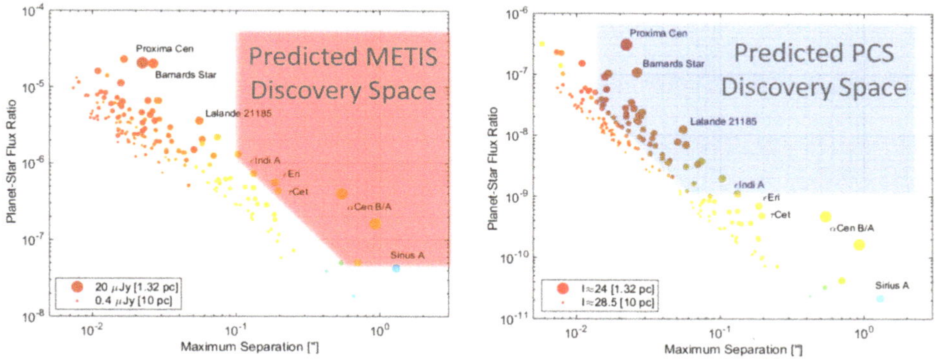


**Figure 5.6.** The expected discovery spaces for a set of both fictitious and actual Earth-sized planets in the habitable zones of all stars within 10pc. Two panels are shown with relative planet/host contrast shown for 10 microns (left panel), and I-band (right panel). Planets associated with M dwarf host stars are shown in red while those for earlier type stars are shown in increasingly bluer colors. The left panel shows the expected discovery space for METIS indicating that a handful of Earth-like planets in the habitable zones orbiting early type stars should be reachable with METIS. In the more distant future, the Planetary Camera and Spectrograph (PCS) should be capable of detecting and characterizing spectroscopically a handful of Earth-like plants around the closest M-dwarfs (e.g., Proxima b, Barnard's star).

are immediately available, those that will be available in the intermediate term (the next five to ten years), and those that will be available in the more distant future.

- **Immediate opportunities:** A wealth of instrumentation currently exists not only for the spectroscopic characterization of the atmospheres of exoplanets, but also for the characterization of their dynamical properties. The SPHERE instrument is finishing its large survey of southern stars, and GPI will migrate to Mauna Kea in the Northern Hemisphere providing access to a completely new set of stars, complementing other Northern Hemisphere efforts such as SCExAO. At the same time, the GRAVITY instrument operating at the VLTI must continue to be exploited for its remarkable astrometric precision at small angular separations (Gravity Collaboration et al. 2019), to characterize the planets detected by other direct (SPHERE and GPI) or indirect (astrometry, RV) techniques. Information about the eccentricities of these orbits may provide vital information on the formation mechanisms of substellar objects. In the near future, much deeper characterization of exoplanet atmospheres should be possible at much higher spectral resolutions using the HiRISE (Vigan et al. 2018; Otten et al. 2021) and KPIC (Mawet et al. 2016) platforms at VLT and Keck, respectively. This much higher spectral resolution will allow not just more precise characterization of the atmospheres of these objects, but potentially also quantities such as their rotation rates (Snellen et al. 2014), and atmospheric dynamics.
- **Intermediate-term opportunities (5–10 years):** In the next 5–10 years, JWST will be fully operational, and an organized effort by the direct imaging community must be initiated to carry out a dedicated survey of nearby, young stars with the goal of identifying the frequency of Saturn/Neptune analogs at

10–1000au. Planning for this survey needs to consider the latest in-flight performance of the observatory as determined by the JWST Early Release Science and GTO programs (Hinkley et al. 2018), but also the updated understandings of the exoplanet SEDs at totally new wavelengths (>5 μm) that these programs will provide. This intermediate timeframe should also witness several more data releases from Gaia, pointing the way to thousands of long-period planets (Sozzetti et al. 2014) with dynamically-measured masses in need of direct characterization with the facilities described in this chapter. In the same timeframe, in the mid/late-2020s the ELT will be fully operational at Cerro Armazones, and the first-generation suite of instruments will have the sensitivity and resolution to not only obtain the first direct images of the Jovian-mass planets detected by the RV technique, but also reach into snow-line separations using advanced coronagraphs and techniques such as Sparse Aperture Masking (Hinkley et al. 2011, 2015), and molecule mapping (Snellen et al. 2010, 2014; Hoeijmakers et al. 2018).

- **Longer term:** Beyond a decade from now, the landscape for exoplanet direct imaging will have changed significantly. JWST will be reaching the end of its mission, dedicated ground-based instruments such as GPI and SPHERE will have easily completed their initial surveys, and the GSMTs will be operational. As discussed above, the GSMTs will possess the sensitivity and resolution necessary to directly image a handful of terrestrial planets orbiting the nearest stars. Specifically, METIS on the ELT will be the first instrument with sensitivity in the thermal infrared, providing sensitivity not only to Jovian-mass exoplanets, but also a small number of temperate, terrestrial planets in the habitable zones of solar-type host stars. Looking forward into the 2030s and beyond, the Planetary Camera and Spectrograph (PCS) will be capable of detecting and characterizing a handful of Earth-like planets around the closest M-dwarfs (Figure 5.6). These two efforts at the ELT will be highly complementary, providing direct imaging access to temperate exoplanets in the habitable zones of a variety of host star spectral types.

## 5.5 Challenges

Despite the very optimistic outlook for the technique of exoplanet direct imaging, there still exist a number of challenges that are fundamental to this technique. Below we highlight a few of those that relate to the points in this article, but the list is by no means exhaustive.

- **Old Versus Young Populations:** Due to the enhanced luminosities of planets at young ages, the direct imaging technique has always had the technical constraint of observing the youngest possible systems, providing a valuable "snapshot" of the properties of nascent exoplanetary systems. However, the population of planets studied by early RV surveys have ages typical of solar-type stars (~1–10 Gyr), and may have had their initial conditions significantly altered by subsequent dynamical processes (e.g., Chatterjee et al. 2008; Scharf & Menou 2009). As discussed previously, upcoming direct imaging instrumentation on the

GSMTs will possess the angular resolution and sensitivity to detect numerous planets initially identified through RV surveys (Quanz et al. 2015). This presents an extraordinary opportunity to compare the demographics (e.g., overall frequencies, orbital distributions, mass distributions, etc.) of these two populations of exoplanets. Thus, any discrepancies that arise when comparing the demographics of these two populations of planets must take into account these differing ages.

- **Spectroscopy of Earth-like Planets:** As discussed above, METIS on the ELT will have the capability to return broadband 10 μm detections of a handful of terrestrial planets in their habitable zones. And while such a single photometric detection would be an unambiguous breakthrough, true *characterization* of these planets will require spectroscopy in the vicinity of 10 μm. Thus, every effort should be made now to equip future GSMT instrumentation operating in the thermal infrared with some spectroscopic capabilities. It is not until PCS sees first-light in the 2030s that true spectroscopic characterization of terrestrial planets at visible wavelengths will be a possibility.
- **Improved Theoretical Atmosphere Models to Derive Intrinsic Planetary Compositions:** Numerous attempts over the last decade have been made to determine the bulk compositions of planets, often using the metric of the carbon-to-oxygen (C/O) ratio, with the goal of gaining insight into the formation history of these objects. However, an ambiguity remains within the modeling community about to what degree the derived compositions from direct spectroscopy truly reflect the intrinsic chemical abundances in the atmosphere, and to what degree these abundances have been affected by inadequate treatments of processes like atmospheric disequilibrium chemistry, or the degree of vertical atmospheric mixing. Processes such as these will have a dramatic impact on the derived atmospheric abundances, and this therefore should form a significant modeling effort going forward.

The enormous gains made within the discipline of exoplanet imaging in the last 20 years have established it as one of the leading techniques for exoplanet characterization. But the next ten years, and subsequent decades, are poised to show astonishing gains within the direct imaging of exoplanets, including: characterizing planets over their full luminous range; establishing the frequency of Saturn and Neptune analogs at wide orbital separations; and potentially exhibiting the first direct detection of a terrestrial planet orbiting a nearby star. Such characterization work will be an essential step on the path to searching for the first unambiguous evidence of biosignatures in a small number of very nearby planets.

## References

Angel, J. R. P. 1994, Natur, 368, 203

Barman, T. S., Macintosh, B., Konopacky, Q. M., & Marois, C. 2011, ApJ, 733, 65

Beuzit, J. L., Vigan, A., Mouillet, D., et al. 2019, A&A, 631, A155

Biller, B. A., Close, L. M., Masciadri, E., et al. 2007, ApJS, 173, 143

Biller, B. A., Liu, M. C., Wahhaj, Z., et al. 2013, ApJ, 777, 160
Bloemhof, E. E., Dekany, R. G., Troy, M., & Oppenheimer, B. R. 2001, ApJ, 558, L71
Bowler, B. P. 2016, PASP, 128, 102001
Bowler, B. P., Blunt, S. C., & Nielsen, E. 2020, AJ, 159, 63
Brandl, B. R., Absil, O., Agócs, T., et al. 2018, Proc. SPIE, 10702, 107021U
Brandt, T. D., Kuzuhara, M., McElwain, M. W., et al. 2014, ApJ, 786, 1
Carson, J. C., Hiner, K. D., Villar, G. G., et al. 2009, AJ, 137, 218
Carter, A. L., Hinkley, S., Bonavita, M., et al. 2021, MNRAS, 501, 1999
Charnay, B., Bézard, B., Baudino, J. L., et al. 2018, ApJ, 854, 172
Chatterjee, S., Ford, E. B., Matsumura, S., & Rasio, F. A. 2008, ApJ, 686, 580
Chauvin, G., Desidera, S., Lagrange, A.-M., et al. 2017, A&A, 605, L9
Chauvin, G., Lagrange, A.-M., Bonavita, M., et al. 2010, A&A, 509, A52
Chauvin, G., Lagrange, A.-M., Dumas, C., et al. 2005, A&A, 438, L25
Dressing, C. D., & Charbonneau, D. 2013, ApJ, 767, 95
Durkan, S., Janson, M., & Carson, J. C. 2016, ApJ, 824, 58
Fernandes, R. B., Mulders, G. D., Pascucci, I., Mordasini, C., & Emsenhuber, A. 2019, AJ, 874, 81
Fortney, J. J., Marley, M. S., Saumon, D., & Lodders, K. 2008, ApJ, 683, 1104
Frelikh, R., Jang, H., Murray-Clay, R. A., & Petrovich, C. 2019, ApJ, 884, L47
Fulton, B., Rosenthal, L., Hirsch, L., et al. 2021, ApJS, 255, 14
Hinkley, S., Carpenter, J. M., Ireland, M. J., & Kraus, A. L. 2011, ApJ, 730, L21
Hinkley, S., Kraus, A. L., Ireland, M. J., et al. 2015, ApJ, 806, L9
Hinkley, S., Oppenheimer, B. R., Soummer, R., et al. 2007, ApJ, 654, 633
Hinkley, S., Oppenheimer, B. R., Zimmerman, N., et al. 2011, PASP, 123, 74
Hinkley, S., Pueyo, L., Faherty, J. K., et al. 2013, ApJ, 779, 153
Hinkley, S., Carter, A., Biller, B., & Skemer, A. 2018, in European Planetary Science Congress, EPSC2018
Hoeijmakers, H. J., Schwarz, H., Snellen, I. A. G., et al. 2018, A&A, 617, A144
Husser, T. O., Wende-von Berg, S., Dreizler, S., et al. 2013, A&A, 553, A6
Janson, M., Quanz, S. P., Carson, J. C., et al. 2015, A&A, 574, A120
Jovanovic, N., Martinache, F., Guyon, O., et al. 2015, PASP, 127, 890
Kasper, M., Apai, D., Janson, M., & Brandner, W. 2007, A&A, 472, 321
Kasper, M., Beuzit, J.-L., Verinaud, C., et al. 2010, Proc. SPIE, 7735, 77352E
Kasper, M., Verinaud, C., & Mawet, D. 2013, in Proc. Third AO4ELT Conf., ed. S. Esposito, & L. Fini, 8
Keppler, M., Benisty, M., Müller, A., et al. 2018, A&A, 617, A44
Konopacky, Q. M., Barman, T. S., Macintosh, B. A., & Marois, C. 2013, Sci, 339, 1398
Labeyrie, A. 1995, A&A, 298, 544
Lacour, S., Nowak, M., et al. 2019, Gravity Collaboration, A&A, 623, L11
Lagrange, A.-M., Bonnefoy, M., Chauvin, G., et al. 2010, Sci, 329, 57
Lowrance, P. J., Becklin, E. E., Schneider, G., et al. 2005, AJ, 130, 1845
Luhman, K. L., McLeod, K. K., & Goldenson, N. 2005, ApJ, 623, 1141
Macintosh, B., Graham, J. R., Barman, T., et al. 2015, Sci, 350, 64
Macintosh, B., Graham, J. R., Ingraham, P., et al. 2014, PNAS, 111, 12661
Marois, C., Lafrenière, D., Doyon, R., Macintosh, B., & Nadeau, D. 2006, ApJ, 641, 556
Marois, C., Macintosh, B., Barman, T., et al. 2008, Sci, 322, 1348

Marois, C., Zuckerman, B., Konopacky, Q. M., Macintosh, B., & Barman, T. 2010, Natur, 468, 1080
Masciadri, E., Mundt, R., Henning, T., Alvarez, C., & Barrado, D. 2005, ApJ, 625, 1004
Mawet, D., Bond, C. Z., Delorme, J. R., et al. 2018, Proc. SPIE, 10703, 1070306
Mawet, D., Wizinowich, P., Dekany, R., et al. 2016, Proc. SPIE, 9909, 99090D
Metchev, S. A., & Hillenbrand, L. A. 2009, ApJS, 181, 62
Mollière, P., Stolker, T., Lacour, S., et al. 2020, A&A, 640, A131
Moses, J. I., Marley, M. S., Zahnle, K., et al. 2016, ApJ, 829, 66
Nakajima, T., Oppenheimer, B. R., Kulkarni, S. R., et al. 1995, Natur, 378, 463
Nielsen, E. L., Close, L. M., Biller, B. A., Masciadri, E., & Lenzen, R. 2008, ApJ, 674, 466
Nielsen, E. L., Rosa, R. J. D., Macintosh, B., et al. 2019, AJ, 158, 13
Nowak, M., Lacour, S., Lagrange, A. M., et al. 2020, A&A, 642, L2
Oppenheimer, B. R., Golimowski, D. A., Kulkarni, S. R., et al. 2001, AJ, 121, 2189
Otten, G. P. P. L., Vigan, A., Muslimov, E., et al. 2021, A&A, 646, A150
Perrin, M. D., Pueyo, L., Van Gorkom, K., et al. 2018, Proc. SPIE, 10698, 1069809
Phillips, M. W., Tremblin, P., Baraffe, I., et al. 2020, A&A, 637, A38
Quanz, S. P., Crossfield, I., Meyer, M. R., Schmalzl, E., & Held, J. 2015, JAsB, 14, 279
Racine, R., Walker, G. A. H., Nadeau, D., Doyon, R., & Marois, C. 1999, PASP, 111, 587
Scharf, C., & Menou, K. 2009, ApJ, 693, L113
Skemer, A. J., Hinz, P. M., Esposito, S., et al. 2012, ApJ, 753, 14
Snellen, I. A. G., Brandl, B. R., de Kok, R. J., et al. 2014, Natur, 509, 63
Snellen, I. A. G., de Kok, R. J., de Mooij, E. J. W., & Albrecht, S. 2010, Natur, 465, 1049
Sozzetti, A., Giacobbe, P., Lattanzi, M. G., et al. 2014, MNRAS, 437, 497
Spiegel, D. S., & Burrows, A. 2012, ApJ, 745, 174
Steinbring, E., Faber, S. M., Hinkley, S., et al. 2002, PASP, 114, 1267
Tamura, M. 2016, PJAB, 92, 45
Tennyson, J., & Yurchenko, S. N. 2012, MNRAS, 425, 21
Vigan, A., Bonavita, M., Biller, B., et al. 2017, A&A, 603, A3
Vigan, A., Fontanive, C., Meyer, M., et al. 2021, A&A, 651, A72
Vigan, A., Otten, G. P. P. L., Muslimov, E., et al. 2018, Proc. SPIE, 10702, 1070236
Vigan, A., Patience, J., Marois, C., et al. 2012, A&A, 544, A9
Zahnle, K. J., & Marley, M. S. 2014, ApJ, 797, 41

# Chapter 6

## Small Star Opportunities

**Amaury H. M. J. Triaud**

### 6.1 Introduction and State of the Art

The *small star opportunity* usually refers to considering low-mass stars of M dwarf spectral type in order to improve the prospect of planet detection. This concept underpinned the design of the MEarth transit survey (Nutzman & Charbonneau 2008). It has since been invoked by other surveys such as SPECULOOS, extending the concept to increasingly smaller stars, reaching late M dwarfs (Delrez et al. 2018) and brown dwarfs (He et al. 2017), but also white dwarfs (Faedi et al. 2011). The small star opportunity is being exploited by Doppler and by transit surveys, which led to spectacular discoveries on solar-neighborhood M dwarfs such as GJ 1214 (Charbonneau et al. 2009), Proxima Centauri b (Anglada-Escudé et al. 2016), Teegarden's star (Zechmeister et al. 2019), and the TRAPPIST-1 system (Gillon et al. 2016, 2017) to name only a few. Investigations for the planets' atmospheres have been immediately launched (Bean et al. 2011; de Wit et al. 2016; Southworth et al. 2017; Benneke et al. 2019) demonstrating the relevance and promise of the *small star opportunity*. Meanwhile the TESS mission has dedicated a large number of its 2 minutes cadence windows to survey M dwarfs. While the *small star opportunity* is often seen as an opportunity of detection, or atmospheric characterization, small stars also offer multiple opportunities to solve other problems, such as planet formation. However, M dwarfs also bring a number of setbacks that need to be overcome to fully exploit the many advantages they present. Despite those, the small star opportunities are a unique chance to understand how rocky worlds form, how frequently they do so, and what range of atmospheric and internal composition rocky worlds have. This paves the way to the detection of an Earth-twin by direct imaging, and the interpretation of its spectrum.

### 6.2 Opportunities and Challenges

The concept of a *small star opportunity* is often confined to the detectability of planets, particularly as a fast-track approach toward the identification of temperate

doi:10.1088/2514-3433/abfa8fch6  

rocky worlds, whose atmospheres can be remotely explored. There is no better way than to use an example to illustrate it: a $1 M_\oplus$, $1 R_\oplus$ planet, orbiting a 0.1 $M_\odot$, 0.1 $R_\odot$, 2500 K star (where $\oplus$ denote Earth, and $\odot$ is the Sun). To receive the same flux as Earth receives form the Sun, that planet has an orbital period of 2.96 days, and a semimajor axis of 0.019 au. Compared to an Earth-twin orbiting a Solar-twin, the small star opportunities consist of:

- Opportunity 1: the planet produces a transit signal 100× larger (allowing ground-based telescopes) because the transit depth scales with $r_\mathrm{p}^2/R_\star^2$.
- Opportunity 2: the planet has an orbital period >100× shorter (more efficient surveys), because the semimajor axis scales as: $a = 1.0\,(T_\star/T_\odot)^2\,(R_\star/R_\odot)$ au.
- Opportunity 3: the transit probability is enhanced by 5× (fewer stars to survey) because Prob $\propto R_\star/a$.
- Opportunity 4: the Doppler signal of the planet is 23× stronger because the semi-amplitude scales as: $K_\star = 9.0\,(P/365\mathrm{d})^{-1/3}\,(m_\mathrm{p}/m_\oplus)\,(M_\star/M_\odot)^{-2/3}$ cm s$^{-1}$.
- Opportunity 5: the transmission spectrum features are 100× stronger because they scale as: $r_\mathrm{p}/R_\star^2$.
- Opportunity 6: the secondary occultation is 200× stronger because, in the Rayleigh–Jeans end of the spectrum, its depth scales as: $(r_\mathrm{p}^2/R_\star^2)\,(T_\mathrm{p}/T_\star)$.
- Opportunity 7: the reflection spectrum is 2500× stronger because it scales as $r_\mathrm{p}^2/a^2$.

Some of these opportunities are unfortunately moderated by the intrinsically low brightness of M dwarfs. Using the example above, the star emits a flux which is 2850× smaller than a Sun-like star however, the Doppler signal of a Earth-like planet is 23× larger. A back of the envelope calculation with simplifying assumptions (constant radial-velocity information with wavelength, and observing the whole spectrum) informs that discovering an Earth-like planet around a 0.1 $M_\odot$ star would require of order 5× more observing time (at least). The small star opportunities therefore arise not from simplicity, but because astronomers are faced with some intrinsic limitations. For over a decade the stability of high-resolution spectrographs was capped at 1 m s$^{-1}$, and ground-based survey photometry could not do better than 500 ppm. The small star opportunities allow to seek Earth-mass planets whose signatures exceed those instrumental constraints.

Probably the most important advantage of the small star opportunity is that temperate orbits are significantly shorter around low-mass stars. Short-period signals are usually easier to extract from noisy data since their coherence can be tested over multiple cycles rapidly. In an important practical reason, short surveys are easier to motivate than multi-decadal efforts to funding agencies. In addition, short orbital periods come at a significant advantage when it comes to the atmospheric characterization, particularly in the case of spacecraft with limited lifetimes and/or no full-time, full-sky coverage such as the James Webb.

Despite obvious advantages, observing low-mass stars remains a difficult task because such stars are faint, particularly in visible wavelengths, and especially for the later spectral types. A case in point is TESS, whose specifications imply that it can detect planets that are either Earth-sized, or within the habitable-zone, but will struggle to identify planets that are Earth-sized *and* in the habitable-zone, on account of its small aperture (Sullivan et al. 2015). M dwarfs of spectral type M6 and later are too faint to achieve the correct precision to reach an Earth-size, except for the very shortest orbits (Vanderspek et al. 2019; Crossfield et al. 2019). SPECULOOS, a ground-based experiment can remain competitive with TESS because it collects 100× more flux (Delrez et al. 2018; Murray et al. 2020).

To benefit from the small star opportunities, many technological developments were needed. Spectrographs have taken a long technological development to achieve 1 m $s^{-1}$ into the near-infrared with CARMENES (Quirrenbach et al. 2018), SPIRou (Artigau et al. 2014), NIRPS (Wildi et al. 2017), and HPF (Mahadevan et al. 2012). Reaching late M dwarfs was possible for photometric transit surveys like SPECULOOS thanks to two important developments, 1- deep-depleted CCDs reaching quantum efficiencies around 90% at 900 nm (also used by TESS), and 2-medium-sized robotic telescopes at acceptable prices (Delrez et al. 2018). Alternatively, near-infrared detectors are needed, (a choice made by ExTrA; Bonfils et al. 2015), or larger telescope mirrors (EDEN[1], Tierras[2]). If instead of M dwarfs the focus is to survey a meaningful number of brown dwarfs, both larger mirrors, and near-infrared detectors are then required (Triaud et al. 2013; Tamburo & Muirhead 2019), or the use of a spacecraft (Triaud et al. 2013).

All opportunities above are purely about detection and characterization. Other opportunities arise thanks to the type and amount of planets discovered orbiting low-mass stars. Those opportunities are aligned with the major goals of the field.

## 6.3 Important Questions and Goals

**Will M dwarfs offer a less biased view of planet demographics?** Most exoplanet surveys, Doppler or transit have so far focused on Sun-like stars (0.9–1.1 $M_\odot$), which represents roughly 15% of the content of our Galaxy (Henry et al. 2018). Since half of those are binaries (Duquennoy & Mayor 1991), which are usually rejected, current surveys really cover only about 8% of stellar types. In comparison, M dwarfs represent 70% of stellar content of our Galaxy, and stars with masses <0.25 $M_\odot$ are 50% of the whole (Henry et al. 2018). M dwarfs have a lower binary fraction (∼30%). Several opportunities arise from this. Planets discovered orbiting stars <0.25 $M_\odot$ are more representative of planet formation processes at large, than planets discovered orbiting single Sun-like stars. In addition, larger number of stars means that late M dwarfs offset in some way the fact that they are intrinsically fainter. Assuming similar occurrence rates, a transiting Earth-like planet will be hosted by a star <0.25 $M_\odot$ that is ∼4× closer to the solar system, than a transiting Earth-twin.

**Are rocky planets are more frequent around low-mass stars?** Doppler (Bonfils et al. 2013) and transit surveys (Dressing & Charbonneau 2015) both indicate that the fraction of stars hosting small planets increases with decreasing mass. Proxima

[1] http://project-eden.space/eden/

[2] https://tsc.mit.edu/talks/JULIANA%20GARCIA-MEJIA.pdf

Centauri and TRAPPIST-1 have stellar properties which are, to a first order, similar. Combining those two detections, it is possible to derive a rough occurrence rate for habitable-zone rocky planets. The geometric probability that the temperate, Earth-mass Proxima Centauri b would transit is 1.5%. This means that only one in 66 very low-mass stars has its planetary orbital plane aligned such that a habitable-zone planet can transit. The RECONS survey counted 93 stars within 10pc, with a mass less than 0.15 $M_{\odot}$ (Henry et al. 2018). This implies of order 160 very low-mass stars within a volume that comprises TRAPPIST-1 (12pc). Depending on whether we assume that TRAPPIST-1e is the only *transiting* temperate planet in the Southern Hemisphere, or the only one in the whole sky (within 12pc), then this transit detection implies that habitable-zone rocky planets likely exist between 40% and 80% of very low-mass stars—a massive opportunity, but also a crucial aspect in need of urgent verification. By combining the high frequency of low-mass stars to the occurrence rate of small planets orbiting them, there might exist $\sim 10\times$ more Earth-size planets at the bottom of the main sequence than around Solar analogs. Studying low-mass stars is to access the most common Earth-sized planets. The access to important number of rocky worlds, combined with the various observing opportunities described above means that we can explore the atmospheric diversity of rocky planet environments with upcoming facilities.

**Do close-in rocky planets orbiting low-mass stars have atmospheres?** Ram pressure from stellar wind, lack of shielding planetary magnetic fields due to synchronous, or near-synchronously rotating planets, cometary impacts, flares, and a Gyr long pre-main sequence phase are all dangers for the presence of atmospheres (Lingam & Loeb 2019). The main promise of the small star opportunity is the chance to remotely-explore alien environments comparable to those of Venus, Earth, and Titan, using JWST and the E-ELT (Morley et al. 2017; Lustig-Yaeger et al. 2019). The lack of atmospheres would prevent this science case. Secondary atmospheres could be reconstructed from interior de-gassing (Gaillard & Scaillet 2014), which in itself would grant astronomers a glimpse of the internal composition of the planets. Cometary impacts have both a destructive and constructive effect (Kral et al. 2018), and can reconstruct a secondary atmosphere too. Finally atmospheric loss can be prevented if the original atmospheric reservoir was sufficiently large to begin with (Bolmont et al. 2017). Whether or not rocky worlds orbiting small stars have atmospheres can only be resolved by observations. The expectation is that the presence of an atmosphere on the TRAPPIST-1 planets can be established rapidly in a handful of JWST transits/occultations. Two rocky planets have had observations with Spitzer, with 55 Cnc e thought to host an atmosphere (Demory et al. 2016), and GJ 1132b interpreted as atmosphere-less (Kreidberg et al. 2019).

**Will planets orbiting low-mass stars help develop a universal law of planet formation?** Understanding the many physical processes behind planet formation and orbital evolution is one of the main drivers behind exoplanet investigations. True insight in physics is when a law applies over order of magnitudes. Therefore any attempt at modeling the planet population found orbiting single Sun-like stars (Mordasini et al. 2009) ought to be applied over orders of magnitude. Since discovering planets orbiting 10 $M_{\odot}$ stars is unlikely due to their high irradiation, short lifetime, and low occurrence rate, we therefore have to explore masses below the Sun's. The discovery of TRAPPIST-1 marked a milestone in being the first transiting system whose host was more than one order of magnitude smaller than the Sun (Gillon et al. 2017). As such ultra-cool dwarfs provide a unique opportunity to explore the outcome of planet formation. Summing the masses of the TRAPPIST-1 planets (Grimm et al. 2018) we obtain $\sim 5.7 M_{\oplus}$, which means there is a ratio $\sim 2 \times 10^{-4}$ between the amount of metals locked into planets and the stellar mass. A similar calculation for the solar system (assuming 10 $M_{\oplus}$ of metals in each of the gas giants), gives a ratio $\sim 1.1 \times 10^{-4}$. This ratio is also similar to the ratio between the total satellite mass around Jupiter, to the mass of Jupiter. This is also seen for Saturn, Uranus, and Neptune. The mass ratio of the TRAPPIST-1 planets to their star is comparable to the mass ratio of super-Earths to their (usually) Sun-like star (Grimm et al. 2018). From the Sun to Neptune, via TRAPPIST-1 and exoplanets we observe a similar fraction of metals locked into orbiting bodies. Additional detections of planets orbiting low-mass stars will verify whether this is a coincidence or related to a fundamental aspect of how disc behave and how orbiting bodies form within them. A ratio of $10^{-4}$ is the number routinely assumed for the dust mass within a protoplanetary disc, itself assumed to be 1% of the stellar mass. This has two corollaries, either planet formation is an extremely efficient process, able to convert of order 100% of dust into planets, or, planet formation happens early in the history of the system, when the disc mass is closer to 10% of its host's mass.

## References

Anglada-Escudé, G., Amado, P. J., Barnes, J., et al. 2016, Natur, 536, 437
Artigau, É., Kouach, D., Donati, J.-F., et al. 2014, Proc. SPIE, 9147, 914715
Bean, J. L., Désert, J.-M., Kabath, P., et al. 2011, ApJ, 743, 92
Benneke, B., Knutson, H. A., Lothringer, J., et al. 2019, NatAs, 3, 813
Bolmont, E., Selsis, F., Owen, J. E., et al. 2017, MNRAS, 464, 3728
Bonfils, X., Almenara, J. M., Jocou, L., et al. 2015, Proc. SPIE, 9605, 96051L
Bonfils, X., Delfosse, X., Udry, S., et al. 2013, A&A, 549, A109
Charbonneau, D., Berta, Z. K., Irwin, J., et al. 2009, Natur, 462, 891
Crossfield, I. J. M., Waalkes, W., Newton, E. R., et al. 2019, ApJL, 883, L16
de Wit, J., Wakeford, H. R., Gillon, M., et al. 2016, Natur, 537, 69
Delrez, L., Gillon, M., Queloz, D., et al. 2018, Proc. SPIE, 10700, 107001I
Demory, B.-O., Gillon, M., Madhusudhan, N., & Queloz, D. 2016, MNRAS, 455, 2018
Dressing, C. D., & Charbonneau, D. 2015, ApJ, 807, 45
Duquennoy, A., & Mayor, M. 1991, A&A, 500, 337
Faedi, F., West, R. G., Burleigh, M. R., Goad, M. R., & Hebb, L. 2011, MNRAS, 410, 899
Gaillard, F., & Scaillet, B. 2014, E&PSL, 403, 307
Gillon, M., Jehin, E., Lederer, S. M., et al. 2016, Natur, 533, 221
Gillon, M., Triaud, A. H. M. J., & Demory, B.-O. 2017, Natur, 542, 456
Grimm, S. L., Demory, B.-O., Gillon, M., et al. 2018, A&A, 613, A68
He, M. Y., Triaud, A. H. M. J., & Gillon, M. 2017, MNRAS, 464, 2687
Henry, T. J., Jao, W.-C., Winters, J. G., et al. 2018, AJ, 155, 265
Kral, Q., Wyatt, M. C., Triaud, A. H. M. J., et al. 2018, MNRAS, 479, 2649

Kreidberg, L., Koll, D. D. B., & Morley, C. 2019, Natur, 573, 87
Lingam, M., & Loeb, A. 2019, RvMP, 91, 021002
Lustig-Yaeger, J., Meadows, V. S., & Lincowski, A. P. 2019, AJ, 158, 27
Mahadevan, S., Ramsey, L., Bender, C., et al. 2012, Proc. SPIE, 8446, 84461S
Mordasini, C., Alibert, Y., & Benz, W. 2009, A&A, 501, 1139
Morley, C. V., Kreidberg, L., Rustamkulov, Z., Robinson, T., & Fortney, J. J. 2017, ApJ, 850, 121
Murray, C. A., Delrez, L., Pedersen, P. P., et al. 2020, MNRAS, 495, 2446
Nutzman, P., & Charbonneau, D. 2008, PASP, 120, 317
Quirrenbach, A., Amado, P. J., Ribas, I., et al. 2018, Proc. SPIE, 10702, 107020W
Southworth, J., Mancini, L., Madhusudhan, N., et al. 2017, AJ, 153, 191
Sullivan, P. W., Winn, J. N., Berta-Thompson, Z. K., et al. 2015, ApJ, 809, 77
Tamburo, P., & Muirhead, P. S. 2019, PASP, 131, 114401
Triaud, A. H. M. J., Gillon, M., Selsis, F., et al. 2013, arXiv:1304.7248
Vanderspek, R., Huang, C. X., Vanderburg, A., et al. 2019, ApJ, 871, L24
Wildi, F., Blind, N., Reshetov, V., et al. 2017, Proc. SPIE, 10400, 1040018
Zechmeister, M., Dreizler, S., Ribas, I., et al. 2019, A&A, 627, A49

# Chapter 7

# Transit Spectroscopy

**Drake Deming, Kevin B. Stevenson and David Ehrenreich**

## 7.1 Introduction

Spectroscopy of light that is transmitted through an exoplanet's atmosphere during transit can reveal the composition of that atmosphere via atomic and molecular absorption features, and cloud/aerosol properties via scattering and/or absorption in the continuum (e.g., Charbonneau et al. 2002; Vidal-Madjar et al. 2003; Deming et al. 2013; Sing et al. 2016). Other atmospheric properties such as temperature can also be deduced from transit spectra. For close-in planets whose rotation is tidally-locked to their orbital period, transit spectra probe the terminator region between day and night. Almost all transiting planets exhibit a secondary eclipse (planet passing behind the star), and observations bracketing the eclipse can reveal the exoplanet's day side thermal emission via photometry (Charbonneau et al. 2005; Deming et al. 2005), and molecular composition via spectroscopy (Grillmair et al. 2008; Todorov et al. 2014). Eclipse spectra probe the day side of a tidally-locked planet, and are more sensitive to the temperature versus height profile in the exoplanetary atmosphere than are transit spectra.

Transit spectroscopy is a useful probe of exoplanetary atmospheres at optical wavelengths where the star is bright, because the star is the illuminating source. Transit spectroscopy can also be favorable at other wavelengths where exoplanetary spectral absorptions are intrinsically strong, such as resonance atomic lines in the ultraviolet (UV), and fundamental vibration–rotation molecular bands in the infrared. Spectroscopy at secondary eclipse is most favorable (in terms of signal-to-noise) at infrared wavelengths approaching the Rayleigh–Jeans limit. In that limit, the Planck function varies linearly with temperature. Linearity of the Planck function (i.e., atmospheric source function in local thermodynamic equilibrium) maximizes the planet-to-star thermal contrast.

doi:10.1088/2514-3433/abfa8fch7  

Transit and eclipse spectroscopy affords us the opportunity to address some of the more prominent questions in exoplanetary science, for example:

- Is there a mass-abundance trend for the atmospheres of exoplanets? We know from masses and radii of transiting planets that less massive planets have higher bulk densities, hence a higher abundance of heavy elements. To what degree that trend extends to their atmospheres and gaseous envelopes is still unclear, and would inform us concerning the nature of planet formation and differentiation.
- What is the diversity of atmospheric properties, processes, and evolution of exoplanets (as a function of their mass)?
- What is the nature and frequency of atmospheres on rocky exoplanets?
- What are the key signatures of important geological (and eventually, biological) processes that we could observe on rocky exoplanets?
- To what degree does liquid water occur on the surfaces of rocky exoplanets?
- What is the impact of the star on the evolution of exoplanets (as a function of their mass and irradiation level)?
- How can we be sure that we have successfully separated exoplanetary signals from stellar activity effects in transit observations?

## 7.2 State of the Art

### 7.2.1 Spectroscopy Using Space-borne Instruments

The first successful application of transit spectroscopy was the detection of atomic sodium in the atmosphere of HD 209458b by Charbonneau et al. (2002). That detection used the Space Telescope Imaging Spectrograph (STIS) on the Hubble Space Telescope (HST). STIS is still an important instrument for transit spectroscopy (e.g., Sing et al. 2016; Alam et al. 2018), but the low-to-medium spectral resolving power of STIS in the optical ($400 \lesssim \lambda/\Delta\lambda \lesssim 8000$) limits the amplitude of the signal in line absorptions (to typically ~0.0001 of the star). In the UV, STIS can provide high spectral resolving power ($10,\ 000 \lesssim \lambda/\Delta\lambda \lesssim 100,\ 000$), allowing in particular to resolve the absorption of neutral hydrogen escaping from hot giant exoplanets (e.g., Vidal-Madjar et al. 2003; des Etangs et al. 2012; Ehrenreich et al. 2015; Bourrier et al. 2018). In spite of limited spectral resolving power at longer wavelengths, space-borne transit spectroscopy using STIS is critical for exoplanetary science because it provides the ability to probe wavelength-dependent scattering due to aerosols.

Spectroscopy of molecules during transit started more slowly than did spectroscopy of strong atomic lines. Water vapor was predicted to be prominent in the spectra of hot Jupiters (Hubbard et al. 2001). Several early detection claims (Barman 2007; Tinetti et al. 2007; Swain et al. 2009) started the observational effort, that expanded greatly when Peter McCullough and John MacKenty implemented spatial scan mode on the Wide Field Camera-3 (WFC3) of HST (see results and technical reports cited in Deming et al. 2013). WFC3 spatial scan has allowed routine measurement of water vapor in hot Jupiters during transit (e.g., Deming et al. 2013; McCullough et al. 2014; Bruno et al. 2018; Kreidberg et al. 2018; Tsiaras et al. 2018; Wakeford et al. 2018). Beyond hot Jupiters, water vapor measurements have been successful for hot Neptunes and sub-Neptunes (Fraine et al. 2014; Wakeford et al. 2018; Benneke et al. 2019), and recently for a temperate sub-Neptune (Benneke et al. 2019; see Figure 7.1).

Because water remains as vapor in the atmospheres of hot planets, and because oxygen is cosmically abundant, water absorption measured using HST/WFC3 can in principle be used as a proxy for the heavy element content (metallicity) of the exoplanetary atmosphere. We expect that, similar to the solar system, atmospheric metallicity will vary inversely with exoplanetary mass. That relation is a signature of planet formation, wherein more massive planets attract and hold more hydrogen during their formation, reducing their atmospheric metallicity. The atmospheric metallicity of about a dozen gaseous exoplanets has been measured, and a tentative relation with mass is shown in Figure 7.2; from Welbanks et al. (2019). They consider the relation separately for water vapor (i.e., oxygen) and sodium and potassium, and compare to the relation for the solar system (based on methane, i.e., carbon). They find a weaker relation for exoplanets using water vapor compared to the solar system; however, some giant exoplanets are enhanced in sodium and potassium, consistent with the solar system trend. However, there are hurdles remaining to overcome in making such measurements, at both transit and secondary eclipse (see Section 7.5 "Challenges").

In addition to spectroscopy using HST, important transit photometry and secondary eclipse photometry/spectroscopy has been performed using Spitzer, either alone or in combination with HST (e.g., Stevenson et al. 2010; Knutson et al. 2011; Todorov et al. 2013; Stevenson et al. 2014; Fraine et al. 2013; Crouzet et al. 2014; McCullough et al. 2014; Fraine et al. 2014; Sing et al. 2015; Kammer et al. 2015; Garhart et al. 2020; Wallack et al. 2019; and many other papers). These studies have inferred atmospheric metallicity values for about a dozen gaseous exoplanets (Figure 7.2).

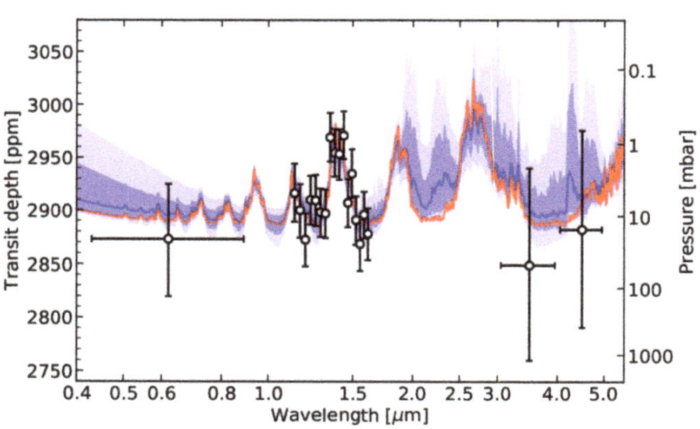


**Figure 7.1.** Transit spectrum of water vapor in the temperate sub-Neptune exoplanet K2-18b. Reproduced from Benneke et al. (2019), © 2019, The American Astronomical Society, All rights reserved. The primary water band accessible to HST/WFC3 peaks at 1.4 μm. Note: absorption by exoplanetary water molecules increases the transit depth, and that is upward on this figure, upside-down from the conventional way that absorption spectra are plotted. Also illustrated on this spectrum is the transit depth from the K2 mission (leftmost point), and transit depths from the two infrared bands of Spitzer (right).

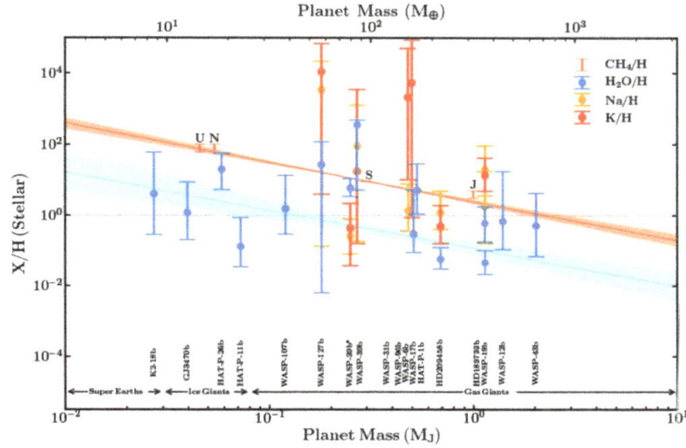


**Figure 7.2.** Relation between exoplanetary mass and atmospheric metallicity (reproduced from Welbanks et al. 2019, © 2019, The American Astronomical Society, All rights reserved, also see Kreidberg et al. 2014; Pinhas et al. 2019; MacDonald & Madhusudhan 2019). For the solar system, there is a strong inverse relation between mass and metallicity. The relation for exoplanets is weaker than the solar system when considering water vapor (i.e., oxygen), but sodium and potassium shown enhancements in exoplanets that are consistent with the trend seen in the solar system.

In the UV, transit spectroscopy has shown that the atmospheres of close-in planets are escaping, both for hot Jupiters (Vidal-Madjar et al. 2003; des Etangs et al. 2012; Fossati et al. 2010; Sing et al. 2019) and for Neptunes (Ehrenreich et al. 2015; Lavie et al. 2017; Bourrier et al. 2018), see Figure 7.3. Observing and understanding the escape process is important for several reasons. The physics of the interaction between the star and the escaping atmosphere is interesting on large scales (tidal stripping, hydrodynamic flow) and micro-scales (charge exchange, non-LTE atomic processes). Moreover, atmospheric escape is believed to play a major role in the evolution of close-in planets, and has specifically shaped the distribution of exoplanetary radii, where gaps appears in the relative frequency of close-in, subjovian planets (the "subjovian (or hot Neptune) evaporation desert") and planets near 1.8 Earth-radii (Fulton et al. 2017; the "super-Earth evaporation valley", Owen 2019). Finally, it is possible that hydrogen-rich exospheres around rocky exoplanets could be used to infer the presence of liquid water on their surfaces, because hydrogen is a product of the photolysis of water vapor (Gómez de Castro et al. 2018; dos Santos et al. 2019; Kislyakova et al. 2019).

### 7.2.2 Spectroscopy Using Ground-based Instruments

Transit spectroscopy from ground-based observatories is complementary to space-borne spectroscopy. Ground-based investigations can utilize larger telescopes to collect more light than is usually possible from space, and ground-based instruments can be very large and capable (e.g., high resolution spectrometers). The high spectral resolving power available in ground-based spectrometers can resolve narrow planetary absorption lines, yielding large amplitude spectral signatures during transit (up to ~0.003 of the star; Nikolov et al. 2018; Seidel et al. 2019). Ground-based spectrometers have also been used for high resolution spectroscopy of hot

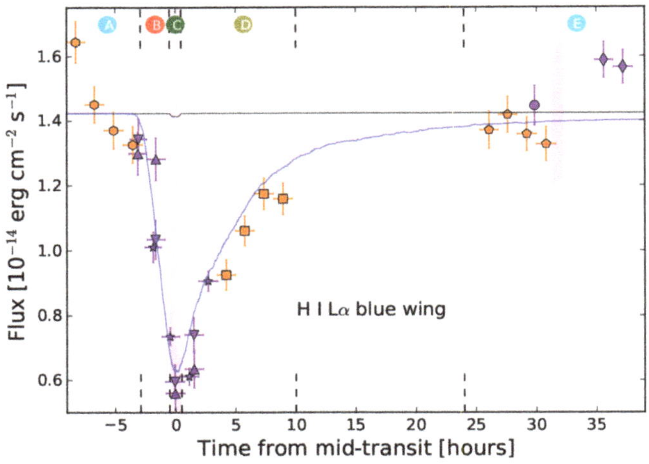


**Figure 7.3.** Absorption in H I Ly$\alpha$ during the transit of the Neptune-sized planet GJ 436b, as observed by Lavie et al. (2017) in the UV using Hubble (reproduced with permission © ESO). Note the large fractional decrease in flux during transit, indicating that the absorbing region is extended well beyond the Roche lobe of the planet. Also note the asymmetric absorption, showing that the escaped hydrogen forms a tail that trails the planet in its orbit, that is shaped by stellar radiation pressure, stellar winds and ionizing radiation (Bourrier et al. 2015, 2016; Vidotto & Bourrier 2017)

Jupiters using the Doppler deconvolution technique (see Deming et al. 2019 for an explanation of that technique).

The new generation of Extremely Large Telescopes (ELTs, Guyon et al. 2012) will have many applications to transit spectroscopy. Combining high resolution spectroscopy with high contrast imaging is a promising method to characterize the atmosphere of the non-transiting (but potentially habitable) planet orbiting the nearby star Proxima Centauri (Lovis et al. 2017). In the more distant future, large ground-based light collectors could arguably detect the biosignature of molecular oxygen in an Earth-like planet transiting a Sun-like star (Snellen et al. 2013).

One prominent exception to the utility of ground-based spectrometers is the ultraviolet domain, mostly inaccessible to ground-based observatories. In addition, the strongest molecular bands in the infrared are often opaque from the ground (e.g., carbon dioxide at 15 μm), requiring that molecular abundances be measured using molecular bands with lower intrinsic strengths, or absorbing from high excitation states (so-called "hot bands"). That can potentially limit ground-based transit spectroscopy to sensing only molecules with high exoplanetary column densities, especially for temperate exoplanets.

## 7.3 Important Questions/Goals

The list of topics in the Introduction are the general and long term questions that we wish to answer. In the immediate term, those long term questions reduce to several specific short-term goals:

1. We need to understand the nature of the potential mass versus atmospheric metallicity relation for gaseous exoplanets (hot Jupiters to Neptunes). The first

consideration is whether the relation (Figure 7.2) is statistically significant, or is a scatter diagram. If it's a scatter diagram, with no statistically significant correlation, our first follow-up hypothesis would be that the metallicities inferred from observations have errors that are larger than expected, possibly due to degeneracies with clouds (see Welbanks & Madhusudhan 2019, for a discussion of cloud-related degeneracies). Aside from observational error, there can be intrinsic scatter in the mass–metallicity relation due to differences in the planet formation process from system to system. That intrinsic scatter is a "signal" that we wish to measure, but it may obscure the reality of the correlation when a group of planets are studied. Understanding the system-to-system differences in planet formation will require improving and thoroughly understanding the observational errors. Forthcoming observations using JWST will be a boon to that effort, especially since the long wavelengths provided by JWST encompass strong molecular bands (e.g., the $\nu$-3 band of methane at 3.3 μm, and the fundamental band of carbon dioxide at 15 μm). JWST may have sufficient signal-to-noise to probe differences in gas-phase and cloud abundances at the morning versus evening terminator (Kempton et al. 2017).

JWST will also provide spectra of the emergent day side radiation of exoplanets, ranging from hot Jupiters potentially down to temperate super-Earths, using secondary eclipse. Secondary eclipses probe the atmosphere using near-vertical radiation paths. Eclipse observations thereby have much less sensitivity to cloud-related degeneracies in interpreting the data than do transit spectra, that are based on long and nearly horizontal slant paths through the atmosphere.

2. We need to understand the thermal structure of hot Jupiter atmospheres, i.e., the relation between temperature and height. That atmospheric structure is determined by the transfer of radiation, hence opacities are central. Opacities per gram of the atmosphere depend on the composition, that in turn can be affected by non-equilibrium processes. Evidence for those non-equilibrium processes includes the puzzling lack of methane in many cool gaseous planets (see below, but also see Guilluy et al. 2019). Garhart et al. (2020) found that Spitzer secondary eclipse photometry using 78 eclipses is inconsistent with current models of hot Jupiter atmospheres, but Baxter et al. (2020) attribute that behavior to the onset of temperature inversions in the hottest planets. The strengths of absorption features in eclipse spectra depend on both the abundance of the species being studied, as well as the temperature gradient in the atmosphere. If we do not understand the temperature gradient, then it will be difficult to derive correct abundances.
3. The nature of atmospheres on hot and temperate super-Earths is largely unknown. There are indications from phase curves (Kreidberg et al. 2019) that the hottest super-Earths are bare rocks, or have at most thin atmospheres. That is not surprising in view of the existence of the evaporation valley (Fulton et al. 2017), and the strong level of irradiation experienced by super-Earths such as LHS3844b. While our information is still very meagre at present, JWST will again enable major progress. JWST has several modes

to detect possible carbon dioxide atmospheres (e.g., Venus-like, or Mars-like), such as using NIRSPEC to observe the strong 4.3 μm band (Deming et al. 2009), or using photometry to measure the fundamental band at 15 μm (Deming et al. 2009; Morley et al. 2017).

4. Where's the methane? One current major puzzle is the apparent absence of methane in gaseous planets with temperatures below ~800K, where thermal equilibrium chemistry predicts that methane is the principal sink for carbon. Observations using Spitzer are currently our strongest constraint on methane, because the strongest methane spectral bands occur at wavelengths longward of Hubble's capability. Moreover, a decade ago, Stevenson et al. (2010) found that the warm Neptune GJ436b lacked methane absorption, using Spitzer photometry at secondary eclipse. Recently, Benneke et al. (2019) found a similar absence in the sub-Neptune GJ3470b, again using Spitzer secondary eclipse photometry (shown as Figure 7.4). The absence of methane also appears very clearly in a large statistical study of Spitzer *transits* (Baxter et al. 2021). Note that an absence of methane in the coolest of the hot Jupiters is consistent with several statistical studies of Spitzer secondary eclipse photometry (Kammer et al. 2015; Garhart et al. 2020; Wallack et al. 2019), and is possibly related to the thermal structure problem mentioned above.

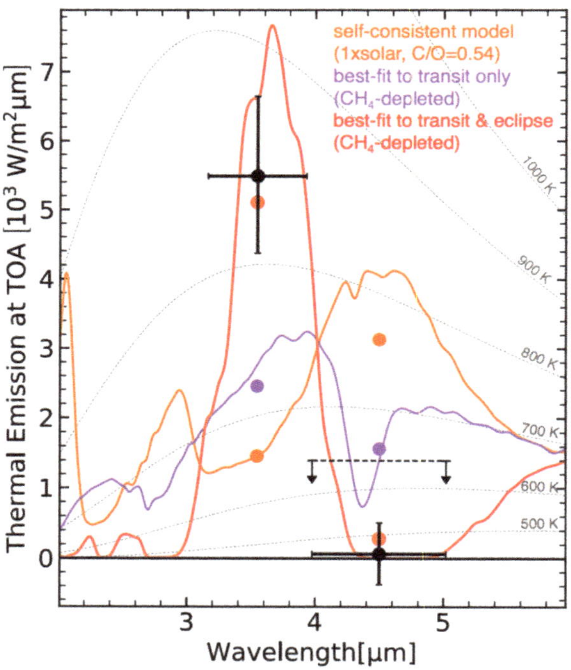


**Figure 7.4.** Spitzer secondary eclipse photometry of the sub-Neptune exoplanet GJ 3470b, reproduced from Benneke et al. (2019), © 2019, The Author(s). Note the large flux at 3.6 μm, relative to 4.5 μm. Spitzer's 3.6 μm bandpass overlaps a strong methane band, so the large emergent flux indicates that methane is not absorbing emergent radiation, i.e., the methane abundance is low.

5. The process of atmospheric escape plays a major role in the evolution of close-in exoplanets. One important constraint on exoplanetary evolution is obtained from statistical studies of the distribution of planet radii (Fulton et al. 2017). However, it is equally important to probe the escape process in detail for individual exoplanets. Transit spectroscopy makes that possible, especially since the highest altitudes do not sustain clouds (rainout is fast), and high altitudes are not obscured by refraction during transit (Bétrémieux & Swain 2018). The UV spectral region is extremely useful for probing atmospheric escape, because many intrinsically strong atomic transitions occur in the UV. Specifically, HI Ly$\alpha$ absorption can reveal hydrogen in the escaping atmosphere, albeit the measurements are rendered difficult by the presence of interstellar and geocoronal hydrogen. Nevertheless, every extra-solar Neptune that has been probed in the UV during transit has been found to have a hugely-extended escaping atmosphere. Extending those measurements to rocky super-Earths is especially important. A rocky planet (lacking a gaseous hydrogen envelope) could have an escaping hydrogen atmosphere if the hydrogen is produced by the photolysis of water. Thus we can potentially probe for the presence of liquid water on rocky planets by detecting abundant hydrogen in their escaping atmospheres.

## 7.4 Opportunities

There are exciting future opportunities to answer the questions discussed above:

1. JWST will be a boon to measuring and understanding abundances in exoplanetary atmospheres. Not only will JWST's aperture collect nearly seven times as many photons as Hubble at a given wavelength, JWST will give us access to the most important wavelengths. The longest wavelengths are important because the planet-to-star contrast is highest at long wavelengths in the Rayleigh–Jeans limit. Spectroscopy at secondary eclipse will make abundant use of JWST's long wavelength coverage, and eclipse spectroscopy has distinct advantages over transit spectroscopy. Those advantages include less sensitivity to interference by clouds, less concern about saturation of strong spectral features, and less interference from stellar activity. With JWST, there exists the potential to detect and possibly characterize atmospheres on rocky exoplanets, such as the TRAPPIST-1 planets (Lustig-Yaeger et al. 2019).
2. JWST will enable us to fully realize the advantages of transit spectroscopy for planets orbiting M-dwarf stars. The fluxes of M-dwarfs peak in the IR spectral region, longward of Hubble's spectroscopic capability. The brightness of the host star is especially important for transit spectroscopy, because the star is the source that illuminates the exoplanetary atmosphere.
3. The forthcoming generation of Extremely Large Telescopes (ELTs) on the ground offer many opportunities for transit spectroscopy. The recent detection of the He I 10,830 Å line, first from space (Spake et al. 2018), then—with exquisite spectral resolution—from the ground (Allart et al. 2018;

Nortmann et al. 2018) is particularly important, because it provides a ground-based probe. Ground-based observations of transits can use large apertures like the ELTs (or existing 10 m telescopes), maximizing the signal-to-noise. Moreover, the He line is not as sensitive to stellar activity as are many UV lines, and it does not suffer from strong interstellar absorption, or geocoronal emission.

4. In the future we may be able to benefit from *stacking*. Stacking refers to the process of combining signals from different planets (that have some common property) to increase the signal-to-noise. Stacking is frequently used in astrophysics (for a recent example, see Elson et al. 2019), but is only beginning to be used for exoplanets (Sheets & Deming 2017; Jansen & Kipping 2018). With a given amount of telescope time, the optimal strategy would always be to obtain a high signal-to-noise spectrum of a single planet, rather than to stack multiple planets. But when archival data become available for multiple planets, stacking offers a way to measure phenomena that would otherwise be inaccessible due to low signal-to-noise. Photometry can be stacked (e.g., to measure secondary eclipses, as per Sheets & Deming 2017), but in principle spectra could also be stacked, for example to measure weak signatures such as HCN (MacDonald & Madhusudhan 2019).
5. Photometry from TESS can help us to understand transiting exoplanetary atmospheres. Already Shporer et al. (2019) have measured a full orbit phase curve of the ultra-hot Jupiter WASP-18b. TESS has a more extended red response (Ricker et al. 2015) than did Kepler, so TESS is sensitive to thermal radiation emitted by the hottest planets. TESS also will measure *transits* to higher precision than do ground-based transit surveys, leading to improved radii in many cases, with implications for atmospheres.
6. Transit spectroscopy has limited access to the lower portions of exoplanetary atmospheres, due to clouds, and also due to refraction. Even in a clear atmosphere, rays that pass through the lower atmosphere are bent by refraction out of our line of sight. However, that refracted light can appear outside of transits, as "shoulders" on transit light curves. Detection of that refracted light can reveal information on the exoplanetary atmosphere, such as the presence or lack of clouds, and the scale height (Sidis & Sari 2010). Although searches for refracted light have been negative to date (Sheets et al. 2018), TESS offers to potential for more sensitive refraction searches for planets transiting bright stars.
7. Although the Hubble Space Telescope is nearing the end of its operational life, its UV capability is critical for probing escaping exoplanetary atmospheres. For example, arguably the best way to infer the presence of liquid water on rocky super-Earths is to detect escaping hydrogen from the photolysis of water vapor.

## 7.5 Challenges

1. Clouds and line-saturation are a potential challenge for transit spectroscopy. The geometry of transits is not ideal for abundance measurements. The long

slant paths that stellar photons follow as they propagate through the exoplanetary atmosphere can produce saturation in individual lines, and saturated lines are not suitable for abundance measurements. Also, the long slant paths make interference by clouds more likely.

2. Measuring the heavy element content of exoplanetary atmospheres has to account for the fraction of elements that are locked up in clouds (Helling et al. 2016). Not only is the cloud inventory difficult to observe, but large aerosol particles may sink beneath the observable atmosphere, making atmospheric metallicities difficult to infer and interpret in terms of the big picture of planet formation and differentiation.
3. We evidently do not understand the thermal structure and chemistry of hot Jupiter atmospheres, and that may make it difficult to derive reliable abundances using secondary eclipse spectroscopy. Studies based on Spitzer secondary eclipses (Garhart et al. 2020; Kammer et al. 2015; Wallack et al. 2019) have found trends in the wavelength variation of hot Jupiter brightness temperatures that are not reproduced by current models.
4. We do not know the noise floor(s) of the JWST instruments, and individual instruments will have their own challenges, i.e., separation of spectral orders could be problematic in NIRISS/SOSS. Spatial scan will apparently not be available with JWST, and many of the spectroscopic modes will produce only spatially narrow spectra. In those cases, detecting weak spectral features may depend on understanding the properties of pixels in small groups, and that may be challenging.
5. JWST measurements of spectral features in small and temperate planets may require averaging the results from multiple transits (e.g., Louie et al. 2018). Multiple transits may not be easy to schedule on JWST, especially if large numbers of transits are required.
6. The larger the column density that we can probe, the greater will be our sensitivity to the composition of exoplanetary atmospheres. However, refraction limits our access to the lower atmosphere during transit (Bétrémieux & Swain 2018). Hence even for clear atmospheres, we may not be able to obtain transit spectra with as great a sensitivity as we would like.
7. Obtaining UV transit spectroscopy after the demise of Hubble will require a new space mission.

## References

Alam, M. K., Nikolov, N., López-Morales, M., et al. 2018, AJ, 156, 298

Allart, R., Bourrier, V., Lovis, C., et al. 2018, Sci, 362, 1384

Barman, T. 2007, ApJ, 661, L191

Baxter, C., Désert, J.-M., Parmentier, V., et al. 2020, A&A, 639, A36

Baxter, C., Désert, J.-M., & Tsai, S.-M. 2021, A&A, 648, A127

Benneke, B., Knutson, H. A., Lothringer, J., et al. 2019, NatAs, 3, 813

Benneke, B., Wong, I., Piaulet, C., et al. 2019, ApJL, 887, L14

Bétrémieux, Y., & Swain, M. R. 2018, ApJ, 865, 12

Bétrémieux, Y., & Swain, M. R. 2018, ApJL, 865, 12

Bourrier, V., Ehrenreich, D., & des Etangs, A. L. 2015, A&A, 582, A65
Bourrier, V., des Etangs, A. L., Ehrenreich, D., et al. 2018, A&A, 620, A147
Bourrier, V., des Etangs, A. L., Ehrenreich, D., Tanaka, Y. A., & Vidotto, A. A. 2016, A&A, 591, A121
Bruno, G., Lewis, N. K., Stevenson, K. B., et al. 2018, AJ, 155, 55
Charbonneau, D., Allen, L. E., Megeath, S. T., et al. 2005, ApJ, 626, 523
Charbonneau, D., Brown, T. M., Noyes, R. W., & Gilliland, R. L. 2002, ApJ, 568, 377
Crouzet, N., McCullough, P. R., Deming, D., & Madhusudhan, N. 2014, ApJ, 795, 166
Deming, D., Seager, S., Winn, J., et al. 2009, PASP, 121, 952
Deming, D., Louie, D., & Sheets, H. 2019, PASP, 131, 013001
Deming, D., Seager, S., Richardson, L. J., & Harrington, J. 2005, Natur, 434, 740
Deming, D., Wilkins, A., McCullough, P., et al. 2013, ApJ, 774, 95
dos Santos, L. A., Bourrier, V., Ehrenreich, D., & Kameda, S. 2019, A&A, 622, A46
Ehrenreich, D., Bourrier, V., Wheatley, P. J., et al. 2015, Natur, 522, 459
Elson, E. C., Baker, A. J., & Blyth, S. L. 2019, MNRAS, 486, 4894
Fossati, L., Haswell, C. A., Froning, C. S., et al. 2010, ApJL, 714, L222
Fraine, J., Deming, D., Benneke, B., et al. 2014, Natur, 513, 526
Fraine, J. D., Deming, D., Gillon, M., et al. 2013, ApJ, 765, 127
Fulton, B. J., Petigura, E. A., Howard, A. W., et al. 2017, AJ, 154, 109
Garhart, E., Deming, D., Mandell, A., et al. 2020, AJ, 159, 137
Gómez de Castro, A. I., Beitia-Antero, L., & Ustamujic, S. 2018, ExA, 45, 147
Grillmair, C. J., Burrows, A., Charbonneau, D., et al. 2008, Natur, 456, 767
Guilluy, G., Sozzetti, A., Brogi, M., et al. 2019, A&A, 625, A107
Guyon, O., Martinache, F., Cady, E. J., et al. 2012, Proc. SPIE, 8447, 84471X
Helling, C., Lee, G., Dobbs-Dixon, I., et al. 2016, MNRAS, 460, 855
Hubbard, W. B., Fortney, J. J., Lunine, J. I., et al. 2001, ApJ, 560, 413
Jansen, T., & Kipping, D. 2018, MNRAS, 478, 3025
Kammer, J. A., Knutson, H. A., Line, M. R., et al. 2015, ApJL, 810, 118
Kempton, E. M. R., Bean, J. L., & Parmentier, V. 2017, ApJL, 845, L20
Kislyakova, K. G., Holmström, M., Odert, P., et al. 2019, A&A, 623, A131
Knutson, H. A., Madhusudhan, N., Cowan, N. B., et al. 2011, ApJ, 735, 27
Kreidberg, L., Bean, J. L., Désert, J.-M., et al. 2014, ApJL, 793, L27
Kreidberg, L., Koll, D. D. B., Morley, C., et al. 2019, Natur, 573, 87
Kreidberg, L., Line, M. R., Parmentier, V., et al. 2018, AJ, 156, 17
Lavie, B., Ehrenreich, D., Bourrier, V., et al. 2017, A&A, 605, L7
Lecavelier des Etangs, A., Bourrier, V., Wheatley, P. J., et al. 2012, A&A, 543, L4
Louie, D. R., Deming, D., Albert, L., et al. 2018, PASP, 130, 044401
Lovis, C., Snellen, I., Mouillet, D., et al. 2017, A&A, 599, A16
Lustig-Yaeger, J., Meadows, V. S., & Lincowski, A. P. 2019, AJ, 158, 27
MacDonald, R. J., & Madhusudhan, N. 2019, MNRAS, 486, 1292
McCullough, P. R., Crouzet, N., Deming, D., & Madhusudhan, N. 2014, ApJ, 791, 55
Morley, C. V., Kreidberg, L., Rustamkulov, Z., Robinson, T., & Fortney, J. J. 2017, ApJ, 850, 121
Nikolov, N., Sing, D. K., Fortney, J. J., et al. 2018, Natur, 557, 526
Nortmann, L., Pallé, E., Salz, M., et al. 2018, Sci, 362, 1388
Owen, J. E. 2019, AREPS, 47, 67
Pinhas, A., Madhusudhan, N., Gandhi, S., & MacDonald, R. 2019, MNRAS, 482, 1485

Ricker, G. R., Winn, J. N., Vanderspek, R., et al. 2015, JATIS, 1, 014003
Seidel, J. V., Ehrenreich, D., Wyttenbach, A., et al. 2019, A&A, 623, A166
Sheets, H. A., & Deming, D. 2017, AJ, 154, 160
Sheets, H. A., Jacob, L., Cowan, N. B., & Deming, D. 2018, RNAAS, 2, 153
Shporer, A., Wong, I., Huang, C. X., et al. 2019, AJ, 157, 178
Sidis, O., & Sari, R. 2010, ApJ, 720, 904
Sing, D. K., Wakeford, H. R., Showman, A. P., et al. 2015, MNRAS, 446, 2428
Sing, D. K., Fortney, J. J., Nikolov, N., et al. 2016, Natur, 529, 59
Sing, D. K., Lavvas, P., Ballester, G. E., et al. 2019, AJ, 158, 91
Snellen, I. A. G., de Kok, R. J., le Poole, R., Brogi, M., & Birkby, J. 2013, ApJ, 764, 182
Spake, J. J., Sing, D. K., Evans, T. M., et al. 2018, Natur, 557, 68
Stevenson, K. B., Bean, J. L., Seifahrt, A., et al. 2014, AJ, 147, 161
Stevenson, K. B., Harrington, J., Nymeyer, S., et al. 2010, Natur, 464, 1161
Swain, M. R., Vasisht, G., Tinetti, G., et al. 2009, ApJ, 690, L114
Tinetti, G., Vidal-Madjar, A., Liang, M.-C., et al. 2007, Natur, 448, 169
Todorov, K. O., Deming, D., Burrows, A., & Grillmair, C. J. 2014, ApJ, 796, 100
Todorov, K. O., Deming, D., Knutson, H. A., et al. 2013, ApJ, 770, 102
Tsiaras, A., Waldmann, I. P., Zingales, T., et al. 2018, AJ, 155, 156
Vidal-Madjar, A., Lecavelier des Etangs, A., Désert, J. M., et al. 2003, Natur, 422, 143
Vidotto, A. A., & Bourrier, V. 2017, MNRAS, 470, 4026
Wakeford, H. R., Sing, D. K., Deming, D., et al. 2018, AJ, 155, 29
Wallack, N. L., Knutson, H. A., Morley, C. V., et al. 2019, AJ, 158, 217
Welbanks, L., & Madhusudhan, N. 2019, AJ, 157, 206
Welbanks, L., Madhusudhan, N., Allard, N. F., et al. 2019, ApJL, 887, L20

AAS | IOP Astronomy

ExoFrontiers
Big questions in exoplanetary science
Nikku Madhusudhan

# Chapter 8

# High-resolution Spectroscopy

**Matteo Brogi and Jayne Birkby**

## 8.1 Introduction

Spectroscopy at high resolving powers (currently $R = 25, 000 - 140, 000$) from the visible to the NIR allows us to (i) resolve molecular bands into the individual lines (see Figure 8.1), and (ii) use the Doppler shift of the planet spectrum to disentangle it from other sources, such as the transmission spectrum of the Earth's atmosphere and the host star spectrum (Snellen et al. 2010). Spectral information can be extracted through cross-correlation with model spectra or templates derived from them, and the best solutions are found by maximizing the cross-correlation. The method excels at identifying species with numerous spectral lines and can be used for transmission, day/night-side emission, and reflected light spectroscopy. In the latter cases, it is also applicable to non-transiting planets (Brogi et al. 2012). Due to the facts that high-resolution spectroscopy (HRS) is currently exclusively available on ground-based facilities, does not need reference stars or telluric standards, and can observe the brightest stars in the sky, it is the natural complement to lower spectral resolution space-based observations.

In this chapter, we consider the important questions and goals for HRS, the opportunities it offers and the challenges it faces. We specifically consider this in the context of detecting molecular species. The main science questions that high-resolution spectroscopy can address (either currently or in the future), are:

- Determining the chemical inventory of exoplanet atmospheres, including elemental abundances and metallicity. This is due to the robustness of cross-correlation at isolating the specific fingerprint of each species, and to the sensitivity of HRS to relative line ratios, which constrain abundances.
- Constraining the overall vertical thermal structure of exoplanets, in particular the presence of thermal inversion layers, which produce emission lines in the planet's thermal spectrum. Cross correlating emission lines with an absorption spectrum produces anti-correlation rather than correlation.

doi:10.1088/2514-3433/abfa8fch8  

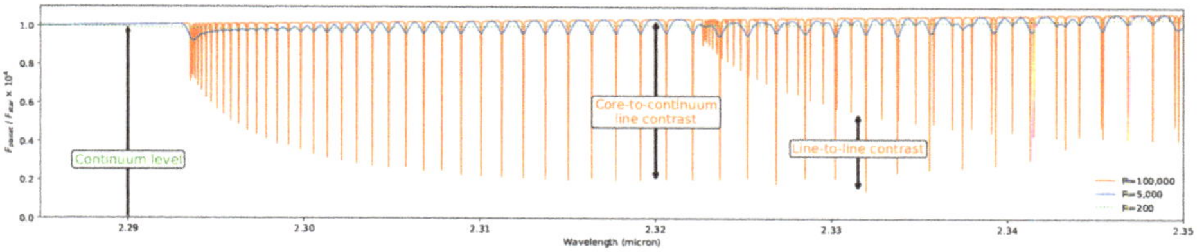


**Figure 8.1.** This figure shows a model spectrum of carbon monoxide in a hot Jupiter atmosphere at different resolving powers. At very low resolution ($R$ = 200, green dotted line), observations measure the absolute planet/star flux as a function of wavelength, marked as "Continuum level" in the figure. For increasing resolving powers, spectra are increasingly resolved into molecular lines. Each resolved line contributes to the total detection strength of the planet signal. The more lines there are, by spectral resolution and/or by wavelength coverage, the stronger the signal. Resolutions at $R$ = 5000 (blue solid line) can result in detections, but in more complex spectra, e.g., water, or in overlapping bands, the dense forest of lines can still present as broad features. Consequently, the higher the resolution the more robust the extraction of different molecular species distinct from other species and the noise. Since the absolute continuum level can be removed in the data analysis process, high-resolution spectroscopy focuses instead on measuring core-to-continuum line depth or the line-to-line depth.

- Constraining the atmospheric circulation and energy redistribution of giant planets by measuring their atmospheric winds and rotation, and mapping their atmospheric features. This is due to the sensitivity of the method to the line shape (broadening, asymmetries, shifts).
- Constraining the nature of the most common exoplanets (with radii between 2 and 4 $R_{\oplus}$) and disentangling atmospheres with strong aerosol opacity from atmospheres with a high metallicity.
- Remotely detecting molecules commonly associated with life, among which are (but not exclusively) molecular oxygen.

## 8.2 State of the Art

The first attempts at using HRS aimed at detecting reflected optical starlight (Cameron et al. 1999; Charbonneau et al. 1999), but only resulted in upper limits. After a decade of further inconclusive attempts at measuring reflected light (Cameron et al. 2002; Leigh et al. 2003; Rodler et al. 2008, 2010) or signatures of molecular species (Wiedemann et al. 2001; Deming et al. 2005), a novel analysis technique was demonstrated on VLT/CRIRES data and detected CO absorption in the transmission spectrum of HD 209458 b around 2.3 μm (Snellen et al. 2010), and $H_2O$ in the atmosphere of HD 189733 b at 3.2 μm (Birkby et al. 2013). Further application to non-transiting hot Jupiters resulted in multiple detection of CO (Brogi et al. 2012; Rodler et al. 2012; Brogi et al. 2013; de Kok et al. 2013; Rodler et al. 2013) and $H_2O$ (Lockwood et al. 2014; Birkby et al. 2017). While these species were routinely found in exo-atmospheres throughout the last decade, detections of additional species were only recently reported for TiO (Nugroho et al. 2017), HCN (Hawker et al. 2018; Cabot et al. 2019), and $CH_4$ (Guilluy et al. 2019). With the discovery of the class of ultra-hot Jupiters with temperatures rivaling those of M- and K-dwarfs, more exotic species were detected at optical wavelengths, among which Fe and Ti, also in their ionized state, were revealed from a cross-correlation spectral atlas (Hoeijmakers et al. 2019). There is also evidence with HRS that some of these species condense out again on the night side (Ehrenreich et al. 2020).

Molecular high-resolution spectroscopy managed to constrain wind patterns in hot Jupiter atmospheres, in particular the presence of day-to-night side winds at high altitude and overall equatorial super-rotation. It measured for the first time the spin of directly-imaged planets (Snellen et al. 2014; Schwarz et al. 2016), confirmed the tidal locking of HD 189733b (Brogi et al. 2016), and compared results of general circulation models to observations (Flowers et al. 2019). Near-infrared spectroscopy has also delivered the first direct measurement of a thermal inversion layer through TiO emission lines (Nugroho et al. 2017), more recently confirmed in the optical as well (Pino et al. 2020).

In spite of the initial efforts to detect stellar reflected light, evidence remains tentative, and with several upper limits (Martins et al. 2015; Hoeijmakers et al. 2018). This is likely due to the overall low albedo of hot Jupiters in the optical.

Currently, high-resolution spectroscopy is undergoing a series of developments, including instrumentation, observational strategy and analysis techniques:

- Radial-velocity machines capable of NIR HRS have been built in the past 5 years and are currently available at 3–4 m class telescopes. Despite the smaller telescope mirror, these spectrographs have a higher throughput, much larger spectral range, and much higher stability compared to the instruments used for previous studies. Early demonstration on these instruments show performances at least comparable if not superior to VLT/CRIRES or Keck/NIRSPEC observations (Brogi et al. 2018; Guilluy et al. 2019; Alonso-Floriano et al. 2019).
- There has been substantial work in modeling and retrieving atmospheric properties at high spectral resolution (Brogi et al. 2017; Pino et al. 2018), which resulted in the first applicable Bayesian retrieval on these data and its possible combination with space spectroscopy (Brogi & Line 2019; Gandhi et al. 2019; Gibson et al. 2020).
- There is ongoing research assessing the performances of HRS when combined with high-contrast imaging (HCI). The principle behind this application is that photon noise from the parent star, which is the dominant source of noise in HRS observations of bright sources, can be significantly reduced by using direct-imaging techniques, i.e., by suppressing the starlight at the spatial location of the planet. Cross-correlation with model templates then uses both spatial and RV separation of the planet and host star to extract the planet spectrum. HRS+HCI can be achieved through long-slit high-resolution spectroscopy fed via an adaptive optic system, as in the case of CRIRES (Snellen et al. 2014; Schwarz et al. 2016). Alternatively, it can be achieved through an integral field unit (IFU) such as that planned for the METIS instrument for the Extremely Large Telescope (Snellen et al. 2015). Early demonstration of IFUs (albeit at lower spectral resolution) have been obtained with OSIRIS on Keck (Konopacky et al. 2013) and with SINFONI at the VLT (Hoeijmakers et al. 2018). Even at spectral resolving

powers of only 2500–5000, it is still possible to apply cross-correlation techniques typical of HRS to filter out the speckle noise from the star (which has a mismatching spectrum) and enhance the signature of the exoplanet.

- In line with similar theoretical development at low resolution (Feng et al. 2016; Parmentier et al. 2016; Taylor et al. 2020), HRS has been used to demonstrate that exoplanet atmospheres are inherently three-dimensional, and that information about their dynamics can be obtained by studying the shape and shift of the spectral lines, as mentioned above. In the infrared, current research is trying to assess the reliability and predictive power of general circulation models, i.e., three-dimensional model of the fluid-dynamics driving the atmospheric flow. Generally, it is observed that GCMs succeed at reproducing the main flow in hot Jupiter atmospheres, especially the day-to-night side flow at low pressure and the overall eastward jet stream at the level of the photosphere (Flowers et al. 2019).
- The exoplanet spectra extracted directly via HRS with current instrumentation are noisy, with typically $S/N < 1$ per line (Brogi et al. 2012). Template-matching, such as cross-correlation, instead combines all of the spectral lines to measure the level of correlation with model spectra. Past work has highlighted that this process is somewhat dependent on the choice of line lists (Brogi et al. 2017; Brogi & Line 2019; Webb et al. 2020), however there have been significant advances in the accuracy of line lists. For HRS application, the most important factor is the knowledge of transition frequencies at the 0.01 $cm^{-1}$ level at least in the NIR. Recent work (Gandhi et al. 2020) has shown that such accuracy has been reached for several molecular species including water vapor, while some uncertainty remains for methane and carbon dioxide.

## 8.3 Important Questions and Goals

HRS has been undergoing major development as a technique for characterizing exoplanet atmospheres over the last decade since its first successful detection in 2010 (Snellen et al. 2010). However, as noted in Section 8.5; there has been only a small community with limited suitable instruments driving the field and consequently, there is still significant room for improvement in understanding the scope and power of the technique, its limitations, and its future potential. This makes for an exciting time, with a number of pertinent questions and goals. The key opportunities are described in Section 8.4, but below we highlight (a non-exhaustive list of) important considerations to address.

- **Absolute abundances**: Unlike low-resolution/broadband secondary eclipse spectra, HRS analysis techniques typically remove the absolute continuum level and thus extract atmospheric information using core-to-continuum line contrast and line-to-line contrast (see Figure 8.1). Initially, there were concerns this would mean only relative abundances would result from HRS observations alone (de Kok et al. 2014), but recent work has shown absolute abundances (i.e., volume mixing ratios) reach the same level of precision and accuracy as their lower resolution counterparts, constraining CO abundance to ± 0.4 dex and C/O ratios to ± 0.2 from emission spectroscopy of a hot Jupiter around a bright host with VLT/CRIRES (Brogi & Line 2019). An important goal is to understand the true limit of precision and accuracy of these measurements, and what is needed to improve them (e.g., observational scope and/or analysis techniques), especially as we consider the physical and chemical diversity of exoplanet atmospheres for both gas giants and terrestrial planets in the future.

- **M-dwarfs**: Do exoplanet atmospheres survive the harsh M-dwarf environment? This is a question that can be answered with HRS and the technique will likely target many nearby planets orbiting potentially very active small stars. Proxima b's host star has significant flares (Howard et al. 2018) and such extreme activity can cause certain lines in the spectra to change rapidly or even saturate, resulting in unusable sections of wavelength and time series during an observing sequence. Understanding the extent of this impact on HRS analysis is a key issue. For example, recent observations of hot pulsating exoplanet host stars have shown that variations in the stellar spectrum, if it overlaps with lines in the planet spectrum used in cross-correlation, results in regions of velocity space that cannot be accessed (Nugroho et al. 2020). M-dwarf spectrum are also more complex compared to hot A-stars and their spectral features can complicate cleaning the data to the photon noise limit. Better understanding of M-dwarf spectra is needed to assist this (see Section 8.5 for more detail), and to date, no exoplanet orbiting an M-dwarf has had molecules detected in is atmosphere with HRS. In addition, when studying exoplanet atmospheres in reflected light with HRS, it is the stellar spectrum itself that is reflected, albeit modulated by planet's albedo as a function of wavelength. In this case, M-dwarf hosts become the optimal target as their spectra have more lines to detect, thus boosting the detection strength of the cross-correlation. However, this again requires accurate models of M-dwarfs spectra, as well as understanding how to extract the albedo information and how to circumvent degeneracy with planet radius given it is likely that the planets will be non-transiting.
- **Necessary resolution**: As typical spectral lines have FWHM below the resolution element of infrared spectrographs, there is a concrete possibility of increasing the cross-correlation signal by increasing the resolving power, possibly up to $R = 500,000$ (López-Morales et al. 2019). This is important both for detecting species with many weak/shallow spectral lines, as well as measuring the exact shape of the lines in the spectrum. However, technical considerations regarding the physical size and throughput of such instrumentation, as well as the consequences of lowering the signal-to-noise per resolution element, would have to be carefully assessed.
- **Detection significance**: Early analyses of HRS used signal-to-noise theory to assess the strength of cross-correlation signals, which was shown to be a good proxy for significance from the statistical analysis of the distribution of cross-correlation values (Brogi et al. 2012). While straightforward if cross-correlation values can be considered independent and Gaussian, this

method can be biased by the presence of residual red noise, aliases of the cross-correlation function in velocity space, and the unavoidable correlation between neighboring velocity lags due to the finite instrumental resolution. The recent introduction of Bayesian techniques to estimate chemical and physical properties opens up to an alternative method to assess the significance of a detection, that is by comparing the Bayesian evidence of models with and without a certain species (Brogi & Line 2019; Gandhi et al. 2019; Gibson et al. 2020). This approach has the advantage of being consistent with the analysis of low-resolution spectroscopy, however it still needs to be demonstrated on a large scale.

## 8.4 Opportunities

HRS offers a cornucopia of opportunities, both near term and long term:

- **TESS and the Sub-Neptunes:** The imminent discovery of a sample of planets orbiting bright stars by the TESS mission, opens up the detailed atmospheric characterization of sub-Neptunes in support of deep HST observations and in advance of JWST observations. This has an enormous potential for delivering informative discoveries with a strong legacy aspect, especially if the information at low and high spectral resolution is routinely combined.
- **Seeing Above the Clouds:** The current availability of high-resolution spectroscopy instrumentation is on medium-class facilities (4–8 m) including some with unprecedented spectral range and throughput. The most performing of these instruments at the moment seems to be CARMENES and SPIRou, with a potential of 3× the S/N per unit observing time when looking for broadband absorbers in the NIR such as water vapor and methane. Such opportunity opens up the possibility of surveying exoplanets that show a flat transmission spectrum at low spectral resolution, e.g., due to aerosols, and to detect the fraction of the spectrum forming above the aerosol deck (Hood et al. 2020; Gandhi et al. 2020).
- **Nearest Neighbors:** Given that HRS does not need a planet to transit in order to characterize its atmosphere, it offers the best chance in the coming decade to remotely study our nearest neighbors, including rocky worlds such as Proxima b (Anglada-Escudé et al. 2016). METIS/ELT is ideally suited to target this planet in the 3–5 μm region (Snellen et al. 2015), alongside GMTIFS and GMTNIRS at GMT, while other optical high-resolution instruments can target oxygen, such as HARMONI and HIRES at ELT, G-CLEF and GMagAOX at GMT. Similar instruments for studying northern hemisphere counterparts include MICHI, MIDHIS, HROS, WFOS, and PSI at TMT. Given the proximity of our nearest neighbor, it can be studied both in Doppler shift and spatial separation at high spectral resolution, and there is even potential to observe it already with a combination of SPHERE +ESPRESSO or other new instruments for the VLT (Lovis et al. 2017).
- **Atmospheric Maps, Biosignatures, and Surfaces:** The increased sensitivity of the ELTs means that hot Jupiter and high signal-to-noise systems observed with

HRS will give the planet spectrum directly, i.e., cross-correlation will not be needed as the spectral lines will stand out above the noise. With METIS/ELT it is expected that individual lines in hot Jupiter spectra will have S/N~10 each. With such exquisite quality of high-resolution exoplanet spectra, we will be able to essentially apply stellar spectroscopy techniques to exoplanets. For Hot Jupiters, that means each lines gives information about a different altitude in the planet. With time resolution, a fully 3D and phase-resolved model of the planet can be built, including measuring wind speeds at different altitudes. Excitingly, the ELTs can monitor the detailed line shape over time to enable Doppler imaging of giant planets (Snellen et al. 2014), in a similar manner that starspots are imaged, or indeed as was used to make the atmospheric map of the brown dwarf Luhman 16 B (Crossfield et al. 2014). For smaller planets, the increased sensitivity of the ELTs suggests that low abundances molecules can also be studied. An interesting avenue for study is the potential to detect biosignatures with no false positive, for example phosphine as has recently been suggested (Sousa-Silva et al. 2020; Greaves et al. 2020). Finally, it is important to investigate how distinct the different types of terrestrial worlds will appear in HRS observations and if their surface process can be studied this way too.

- **HRS in Space:** There is a reasonable case for space-based high-resolution spectroscopy in order to overcome the high thermal background noise and instability of the Earth's atmosphere, and unlock precious observable windows targeting the main water vapor bands and much further into the infrared where spectral features can be larger, less obscured by clouds, and contain additional biosignature features. This does not necessarily need to be a flagship mission, but alternative routes could be exploited such as the use of 24U CubeSats with foldable components (see Chapter 10), however considerations need to be made for pointing stability and cooling.
- **Combining ELTs with JWST and the Next Great Observatory**: As discussed, the combination of HRS and lower resolution spectra is a power method for building a complete, accurate, and precise view of an exoplanet atmosphere. There is a special opportunity then to combine observations with HRS instruments on ELTs with space-based exoplanet spectra from initially JWST and HST, and eventually the Next Great Observatory (e.g., LUVOIR, OST, HabEx), with even the option for simultaneous monitoring. This is a possible long term avenue for the detailed study of nearby Earth-like exoplanets.

## 8.5 Challenges

In order to capitalize on the opportunities offered by HRS in the study of exoplanet atmospheres, there are several key challenges to overcome.

- **Stellar Spectra and 3D:** The stellar spectrum is a non-negligible contaminant when looking for species simultaneously present in the stellar and planet atmospheres. For transmission/emission, these are currently the atomic lines from Na and potentially He, molecular lines of CO for G and K-type stars, TiO and $H_2O$ for M-dwarfs. For reflection, the entire stellar spectrum is

reflected, albeit at a different Doppler shift. Stellar lines are non-stationary in the observer's reference frame mostly due to the combined rotational and orbital motion of the Earth, and during transit stellar lines are distorted by Rossiter–McLaughlin effect and center-to-limb variations. One-dimensional stellar models applied to a proper simulation of transit geometry can mitigate these effects, and excellent correction is achieved for photospheric lines (e.g., CO) in main-sequence F to K stars when applying 3D stellar modeling (Chiavassa & Brogi 2019). However, uncertainties in molecular opacities remain an issue for proper modeling and correction of stellar lines, furthermore non-photospheric lines and effects of starspots are still challenging to model, with negative impact especially on the correction of M-dwarf spectra. Additional stellar activity in the form of flares can further influence HRS observations and its impact still needs to be quantified.
- **Accuracy and Completeness of Line Lists:** Uncertainties in molecular line lists at high spectral resolution will still affect accurate modeling of exoplanet spectra at least for a few years to come. These uncertainties come in two main flavors: completeness and accuracy. While the level of completeness is somewhat secondary for detecting species through cross-correlation, it is still important when retrieving abundances and temperatures, because the myriad of weak molecular lines form a pseudo-continuum capable of altering the line ratios. Accuracy refers to the exact determination of the frequency and intensity of a certain transition, and it is the most crucial aspect of current analysis of high-resolution spectra. In particular, inaccurate transition frequencies redistribute power from the peak to the wings of the cross-correlation function, with damaging effects on the detectability of species, especially at the relatively low S/N of HRS observations. While great progress has been made on some species such as CO, $H_2O$, TiO, and HCN, other key molecules such as $CH_4$ and $CO_2$ still lack the accuracy and completeness sufficient at resolving powers close to $10^5$ (Gandhi et al. 2020).
- **Community Size and Instrumentation:** The community of HRS users is still too small to process even archival data. This is especially true for spectrographs mounted at smaller telescopes (CARMENES, GIANO, SPIRou), due to the significant amount of telescope time invested by small teams with proprietary access. While instrumentation has begun to ramp up in recent years, especially in the optical, and the imminent return of CRIRES+/VLT remains highly anticipated. One main obstacle to create a wider community is the lack of standardized analysis techniques that are applicable on large scale and over heterogeneous data sets. Currently, the analysis is often optimized on a case-by-case basis and with proprietary codes of each research group. Specialist conferences and data challenges are still rare and do not offer enough support for young academics and group leaders to get involved. It is especially important for the community to explore different avenues within the field of HRS which is still in its relative infancy, but also to consolidate its findings on the best approaches for analysis.

## Acknowledgements

JB acknowledges funding from the European Research Council (ERC) under the European Union's Horizon 2020 research and innovation program under grant agreement No 805445.

## References

Alonso-Floriano, F. J., Sánchez-López, A., Snellen, I. A. G., et al. 2019, A&A, 621, A74
Anglada-Escudé, G., Amado, P. J., Barnes, J., et al. 2016, Natur, 536, 437
Birkby, J. L., de Kok, R. J., Brogi, M., et al. 2013, MNRAS, 436, L35
Birkby, J. L., de Kok, R. J., Brogi, M., Schwarz, H., & Snellen, I. A. G. 2017, AJ, 153, 138
Brogi, M., de Kok, R. J., Albrecht, S., et al. 2016, ApJ, 817, 106
Brogi, M., Giacobbe, P., Guilluy, G., et al. 2018, A&A, 615, A16
Brogi, M., Line, M., Bean, J., Désert, J. M., & Schwarz, H. 2017, ApJL, 839, L2
Brogi, M., Snellen, I. A. G., de Kok, R. J., et al. 2013, ApJ, 767, 27
Brogi, M., & Line, M. R. 2019, AJ, 157, 114
Brogi, M., Snellen, I. A. G., de Kok, R. J., et al. 2012, Natur, 486, 502
Cabot, S. H. C., Madhusudhan, N., Hawker, G. A., & Gandhi, S. 2019, MNRAS, 482, 4422
Charbonneau, D., Noyes, R. W., Korzennik, S. G., et al. 1999, ApJL, 522, L145
Chiavassa, A., & Brogi, M. 2019, A&A, 631, A100
Cameron, A. C., Horne, K., Penny, A., & James, D. 1999, Natur, 402, 751
Cameron, A. C., Horne, K., Penny, A., & Leigh, C. 2002, MNRAS, 330, 187
Crossfield, I. J. M., Biller, B., Schlieder, J. E., et al. 2014, Natur, 505, 654
de Kok, R. J., Birkby, J., Brogi, M., et al. 2014, A&A, 561, A150
de Kok, R. J., Brogi, M., Snellen, I. A. G., et al. 2013, A&A, 554, A82
Deming, D., Brown, T. M., Charbonneau, D., Harrington, J., & Richardson, L. J. 2005, ApJ, 622, 1149
Ehrenreich, D., Lovis, C., Allart, R., et al. 2020, Natur, 580, 597
Feng, Y. K., Line, M. R., Fortney, J. J., et al. 2016, ApJ, 829, 52
Flowers, E., Brogi, M., Rauscher, E., Kempton, E. M. R., & Chiavassa, A. 2019, AJ, 157, 209
Gandhi, S., Brogi, M., & Webb, R. K. 2020, MNRAS, 498, 194
Gandhi, S., Brogi, M., Yurchenko, S. N., et al. 2020, MNRAS, 495, 224
Gandhi, S., Madhusudhan, N., Hawker, G., & Piette, A. 2019, AJ, 158, 228
Gibson, N. P., Merritt, S., Nugroho, S. K., et al. 2020, MNRAS, 493, 2215
Greaves, J. S., Richards, A. M. S., Bains, W., et al. 2020, arXiv:2009.06593
Guilluy, G., Sozzetti, A., Brogi, M., et al. 2019, A&A, 625, A107
Hawker, G. A., Madhusudhan, N., Cabot, S. H. C., & Gandhi, S. 2018, ApJL, 863, L11
Hoeijmakers, H. J., Ehrenreich, D., Kitzmann, D., et al. 2019, A&A, 627, A165
Hoeijmakers, H. J., Schwarz, H., Snellen, I. A. G., et al. 2018, A&A, 617, A144
Hoeijmakers, H. J., Snellen, I. A. G., & van Terwisga, S. E. 2018, A&A, 610, A47
Hood, C. E., Fortney, J. J., Line, M. R., et al. 2020, AJ, 160, 198
Howard, W. S., Tilley, M. A., Corbett, H., et al. 2018, ApJL, 860, L30
Konopacky, Q. M., Barman, T. S., Macintosh, B. A., & Marois, C. 2013, Sci, 339, 1398
Leigh, C., Cameron, A. C., Horne, K., Penny, A., & James, D. 2003, MNRAS, 344, 1271
Lockwood, A. C., Johnson, J. A., Bender, C. F., et al. 2014, ApJL, 783, L29
López-Morales, M., Ben-Ami, S., Gonzalez-Abad, G., et al. 2019, AJ, 158, 24

Lovis, C., Snellen, I., Mouillet, D., et al. 2017, A&A, 599, A16
Martins, J. H. C., Santos, N. C., Figueira, P., et al. 2015, A&A, 576, A134
Nugroho, S. K., Gibson, N. P., de Mooij, E. J. W., et al. 2020, ApJL, 898, L31
Nugroho, S. K., Kawahara, H., Masuda, K., et al. 2017, AJ, 154, 221
Parmentier, V., Fortney, J. J., Showman, A. P., Morley, C., & Marley, M. S. 2016, ApJ, 828, 22
Pino, L., Désert, J.-M., Brogi, M., et al. 2020, ApJL, 894, L27
Pino, L., Ehrenreich, D., Allart, R., et al. 2018, A&A, 619, A3
Rodler, F., Kürster, M., & Barnes, J. R. 2013, MNRAS, 432, 1980
Rodler, F., Kürster, M., & Henning, T. 2008, A&A, 485, 859
Rodler, F., Kürster, M., & Henning, T. 2010, A&A, 514, A23
Rodler, F., Lopez-Morales, M., & Ribas, I. 2012, ApJL, 753, L25
Schwarz, H., Ginski, C., de Kok, R. J., et al. 2016, A&A, 593, A74
Snellen, I., de Kok, R., Birkby, J. L., et al. 2015, A&A, 576, A59
Snellen, I. A. G., Brandl, B. R., de Kok, R. J., et al. 2014, Natur, 509, 63
Snellen, I. A. G., de Kok, R. J., de Mooij, E. J. W., & Albrecht, S. 2010, Natur, 465, 1049
Sousa-Silva, C., Seager, S., Ranjan, S., et al. 2020, AsBio, 20, 235
Taylor, J., Parmentier, V., Irwin, P. G. J., et al. 2020, MNRAS, 493, 4342
Webb, R. K., Brogi, M., Gandhi, S., et al. 2020, MNRAS, 494, 108
Wiedemann, G., Deming, D., & Bjoraker, G. 2001, ApJ, 546, 1068

# Chapter 9

# Space Missions for Exoplanet Science

**Karl Stapelfeldt, Heike Rauer, Aki Roberge and Avi Mandell**

## 9.1 Introduction

Exoplanets are most often detected through precision observations in the time domain. By measuring small changes in a star's brightness, its spectrum, its radial velocity, or its photocenter—and distinguishing these from unrelated astrophysical effects—the perturbations induced by an exoplanet on its star can be used to infer the properties of the unseen planet. These effects have small amplitudes and frequencies ($\leqslant$0.01%–1% in total brightness for transits, from 1 mas to $\leqslant$1 μas for astrometry, and two events per 100,000 stars per year for microlensing) and are therefore most effectively measured from the stable space environment. Of the more than 4500 confirmed exoplanets known as of September 2021, more than 3000 were discovered in dedicated space-based transit surveys executed with the Kepler Space Telescope (Kepler) and Transiting Exoplanet Survey Satellite (TESS) missions. The initial focus for exoplanet observations from space has been to develop the population statistics for short-period transiting planets orbiting stars with radii $\leqslant 1\ R_\odot$, especially the distributions of exoplanet orbits and radii.

The effort to characterize the atmospheres and surfaces of detected transiting planets has begun alongside the search for new worlds. Due to the absence of interference by absorption from Earth's atmosphere, the vast majority of exoplanet atmosphere measurements to date have been achieved from space using existing large observatories such as the Hubble and Spitzer Space Telescopes. These missions were not originally designed to achieve high-precision spectrophotometric time series measurements of atmospheric absorption during a planetary transit or disappearance of atmospheric emission during an eclipse. However, painstaking work to characterize the instrumental response behaviors of both observatories has allowed the needed precision to be achieved. This has enabled measurements of the atmospheric constituents for hot exoplanets ranging from Jupiter-size down to sub-Neptune-size.

doi:10.1088/2514-3433/abfa8fch9  

The importance of space-based measurements to the exoplanet field will accelerate in the 2020s as the TESS, Gaia, Nancy Grace Roman Space Telescope (Roman; formerly known as WFIRST), PLAnetary Transits and Oscillations of stars (PLATO), and Atmospheric Remote-sensing Infrared Exoplanet Large-survey (ARIEL) missions all deliver large statistical samples of exoplanet discovery and characterization. JWST will provide deep spectroscopy of dozens of transiting planets down to Earth-size around red dwarf stars. There is also the exciting possibility that development of a major new observatory capable of directly imaging and spectrally characterizing rocky exoplanets in the habitable zones of nearby Sun-like stars could begin in the 2020s. A timeline of these current, upcoming, and potential facilities is shown in Figure 9.1.

## 9.2 Important Questions and Goals

*How do we determine the populations of different types of planets at larger orbital distances, including the habitable zone?*

No one exoplanet detection method has the practical capability to detect exoplanets across their full range of masses, sizes, and orbital periods. Transit and radial velocity methods have produced the vast majority of exoplanet discoveries to date, but are biased toward planets with shorter orbital periods and larger sizes/masses relative to their host star. Astrometry has a very meager track record of exoplanet detections, is poised for breakout success with the Gaia mission, but biased toward detection of massive planets around nearby stars. The microlensing technique will overcome some of these limitations through its sensitivity to more widely-separated planets beyond the ice line, and to smaller planet masses down to that of Mars. But like the transit method, microlensing is only sensitive to a fraction of the possible

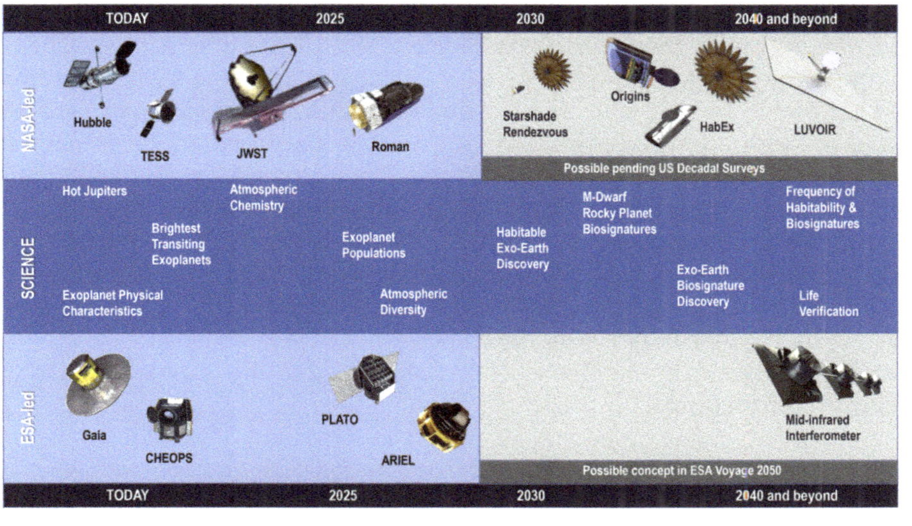


**Figure 9.1.** Timeline of current, planned, and potential space missions for exoplanet research. Adapted from a concept originated by Brendan Crill (JPL/Caltech).

orbital phase space and thus cannot provide a complete inventory of exoplanets in any particular system. Direct imaging overcomes these many limitations by being sensitive to a wide range of planet sizes, orbital inclinations, and orbital distances, and unlike the other methods can achieve this in just a few epochs of observation. Exoplanet direct imaging is however limited to relatively nearby stars where the exoplanetary system has sufficient angular extent to be resolved. A modest number of young, giant exoplanets that retain their heat of formation have been imaged to date in thermal emission. No exoplanets have yet been imaged so far in reflected starlight. Improvements in stellar reflex motion sensitivity and duration, transit survey sensitivity and duration, microlensing cadence and precision, and image contrast will all be needed to build a complete picture of planetary system architectures and populations.

*How do we characterize planetary atmospheres across a range of host stars, planet masses, and planet temperatures?*
Pushing our knowledge of the character of exoplanets from the exotic surprises orbiting close to their host stars to cooler worlds more analogous to those of the solar system will demand different techniques from those used today. Spectroscopy during primary transit, which has been used to obtain the large majority of atmospheric characterization measurements to date, is most sensitive to an exoplanet's upper atmosphere and more easily detects spectral signatures for planets with larger atmospheric scale heights orbiting smaller stars. Since the atmospheric scale height ($H = kT/\mu g$) depends on planet surface gravity ($g$), temperature ($T$), and mean molecular weight ($\mu$), planets become increasingly harder to characterize with primary transit spectroscopy at smaller sizes, cooler temperatures, and denser atmospheric compositions. Furthermore, at cooler temperatures and higher mean molecular weights, the condensation of high-altitude clouds may mask molecular absorption features in transit spectra.

Future exoplanet characterization missions must employ new methods to push exploration to cooler and smaller planets. Direct spectroscopy, where light reflected from or emitted by the planet's disk is observed rather than light from the host star passing through a thin layer of planet's upper atmosphere, can probe deeper into the atmosphere and all the way to the solid surface of a terrestrial planet under the right conditions. Combined-light measurements of thermal emission, such as dayside measurements during secondary eclipse and/or phase-dependent measurements over the full orbit of a planet, have been demonstrated for hot and large planets and could be extended to cooler and smaller planets by utilizing longer IR wavelengths and improved measurement precision. But in the general case, the light of the central star must be blocked from falling on the detector, allowing exoplanet light to be separated and analyzed.

*How do we assess the habitability of terrestrial exoplanets and search them for life?*
One of the most important long-term goals of exoplanet exploration is to characterize potentially habitable exoplanets and probe them for signs of life. However, the

requirements for discovering biological activity go far beyond finding a planet of approximately the right temperature and detecting one or two molecules produced by Earth-based biology (biosignatures). Any molecule may or may not be a biosignature depending on the atmospheric context. The requirements for understanding that context include: knowledge of the planetary size and mass; abundance measurements of major atmosphere constituents (e.g., $H_2O$, $CO_2$); abundance measurements for potential biosignature gases (e.g., $O_2$, $O_3$, $CH_4$); and knowledge of the incident stellar spectrum, especially at the ultraviolet wavelengths that drive photochemistry.

Probing the atmospheres of an Earth-sized planet orbiting a Sun-like star with transit spectroscopy alone may be infeasible (e.g., Kaltenegger & Traub 2009). The transit depth of the whole planet in this case is only ~85 ppm, and thin transmitting layer will induce a change in transit depth of only a few ppm. Detecting this signal would demand spectra with very high signal-to-noise ratios (S/Ns) and vanishingly small systematic errors; thus, the 2018 NAS Exoplanet Science Strategy report[1] concluded that direct spectroscopy is needed to achieve this goal. The extreme planet-to-star contrast ratios ($\sim 10^{-10}$ in the visible, $\sim 10^{-7}$ in the mid-IR) needed for this requires an observatory in space, where the stability of the wavefront entering the telescope can enable ultra-high performance starlight suppression (Wang et al. 2017). Observatory architectures that could execute that starlight suppression include an optical/near-IR single-aperture telescope coupled to a coronagraph or starshade, or a mid-IR interferometer composed of multiple telescopes that combine signals while nulling out the central star.

However, rocky exoplanets orbiting M dwarf stars present an opportunity to use transit spectroscopy to investigate exoplanet habitability in the relatively near-term. The stellar radius for a M1V star is 50% that of the Sun, and for a M7V star it is only 12%; therefore the transit depth for an Earth-sized planet around a M7V star similar to TRAPPIST-1 is 70× deeper than it would be around a Sun-like star. The depth of atmospheric absorption from $CO_2$ without dilution by clouds would therefore be ~70 ppm (Fauchez et al. 2019), potentially detectable within currently achievable measurement uncertainties (Kreidberg et al. 2014; Benneke et al. 2019). The actual strength of habitability indicators (e.g., $H_2O$) as well as biosignature gases will depend on the presence of clouds and other aspects of atmospheric structure and composition.

## 9.3 State of the Art

The study of exoplanets from space observatories has developed in two parallel ways. The first was the application of existing observatories and instruments to characterize transiting exoplanets and microlensing events, with ground-based observatories providing crucial information on when and where to look. New observing modes and calibration procedures had to be developed to get the most out of instruments (Spitzer/IRAC, HST/WFC3) that had not been designed for

[1] https://www.nap.edu/catalog/25187/exoplanet-science-strategy

exoplanet work. The second has been the progressive deployment of wide-field optical imaging capabilities on space telescopes, pushing space astronomy into the forefront of exoplanet demographic studies and precision astrometry. These adaptive and evolutionary approaches have produced a rich harvest of exoplanet science over the past 15–20 years.

### 9.3.1 From CoRoT to Kepler/K2 and TESS

CoRoT, the "Convection, Rotation et Transits plantaires" mission led by the French space agency CNES, was the first space mission to conduct exoplanet transit searches (Deleuil et al. 2018; Auvergne et al. 2009). Observing from a polar low-Earth orbit from 2006 to 2012, CoRoT consisted of a 27 cm telescope with CCD detectors covering a ~10 square degree field of view. It discovered 34 transiting exoplanets including CoRoT-7b, the first rocky exoplanet. The mission helped show the importance of confirming transits with radial velocity follow-up to get the masses, and of asteroseismology to characterize the exoplanet host stars. CoRoT pioneered methods that continue to be used today to detect and characterize exoplanets.

Following in the path of CoRoT, NASA's Kepler mission and its K2 reincarnation have together discovered more exoplanets than any other facility. Kepler employed a 0.95 m effective aperture to direct starlight onto 84 CCD detectors covering 113 square degrees of sky. 170,000 stars were monitored in a broad visual bandpass at 30 minute cadence with 20 ppm photometric precision for $R = 12$, in pursuit of the goal of detecting transits of Earth-sized planets in or near the habitable zones of Sun-like stars (Borucki 2018). Conceived ahead of its time, five iterations of the mission proposal, and the ground-based discovery of the first transiting exoplanet, were required before Kepler was selected for development under the NASA Discovery Program. The mission launched in 2009 into an Earth-trailing orbit and stared at a single field in Cygnus-Lyra for 3.5 years. Kepler achieved its statistical goals by showing that small planets are abundant at orbital periods <200 days, with planets intermediate in size between the Earth and Neptune dominating the detected population. After the failure of two reaction wheels in 2013 the mission could no longer point at its prime target field and was redirected to observing ecliptic fields as the K2 mission (Howell et al. 2014). The mission ended in late 2018 with the exhaustion of its reaction control gas. As of summer 2020 the facility is credited with discovering 2751 confirmed planets and 3307 additional candidates. Analysis of this rich data set will continue for many years to come. Kepler's results give confidence that small planets will be found orbiting nearby stars whenever we choose to begin their exploration.

While Kepler was hugely successful at establishing the exoplanet radius distribution at short orbital periods, too few of its target stars were bright enough for dynamical mass determinations using the radial velocity technique or exoplanet spectral characterization with the Hubble, Spitzer, or upcoming James Webb Space Telescopes. The TESS mission (Ricker et al. 2015) was developed to address those deficiencies. TESS was launched in spring 2018 into an unusual high Earth orbit. From there it uses four cameras with 10.5 cm apertures and optical CCD detectors to step-wise survey almost the entire sky in 26 sectors during its two-year prime mission. TESS will find the short-period transiting planets of bright stars all over the sky, enabling follow-up mass determinations and transit spectroscopy. As of September 2021 TESS has discovered 155 confirmed planets and about 4500 candidates. It is now in an extended mission which could potentially last many more years, if the spacecraft remains healthy and funding support continues.

### 9.3.2 Hubble and Spitzer

The bulk of atmospheric characterization for transiting exoplanets has been accomplished using the Hubble and Spitzer Space Telescopes. Though their apertures are modest compared with ground-based telescopes, the twin benefits of being above the Earth's atmospheric absorption and the ability to remove instrumental systematic effects to high precision has enabled a quality of transit and eclipse measurements unachievable from the ground.

The workhorse instruments for transit spectroscopy with Hubble have been the Space Telescope Imaging Spectrograph (STIS) and the Wide-Field Camera 3 (WFC3). STIS visible transit spectroscopy is typically done with two grisms, G430L and G750L, to cover a wavelength range 0.3–1.0 μm with a spectral resolving power of $R \sim 500$. Such observations probe the short-wavelength spectral slope that is sensitive to both absorption from clouds and hazes (Sing et al. 2016), as well as high-temperature atomic species such as Na and Fe and diatomic molecular absorbers such as TiO. STIS can also probe absorption from neutral H in the Ly$\alpha$ transition using the G140M grating, providing a window into atmospheric escape in highly irradiated atmospheres (Vidal-Madjar et al. 2003). The WFC3 instrument IR channel probes the presence of absorption from $H_2O$ through the 1.15 and 1.4 μm bands, providing a secure detection of water vapor when both are seen (Deming et al. 2013). WFC3 has also been used to measure exoplanet thermal emission through secondary eclipse and phase-curve measurements (Stevenson et al. 2014), but the 1.6 μm wavelength cut-off limited these measurements to the hottest and largest planets.

Spitzer operated from 2003 to 2020 and served as one of the key platforms for transiting exoplanet characterization. The vast majority of Spitzer exoplanet measurements utilized the four broad IRAC photometric bands covering 3.6–8 μm. These measurements proved to be quite powerful as a rough diagnostic of planetary temperature and composition, even after the depletion of Spitzer's cryogen left only the two shortest wavelength bands (3.6 and 4.5 μm) operational after 2009. Spitzer/IRAC provided sensitive measurements of a planet's thermal flux, probed the transit depth at long wavelengths, with some discrimination of molecular absorption features such as $H_2O$, CO, $CO_2$, and $CH_4$. The range of discoveries enabled by combining measurements with one or more of these bands was extraordinary (see Deming & Knutson 2020 for a review). The combination of Hubble spectra and Spitzer photometry has been especially successful at placing constraints on the impact of aerosols on the spectral continuum, due to the possibility of wide wavelength coverage extending from 0.3 to 8 μm.

Photochemistry in planetary atmospheres is primarily driven by stellar UV photons. The spectral distribution of far-UV and near-UV photons affects the production and destruction rates of key molecules, and therefore their equilibrium concentrations. High-quality UV spectra of exoplanet host stars are necessary inputs to current and future atmosphere modeling efforts. This is especially true for the planets of M dwarf stars, which are exposed to higher levels of activity and different spectral distributions of UV radiation than the better-studied solar-type stars. A series of Hubble observing programs (e.g., France et al. 2016, Froning et al. 2019) have acquired time-resolved UV spectra of low-mass stars, in combination with observations at X-ray and optical wavelengths, finding that M dwarfs which appear relatively inactive at optical wavelengths can display strong signs of activity at UV wavelengths (France et al. 2016).

### 9.3.3 CHEOPS and Gaia

ESA's first S-class mission CHEOPS (CHaracterising ExOPlanet Satellite; Benz et al. 2021) was launched December 2019 into a Sun-synchronous low-Earth orbit. The observatory consists of a 33.5 cm aperture telescope capable of ultra-high precision white light (0.4–1.0 μm) photometry. Its primary science goal is to measure highly precise radii of transiting exoplanets and, in combination with available masses ($m_p \sin i$) from radial-velocity data, the bulk densities of known super-Earth- and Neptune-mass planets. After successful completion of its commissioning phase, CHEOPS entered into science operations in late March 2020; the mission duration is expected to be 3.5 years. The first scientific publication of the CHEOPS mission demonstrates its high photometric stability by analyzing the asymmetric transit and occultation lightcurves of WASP-189b (Lendl et al. 2020).

ESA's Gaia mission is performing an all-sky astrometric, photometric and spectroscopic survey of our galaxy with the goals to reveal its dynamics and formation history. It measures the positions, distances, motions and physical parameters of stars. The astrometric precision is around 10 microarcseconds, with more than 20,000 exoplanets expected to be detected (Perryman et al. 2014). Gaia is mainly sensitive to gas giant planets (1–15 $M_{Jup}$) in intermediate orbital periods of about 2–3 years (up to 10 years), for stars up to 500 pc distance from the Sun. However, the full precision will only be reached with the final release of the Gaia catalogue, expected no earlier than 2022. First results on exoplanets with Gaia include an improved mass determination of Beta Pictoris b (Snellen & Brown 2018). In addition to resolving the $\sin i$ ambiguity in the masses of radial velocity planets, Gaia has the potential to discover a new population of giant planets orbiting early-type stars that are unsuitable for detection via radial velocity measurements.

## 9.4 Opportunities

We now turn to the opportunities for exoplanet science offered by upcoming missions that are in development, planned for launch in the coming decade. The largest of these missions—JWST—is a powerful but general-purpose observatory that was not conceived as an exoplanet facility. But we will also see the launch of the first large observatory designed with exoplanet observations in mind (Roman), as well as at least two medium-scale missions. Together, these facilities will fill out our knowledge of exoplanet demographics and usher in a major transition towards exoplanet atmospheric characterization.

### 9.4.1 Upcoming Missions

#### *9.4.1.1 James Webb Space Telescope*

The James Webb Space Telescope (JWST) will provide a great leap in exoplanet atmospheric characterization, extending the initial reconnaissance of the largest and hottest transiting planets accomplished with Hubble and Spitzer down to cooler temperatures and smaller planetary radii. JWST will also provide much more extensive wavelength coverage compared with existing facilities, as well as spectra with higher spectral resolving power and signal to noise. The JWST instrument suite includes four instruments—NIRISS, NIRSpec, NIRCam, and MIRI—spanning the wavelength range between 0.6 and 28 μm. Though none of the instruments were originally designed for exoplanet observations, they have all been modified or optimized to better address the needs of both transiting exoplanet characterization and direct imaging of bright, self-luminous young planets (Figure 9.2).

The NIRISS instrument covers the wavelengths 0.6–2.5 μm using a grism with a spectral resolution of $R \sim 300$–800. A cylindrical lens will broaden the detector footprint of the spectrum to reduce interpixel noise and improve the bright-star limit to $J \sim 7$–8. Both the NIRSpec and NIRCam instruments cover the spectral range from 0.7 to 5 μm, but NIRSpec has a number of spectral resolution options ranging from $R \sim 100$ with the prism to $R = 1000$ and 2700 with gratings. NIRCam is optimized for observing very bright stars ($K \sim 4$–6) and provides photometry, coronagraphy, and low-resolution grism spectroscopy (from 2.4 to 5 μm for spectroscopy). Finally, the MIRI instrument provides photometric, coronagraphic, and spectroscopic capabilities between 5 and 28 μm, with the primary spectroscopic mode for exoplanet transits operating from 5 to 11.5 μm.

With such a broad range of capabilities, JWST will revolutionize many areas of exoplanet science. In particular, transit and eclipse spectroscopy of hot and warm giant planets will provide answers to fundamental questions about the composition and structure of gas and ice giant planets, and address theories of their formation and evolution by examining correlations between planet characteristics. Key questions that JWST will hope to answer include the atmospheric composition of giant planets as a function of system architecture, stellar type and age, and the presence and evolution of secondary outgassed atmospheres on rocky planets as a function of stellar properties.

#### *9.4.1.2 Nancy Grace Roman Space Telescope and Euclid*

The NASA-led Roman observatory will use the microlensing technique to extend our knowledge of exoplanet demographics to greater orbital distances and smaller masses. For red dwarf stars Roman will measure planet-to-host star mass ratios for planets as small as about 0.1 $M_\oplus$ at orbital separations from the ice line all the way

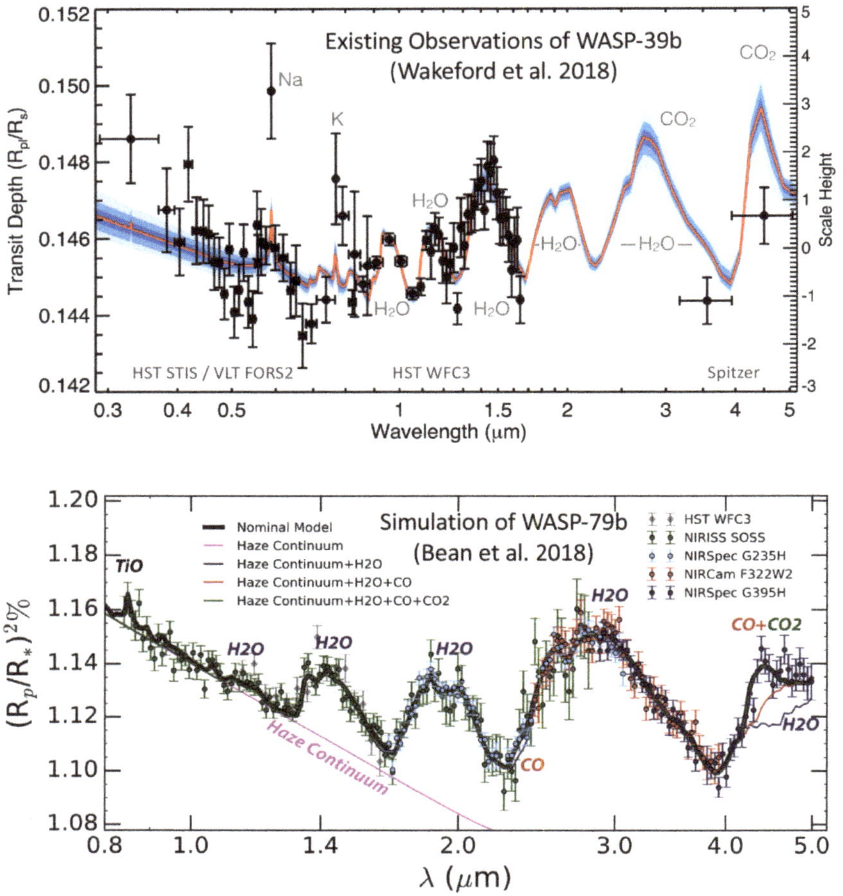


**Figure 9.2.** (Top) A combined multi-instrument spectrum of the hot Saturn-mass planet WASP-39b. Reproduced from Wakeford et al. (2018).  The data available for this planet are representative of the current state-of-the-art for broad-wavelength transmission spectroscopy measurements. (Bottom) A simulation of JWST spectra of WASP-79b showing the greater precision, resolution, and wavelength coverage that will be provided. Reproduced from Bean et al. (2018). 

to free-floating planets (Penny et al. 2019; Johnson et al. 2020). For G dwarfs, the detected planets will extend inward to the habitable zone. The Roman microlensing survey is therefore an important complement to Kepler, which was limited to planets orbiting within 1 AU of their host star.

Due to Roman's diffraction-limited angular resolution, unrelated stars that are blended with the lens and/or source in typical ground-based seeing-limited images will be resolved, enabling clean flux measurements for most bright lens stars. When combined with second-order effects that will be observable in most planetary event lightcurves from Roman, the absolute masses of the host stars and their planets will be measurable for the majority of detected systems (Bennett et al. 2007). It will also be possible to measure the projected separation of the planet from its host star and the distance to the planetary system.

The ESA-led Euclid dark energy mission shares wide-field imaging capabilities with Roman. This is a key capability for an effective microlensing survey, since lensing events are infrequent and many sources must be monitored. Euclid exoplanet microlensing surveys are not part of the prime mission but could potentially occur during an extended mission (Penny et al. 2013).

#### *9.4.1.3 PLATO*

The PLATO (PLAnetary Transits and Oscillations of stars) mission (Rauer et al. 2014, 2016) will detect and characterize a large number of extrasolar planetary systems via the transit method, with an emphasis on terrestrial planets in the habitable zone around solar-type stars. For the challenging benchmark case of an Earth around a Sun, the goal is to determine planetary radii to 3% precision for bright stars. Planetary masses with up to 10% precision will be obtained from ground-based radial-velocity follow-up data. Precise characterization of exoplanet host stars will be achieved from the PLATO asteroseismology program in combination with results from the Gaia mission, allowing the determination of stellar ages to ~10% precision for Sun-like stars in its survey fields. Such well-constrained ages for a large number of planetary systems will eventually allow us to better understand the evolution of planetary systems, similar to how we understand the evolution of stars by observing them in different phases of their life cycle. The resulting large data base of precise planet and stellar parameters will help us to track how planets form and evolve and to identify those objects which could currently host habitable conditions.

PLATO is ESA's M3 mission, with launch into orbit around the L2 Lagrangian point planned for the end of 2026. The payload consists of 26 cameras with a 12 cm aperture each, mounted on a common optical bench and operating in white light (0.5–1.0 μm). 24 cameras are grouped into four sub-groups with overlapping field-of-views (FoVs). This multi-telescope approach provides a total FoV of 2132 $deg^2$ combined with a large dynamic range for photometry. Two cameras are responsible for observatory fine-pointing. Within the nominal mission duration PLATO will observe two target fields for two years each, but has the capability to extend observations up to 8 years total and therefore could perform longer pointings as well as shorter step-and-stare sequences.

#### *9.4.1.4 ARIEL*

ARIEL (Atmospheric Remote-sensing Infrared Exoplanet Large-survey) is ESA's M4 mission, expected to launch in 2029 into the L2 Lagrangian point. ARIEL will perform transit spectroscopy of hundreds of extrasolar planets at visible and infrared wavelengths to characterize their atmospheric composition (Tinetti et al. 2018). The mission will focus on hot-to-warm known extrasolar planets, including gas giants, Neptune-mass and Super-Earth-mass planets. ARIEL will obtain spectra of their atmospheres in the 1.25–7.8 μm spectral range as well as multiple narrow-band photometric points in the optical range. The mission will determine elemental

**Table 9.1.** Key Parameters for the HabEx, LUVOIR, and Origins Mission Concepts.

| | HabEx | LUVOIR | Origins |
|---|---|---|---|
| Proposed launch date | 2036 | 2039 | 2035 |
| Telescope diameter | 4 m | 15 m (LUVOIR-A)<br>8 m (LUVOIR-B) | 5.9 m |
| # of instruments | 4 | 4 (LUVOIR-A)<br>3 (LUVOIR-B) | 3 |
| Total wavelength range | 0.115–1.8 μm | 0.100–2.5 μm | 2.8–588 μm |
| Available exoplanet observing techniques | Direct imaging<br>direct spectroscopy<br>transit spectroscopy | Direct imaging<br>direct spectroscopy<br>transit spectroscopy<br>astrometry | Transit photometry<br>transit spectroscopy<br>thermal phase curves |

compositions (especially the abundances of C, O, N, S, and Si) through the measurement of important atmospheric absorbing species, e.g., $H_2O$, $CO_2$, $CH_4$, $NH_3$, HCN, $H_2S$. Spectra will be obtained in both transit and eclipse geometry. In addition, planetary phase curves will be observed. The targeted warm and hot planets have atmospheres which are more representative to gaseous planetary bulk compositions and can therefore provide new insights into the evolution of such planets in context of their stellar environment. ARIEL consists of an off-axis Cassegrain telescope with an effective collecting area of 0.64 $m^2$. A U.S. hardware contribution "CASE" will extend the simultaneous wavelength coverage down to 0.5 μm. ARIEL will have a 4-year nominal mission lifetime, with a potential to extend for two more years.

### 9.4.2 Future Mission Concepts

#### *9.4.2.1 HabEx, LUVOIR, and Origins*

In January 2016, NASA initiated four concept studies of large space telescopes in preparation for the 2020 Astrophysics Decadal Survey (Astro2020). While all four have exoplanet applications, three are of particular interest: the Habitable Exoplanet Observatory (HabEx),[2] the Large UV/Optical/Infrared Surveyor (LUVOIR),[3] and the Origins Space Telescope (Origins).[4] These missions would all greatly advance atmospheric characterization for a wide range of exoplanets and host stars. HabEx and LUVOIR were specifically designed to tackle habitability and life for Earth-size exoplanets around solar-type stars using direct imaging and spectroscopy of reflected light, while Origins would investigate habitability for planets around M dwarf stars primarily with transit spectroscopy.

[2] HabEx Final Report at https://www.jpl.nasa.gov/habex/documents/.
[3] LUVOIR Final Report at https://asd.gsfc.nasa.gov/luvoir/reports/.
[4] Origins Final Report at https://asd.gsfc.nasa.gov/firs/docs/.

A summary of the basic characteristics of the three observatories appears in Table 9.1. All three missions have a 5-year prime mission lifetime, with 10 years of on-board consumables, and could be serviced in space. HabEx's preferred "hybrid" architecture features both a coronagraph and starshade to provide starlight suppression for near-UV/optical/near-IR high-contrast direct observations. HabEx would use its coronagraph to search the nearest solar-type stars for habitable planet candidates and use the starshade to spectrally probe them for atmospheric biosignatures, while producing a complete "family portrait" of all the planets and planetesimal belts in each system. For nominal assumptions on the frequency of temperate rocky planets and exozodi levels, HabEx would expect to find approximately eight of them. Figure 9.3 shows a simulated spectrum of a modern Earth-analog exoplanet that could be obtained with HabEx or LUVOIR (with differing exposure times). Enabling a broad range of general astrophysics and solar system studies are also important goals for HabEx.

LUVOIR also focuses on habitability and biosignatures for planets around solar-type stars, but has the more ambitious goal of measuring (or constraining) the frequencies of Earth-like planets and global biospheres by searching hundreds of stars to discover and study dozens of habitable planet candidates. Exoplanet science and general astrophysics are on a more equal footing as science goals for LUVOIR; observations of solar system bodies were also prioritized. Two variants of LUVOIR were designed: one with a 15-m obscured primary mirror (LUVOIR-A) and one with an 8-m unobscured primary mirror (LUVOIR-B). Both variants use a coronagraph to perform near-UV/optical/near-IR high-contrast direct observations. Both HabEx and LUVOIR could obtain high-quality UV/optical/near-IR transit spectroscopy for a range of planets, and could perform the complete UV spectral characterization of host stars that is critical for modeling atmospheric photochemistry.

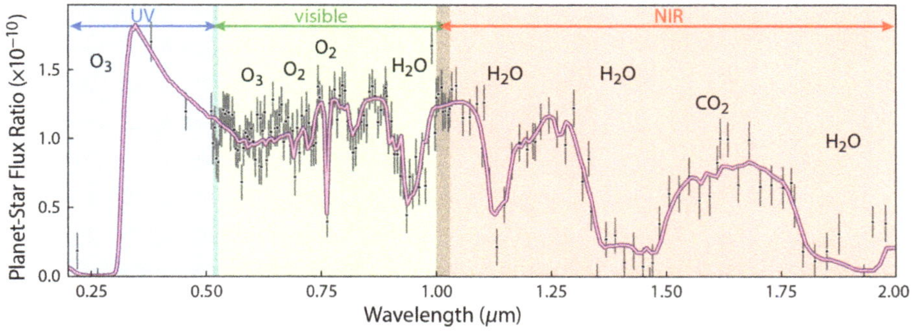


**Figure 9.3.** Simulated spectrum of a modern Earth-twin orbiting a Sun-like star, which could be obtained using HabEx or LUVOIR with differing exposure times. These spectra have S/N = 8.5 in the continuum across the whole bandpass, sufficient for measurement of key molecular abundances in the planet's atmosphere. With either mission concept, the total time to acquire this spectrum varies dramatically depending on the distance of the target system from the Sun. The shaded background regions indicate the near-UV (blue), visible (green), and NIR channels. Based upon Figures 1–10 from the LUVOIR Final Report. Credit: J. Lustig-Yeager (JHU-APL)/T. Griswold (NASA GSFC).

Finally, Origins is a concept for a cooled (4.5 K) telescope equipped for observations from mid-IR to far-IR wavelengths. It would investigate the habitability of exoplanets around M dwarfs using mid-IR transit spectroscopy and obtain phase curves of exoplanet thermal emission. Advanced detectors would allow Origins to obtain higher quality exoplanet transit spectra than expected with JWST. Origins also has a primary science goal of understanding how habitable conditions arise during planet formation by studying water and deuterated molecules in protoplanetary disks, debris disks, and solar system comets. All of these mission concepts were presented to Astro2020 in late 2019 and are currently under consideration for possible prioritization as NASA's next Great Observatory. Representatives of several international space agencies participated in these mission concept studies, and ESA has expressed interest in partnering on whichever mission is prioritized.

#### *9.4.2.2 Darwin/LIFE*

The Darwin mission concept was extensively studied by ESA during the first decade of the 2000s, complementing the Terrestrial Planet Finder (TPF) mission studies done by NASA at the time. Darwin was a mid-IR, space-based nulling interferometer aiming to directly detect and characterize terrestrial exoplanets around nearby stars in the wavelengths range 5–20 μm. Its main goal was to obtain infrared spectra that could detect signs of life (e.g., the ozone band at 9.6 μm), as well as $H_2O$, $CO_2$, $CH_4$, and $NH_3$. The technical challenges were however very large for the time and the Darwin study was eventually stopped. With advances in extrasolar planet science over the last decades, detection of biosignatures on terrestrial planets in the habitable zones of solar-type stars again became a driver for space mission concepts in Europe. The Large Interferometer for Exoplanets (LIFE) initiative was kicked-off in 2018 as a community effort. LIFE is also a concept for a space-based, formation-flying, mid-IR nulling interferometer—building from the heritage of Darwin but adding major scientific and technological developments from the last ten years. The goal of LIFE is to develop a roadmap for a space mission that will be able to detect and characterize the atmospheres of dozens of warm, terrestrial extrasolar planets at mid-IR wavelengths. An initial version of the roadmap was submitted to ESA's call for White Papers for the Voyage 2050 long-term plan (Quanz et al. 2019).

#### *9.4.2.3 Medium-Class Mission Concepts*

ESA is already developing the medium-class PLATO and ARIEL missions. If the U.S. 2020 Decadal Survey decides not to select a large mission with an exoplanet science focus, then a medium-scale "probe" mission would provide a more affordable (~$1 B) option. Three exoplanet mission concepts in this class were extensively studied by NASA during 2013–2018: "Exo-S",[5] "Exo-C" (Stapelfeldt et al. 2015), and "Earthfinder".[6] Even smaller missions at or below the Explorer

[5] Starshade Rendezvous Probe Final Report at https://science.nasa.gov/astrophysics/2020-decadal-survey-planning.

[6] Earthfinder Final Report at https://science.nasa.gov/astrophysics/2020-decadal-survey-planning.

class have been proposed to conduct a TESS-like all-sky transit survey in the near-infrared, or to focus on the $\alpha$ Centauri system with imaging or astrometry (Belikov et al. 2015).

Among these Exo-S stands out for its distinctive approach of flying an external occulter to rendezvous with the Roman Space Telescope (Seager et al. 2015). By aligning this starshade ~25,000 km from Roman along the direction to a target star, this concept would provide starlight suppression of $10^{-9}$ and enable detection of rocky exoplanets in the habitable zones of roughly two dozen nearest neighbor stars. Roman's existing CGI instrument would serve as the camera and spectrograph working with the starshade. Spectral characterization of rocky exoplanets in the HZ of 1–2 such targets is projected, if exozodiacal dust levels are not too high, while also characterizing larger outer planets in the target systems.

## 9.5 Challenges

The next few decades have the potential for revolutionary discoveries regarding the composition of exoplanetary atmospheres and the population of planets with habitable surface conditions around nearby stars. To reach these goals, a number of challenging technological hurdles must be overcome in observatory and instrument performance. We briefly review these below.

### 9.5.1 Technology Development for Direct Imaging and Spectroscopy

#### *9.5.1.1 Roman Coronagraph Instrument*

Achieving the $\leqslant 10^{-10}$ contrast ratio needed for spectroscopy of Earth-like exoplanets around solar-type stars is the major technological challenge for future missions like HabEx and LUVOIR. The highest contrast measured to date in a testbed using 10% broadband light is $2 \times 10^{-9}$ with an obscured primary aperture;[7] Roman and LUVOIR-A have obscured apertures. For an unobscured aperture, broadband contrast of $4 \times 10^{-10}$ has been achieved in a testbed; HabEx and LUVOIR-B have unobscured apertures. In general, current coronagraph designs perform better with unobscured apertures, incentivizing off-axis telescope designs for exoplanet direct imaging mission concepts. For all cases, additional performance improvements beyond contrast are needed (e.g., throughput, inner working angle).

Fortunately, major progress is being made via the Coronagraph Instrument (CGI) on the Roman Space Telescope, a technology development effort partnered with a science mission. CGI will be the first coronagraph with active wavefront control to fly in space. While CGI's required raw contrast performance is only $10^{-7}$, a factor of 10 or more improvement over this may be achieved. The level of low-order wavefront control needed while coupled to the Roman telescope, which is not at all optimized for high-contrast observations, is quite close to the requirements for the more optimal HabEx and LUVOIR-B designs (Mennesson et al. 2020). The CGI will therefore provide a valuable system-level technology demonstration and take a

[7] https://exoplanets.nasa.gov/internal_resources/1200/

large step toward making the goal of finding habitable—or inhabited—exoplanets an observational reality.

*9.5.1.2 Other Technology Developments for Direct Observations*

To meet HabEx requirements, coronagraph performance in the laboratory needs to improve by a factor of 4 in contrast, 2 in spectral bandwidth, and 3 in outer working angle (Patterson et al. 2019; Seo et al. 2019). Laboratory demonstrations to close these gaps are in the planning stages and should take place in the next year or two through upgrades to testbed facilities and advances in coronagraph architectures, masks, and wavefront control techniques. Demonstrations of the needed performance with segmented telescope pupils have received less investment to date and thus will take longer to realize. Relative to LUVOIR-B requirements, recent lab demonstrations (Llop-Sayson et al. 2020) are a factor of 400 away from the needed contrast, a factor $\geqslant 3$ away from the needed outer working angle, and do not yet treat the issue of surface discontinuities from residual segment mis-alignments. In both cases, lab demonstrations have been at the instrument level in a static environment. Coronagraph testbed facilities capable of simulating the full-scale surface errors and thermal drift of a large telescope are in development.[8]

A free-flying starshade provides a separate approach to starlight suppression that does not require exceptional wavefront control or telescope stability. Laboratory tests of a rigid, undeployed sub-scale starshade have shown that $10^{-10}$ contrast can be achieved (Harness et al. 2020). While a system-level test of a flight-scale starshade is not practical, due to its large size (tens of meters) and large distance from the telescope (tens of thousands of km), work on subsystem-level hardware is currently going on under NASA's starshade technology project "S5" (Short et al. 2019). Issues such as the precise deployment of a 26 m diameter starshade, formation flying of the starshade and telescope, and management of the level of sunlight scattered from starshade petal edges are being worked with the goal of achieving technical readiness for a starshade mission by the mid-2020s.

### 9.5.2 Technology Development for Transits

The next generation of infrared space observatories will focus on the goal of detecting and measuring the spectral signatures of biosignature gases such as $H_2O$, $CO_2$ and $CH_4$ in the atmospheres of potentially habitable rocky exoplanets orbiting cool stars. However, as stated in question 3 of Section 9.2, the signal of atmospheric absorption or emission from a cool rocky planet transiting a late M dwarf will be 10–50 parts per million. To make these detections, future near-IR and mid-IR detectors need to have high pixel measurement stability of the order of ~5 ppm over the timescale of a few hours (transit durations) to days (phase-curve durations).

Current state-of-the-art detector arrays are not able to achieve this stability—the expected performance based on Hubble and Spitzer observations is ~5–10× worse than what is needed to detect biosignatures. However, both the HgCdTe and Si:As

[8] https://exoplanets.nasa.gov/internal_resources/1170/

arrays to fly on JWST, as well as a newer technology called transition edge sensor (TES) bolometer arrays, are currently being developed to meet the required performance. Potentially enabling technologies for reaching the required measurement stability include continuous on-orbit flux calibration and/or the use of a lenslet array known as a densified pupil (Matsuo et al. 2016).

### 9.5.3 Supporting Observations

Exoplanets studied via transit or imaging observations require supporting dynamical measurements to determine the planet mass, a fundamental astrophysical parameter. Precision radial velocity measurements from ground-based telescopes can currently detect stellar reflex motions as small as $\sim$1 m s$^{-1}$ for stars with $V < 12$. A factor of at least 10 improvement will be necessary to detect the RV signal of an Earth analog orbiting a Solar analog. Investments in stellar jitter mitigation, instrument stabilization and calibration, and dedicated high-cadence observing facilities will be needed to establish whether the state of the art can be advanced to the needed level. If not, a space-based astrometry capability may prove to be necessary at substantially greater cost.

While an exoplanet's radius is reasonably well-defined by a transit detection, it is ambiguous when an exoplanet is detected in reflected light. The albedo of a terrestrial planet strongly depends on whether the surface is directly visible or covered by clouds or ice; Venus has almost 3x the albedo of Mars. This albedo uncertainty propagates into as much as a factor of two uncertainty in an exoplanet's radius, and also impacts estimates of its effective temperature. To establish the radius and $T_{\rm eff}$ of non-transiting exoplanets, observations of planetary thermal emission will be needed. The LIFE mission concept would be capable of making these measurements.

## 9.6 Conclusions

At least ten space missions from small-sats to large flagships have produced a flood of exoplanet scientific literature since HST was first used to observe HD 209548 b in 2001. While this work has been exciting and transformative, it has primarily been based on just two observational capabilities: (1) precision broadband, wide-field photometry; and (2) precision spectrophotometry and low-resolution spectroscopy. The upcoming PLATO, Roman, and ARIEL missions will be the penultimate expressions of these techniques. In addition, we can look forward to two new observational capabilities becoming available over the next few years which should transform exoplanet space astronomy: the precision astrometry results of Gaia, detecting thousands of giant planets through their stellar reflex motions and opening a new window on exoplanet demographics; and the medium-resolution mid-IR spectroscopy capabilities of JWST, providing access to new molecular probes of exoplanet atmospheres down into the rocky planet regime. The 2020s should be an exciting decade for exoplanet space astronomy.

On the horizon is the capability to suppress bright starlight and explore the planetary systems of nearby stars via high-contrast direct imaging. A large space telescope equipped with advanced starlight suppression technologies offers the only clear path to discovering and characterizing the temperate rocky planets orbiting the nearest Sun-like stars. Can exoplanet astronomers rally their colleagues in other disciplines to support such a bold and expensive project? Can the space agencies of the world come together to plan the mission, and win the support of their governments and populaces? The possibilities for finding other life in the Universe, and for finding destinations for future exploration, are exciting enough that the answers must eventually be yes.

## References

Auvergne, M., Bodin, P., Boisnard, L., et al. 2009, A&A, 506, 411
Bean, J. L., Stevenson, K. B., Batalha, N. M., et al. 2018, PASP, 130, 114402
Belikov, R., Bendek, E., Thomas, S., Males, J., & Lozi, J. 2015, Proc. SPIE, 9605, 960517
Benneke, B., Wong, I., Piaulet, C., et al. 2019, ApJL, 887, L14
Bennett, D. P., Anderson, J., & Gaudi, B. S. 2007, ApJ, 660, 781
Benz, W., Broeg, C., Fortier, A., et al. 2021, ExA, 51, 109
Borucki, W. J. 2018, in Space Missions for Exoplanet Science: Kepler/K2 (Berlin: Springer), 80
Deleuil, M., Aigrain, S., Moutou, C., et al. 2018, A&A, 619, A97
Deming, D., & Knutson, H. A. 2020, NatAs, 4, 453
Deming, D., Wilkins, A., McCullough, P., et al. 2013, ApJ, 774, 95
Fauchez, T. J., Turbet, M., Villanueva, G. L., et al. 2019, ApJ, 887, 194
France, K., Loyd, R. O. P., Youngblood, A., et al. 2016, ApJ, 820, 89
Froning, C. S., Kowalski, A., France, K., et al. 2019, ApJL, 871, L26
Harness, A., Shaklan, S., Kasdin, N., Willems, P., & Balasubramanian, K. 2020, AAS Meeting Abstracts, 373.10
Howell, S. B., Sobeck, C., Haas, M., et al. 2014, PASP, 126, 398
Johnson, S. A., Penny, M. T., Gaudi, B. S., et al. 2020, AJ, 160, 123
Kaltenegger, L., & Traub, W. A. 2009, ApJ, 698, 519
Kreidberg, L., Bean, J. L., Désert, J.-M., et al. 2014, Natur, 505, 69
Lendl, M., Csizmadia, S., Deline, A., et al. 2020, arXiv:2009.13403
Llop-Sayson, J., Ruane, G., Mawet, D., et al. 2020, AJ, 159, 79
Matsuo, T., Itoh, S., Shibai, H., Sumi, T., & Yamamuro, T. 2016, ApJ, 823, 139
Mennesson, B., Juanola-Parramon, R., Nemati, B., et al. 2020, A&A, 643, A94
Patterson, K. D., Seo, B.-J., Balasubramanian, K., et al. 2019, Proc. SPIE, 11117, 111171U
Penny, M. T., Kerins, E., Rattenbury, N., et al. 2013, MNRAS, 434, 2
Penny, M. T., Gaudi, B. S., Kerins, E., et al. 2019, ApJS, 241, 3
Perryman, M., Hartman, J., Bakos, G., & Lindegren, L. 2014, ApJ, 797, 14
Quanz, S. P., Absil, O., Angerhausen, D., et al. 2019, arXiv:1908.01316
Rauer, H., Aerts, C., Cabrera, J., & PLATO Team, 2016, AN, 337, 961
Rauer, H., Catala, C., Aerts, C., et al. 2014, ExA, 38, 249
Ricker, G. R., Winn, J. N., Vanderspek, R., et al. 2015, JATIS, 1, 014003
Seager, S., Turnbull, M., Sparks, W., et al. 2015, Proc. SPIE, 9605, 96050W
Seo, B.-J., Patterson, K., Balasubramanian, K., et al. 2019, Proc. SPIE, 11117, 111171V
Short, K., Willems, P., Lisman, D., et al. 2019, BAAS, 51, 190
Snellen, I. A. G., & Brown, A. G. A. 2018, NatAs, 2, 883
Stapelfeldt, K. R., Dekens, F. G., Brenner, M. P., et al. 2015, Proc. SPIE, 9605, 96050T

Stevenson, K. B., Désert, J.-M., Line, M. R., et al. 2014, Sci, 346, 838
Tinetti, G., Drossart, P., Eccleston, P., et al. 2018, ExA, 46, 135
Vidal-Madjar, A., des Etangs, A. L., Désert, J. M., et al. 2003, Natur, 422, 143
Wakeford, H. R., Sing, D. K., Deming, D., et al. 2018, AJ, 155, 29
Wang, J., Mawet, D., Ruane, G., Hu, R., & Benneke, B. 2017, AJ, 153, 183

# Chapter 10

# Small Satellites for Exoplanet Science

**Luca Fossati, Peter R. McCullough and Ian R. Parry**

## 10.1 Introduction

Opportunities provided by Small Satellites (SmallSats) have quickly become clear to the astrophysical community, and in particular to the exoplanet community, which is in continuous need of observational facilities (Engineering National Academies of Sciences & Medicine 2016; Fleming et al. 2018; Shkolnik 2018). The great interest of the exoplanet community in SmallSats, and in CubeSats particularly, is mainly driven by expectations that such a young field will advance especially via new observations.

CubeSats are a subset of SmallSats that carry a science payload and the service components in a volume typically contained within 30 × 30 × 30 cm and for a total mass of less than 40 kg (Helvajian & Janson 2008). In particular, the volume of CubeSats is standardized into units (U) of 10 × 10 × 10cm that can be adjoined together to form larger satellites. Typical CubeSat sizes are of 1 U, 2 U, 3 U, 6 U, 12 U, or 24 U.[1] The small mass, compact size, and standardized shape have enabled a substantial decrease in launch and deployment costs, fostering the success of CubeSats, and of SmallSat platforms more generally, for educational, technological, commercial, defense, and scientific purposes at low cost. As a matter of fact, the first CubeSats started off as educational projects. The small volume and low launch costs have created the conditions for shrinking development time and for using off-the-shelves components, hence reducing costs.

Despite the limited observing capabilities of SmallSats, their typically small sizes and data downlink rates, there are specific and important scientific questions that can be best answered by SmallSats, such as:

- Which very bright stars have multiple transiting planets?

[1] https://www.cubesat.org/

doi:10.1088/2514-3433/abfa8fch10  

- Are exoplanetary atmospheric spectra (in transmission and/or emission) time variable? If so, how does this variability behave? Is the variability related to (or not to) the intrinsic variability of the host star?

Furthermore, the small and faint nature of planets make the use of SmallSats for exoplanet research challenging, which leads to a number of relevant technical and programmatic questions, such as:

- How can we best use SmallSats to address big questions in exoplanet science?
- How can SmallSats complement capabilities of larger facilities from space and ground?
- Can we push both pointing stability and instrumental sensitivity to enable studying a significant number of planets smaller than Jupiter?

SmallSats have the potential to revolutionize exoplanet science with a flood of unique data focusing on very specific scientific questions, therefore giving us the best chances to get firm and interesting answers.

## 10.2 State of the Art

SmallSats for exoplanet research have only very recently started development and operation, but SmallSats for Astrophysical research have a relatively long history, which precedes the CubeSat era. The asteroseismic community was among the first to realize that SmallSats may be the key to obtain unique data sets. For example, a dedicated SmallSat could enable high-quality observations for bright stars for which high-quality complementary spectroscopic observations either were already available or could be obtained. More importantly, these data sets would be (almost) uninterrupted, unprecedentedly long, and unaffected by seeing, clouds, and, very importantly for some kind of pulsating stars (i.e., $\gamma$ Doradus stars), the day–night cycle of Earth-bound observations. One such mission was the Canadian Microvariability and Oscillations of Stars Telescope (MOST; Walker et al. 2003), which started operations in 2003 and stopped in 2019 as a result of a power failure. Among many contributions to exoplanet science, MOST observed the transit of a small planet, 55 Cnc e, across a star so bright it is visible to the naked eye (Winn et al. 2011). The concept of MOST has been brought forward by the BRITE constellation (Weiss et al. 2014), which is a set of $\approx$4 U SmallSats developed by Austria, Canada, and Poland,[2] which have been launched between 2013 and 2014. Five satellites are currently fully operational. These first SmallSats were enabled by the development of miniaturized reaction wheels that permitted pointing stability sufficient for astrophysical photometric observations.

One SmallSat launched specifically for exoplanet observations is ASTERIA (Knapp et al. 2020). It performs photometry from a 6 U platform and has the primary aim of testing the capabilities and limitations of high-precision pointing on

[2] Originally, the constellation comprised six satellites, but one of the Canadian satellites did not detach from the upper stage of the rocket and was therefore lost.

a small, light payload in space by observing transits of close-in planets orbiting bright stars. ASTERIA employs the tracking system developed by Blue Canyon Technologies that was previously demonstrated on the 3U MinXSS SmallSat (Mason et al. 2017). ASTERIA's photometer is used to provide additional pointing information to the attitude control system (i.e., "payload in the control loop"). In this way, ASTERIA has achieved sub-arcsecond pointing stability (Knapp et al. 2020). In addition to ASTERIA, there was PicSat (David et al. 2018), which aimed at attempting the detection of the possible transit of the Jupiter-like planet $\beta$ Pic b, which had been detected through direct imaging. Unfortunately, the mission failed during commissioning, possibly due to a failure of its reaction wheels.

Inspired by the success of these missions and by the demonstrated pointing capabilities of SmallSats, the exoplanet community has started the development of a few new missions, some of them already in the assembly phase. The next one to be launched (late 2021) is the Colorado Ultraviolet Transit Experiment (CUTE; Fleming et al. 2018; Sreejith et al. 2019), which will observe exoplanets transiting bright, nearby stars to obtain near-ultraviolet transmission spectra. The data will be used to observationally characterize the upper layers of planetary atmospheres to study atmospheric escape, atmospheric evolution, and star-planet interactions. Figure 10.1 illustrates the 6U spacecraft built by Blue Canyon Technologies, the 20×8cm rectangular Cassegrain telescope, and the low-resolution spectrograph with an ion-etched holographic grating operating from 2500 Å to 3300 Å (comparable to that of HST COS G230L).

The year after CUTE was approved by NASA, the agency gave the green light to the development of the Star-Planet Activity Research CubeSat (SPARCS; Shkolnik et al. 2018), which will stare at about a dozen stars. SPARCS consists of a 9 cm reflector telescope attached to a pair of CCD detectors within a 6 U spacecraft. The mission will focus on characterizing the ultraviolet variability of low-mass stars, because of their relevance in the search for habitable-zone planets, in particular to improve our understanding of their flare and quiescent ultraviolet emission. This is relevant for understanding planetary atmospheres as ultraviolet radiation drives photochemistry and escape (e.g., Yelle 2004; France et al. 2016), hence it will have a significant impact on planetary atmospheric evolution (e.g., Kubyshkina et al. 2018).

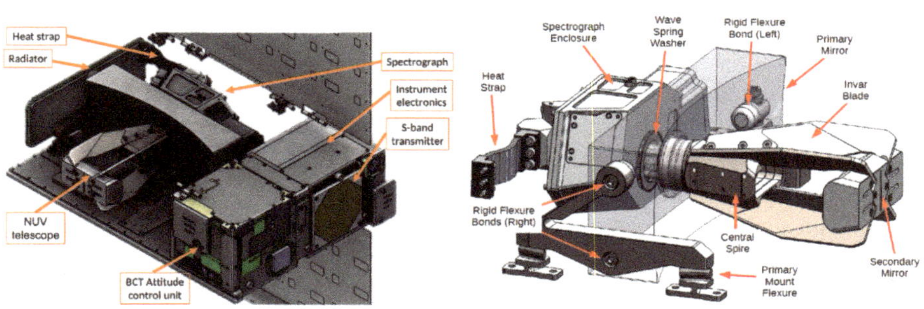


**Figure 10.1.** Schematic of the CUTE satellite (left) and of the CUTE instrument (i.e., telescope and spectrograph; right). Main components are labeled. Right: reproduced from Egan et al. (2018). © (2018) Society of Photo-Optical Instrumentation Engineers (SPIE).

The proposed Skyhopper 12U CubeSat mission (PI: M. Trenti, University of Melbourne; skyhopper; Mearns & Mearns 2018) aims to observe exoplanet transits in the near-infrared (see https://icubesat.files.wordpress.com/2017/05/2017-b-1-2-20170530174534-skyhopper.pdf). It will be very sensitive to water features in the exoplanet atmosphere because it will be unaffected by telluric contamination. It will also search for exoplanet transits of ultra-low mass stars and brown dwarfs, which are very faint at optical wavelengths.

## 10.3 Important Questions/Goals

A SmallSat characteristically is designed to answer robustly one or two very specific scientific questions. In principle this bespoke nature of their design can be an advantage. Also, due to the fast technological development of SmallSats, the major questions and goals related to SmallSats are both scientific and technological.

The need for space-based facilities, typically arises because of the need for observations at wavelengths unreachable from the ground (e.g., x-ray, ultraviolet, and particular infrared wavelengths) or because of the need of reaching a certain instrumental and/or photometric stability required to observe astrophysical time variability. Specifically, SmallSats may be ideally suited to determine whether specific atmospheric features observed in transmission and/or emission spectra of exoplanets are time variable. Furthermore, if any variability is indeed detected, SmallSats will be able to characterize the behavior of the variability and, more importantly, identify whether it is exclusive to the planetary atmosphere or it is connected to the intrinsic variability of the host star. SmallSats are in this unique position because the space-based platform can enable continuous observations uninterrupted by the day–night cycle. Furthermore, because of the dedicated nature of these missions, SmallSats enable monitoring, which is impractical on a shared-used, large facility, such as HST.

One important scientific question in the exoplanet field that SmallSats are well suited to tackle is which of the brightest stars host small, transiting planets and particularly which host multi-planet systems. Because of the lack of suitable comparison stars, photometric monitoring of very bright stars from the ground is very difficult, if not impossible (Weiss et al. 2014). In addition, the small apertures of SmallSats enables observing the brightest stars without saturation, which can occur with larger-aperture telescopes in some observational modes.

On the other hand, the small apertures of SmallSats are also a disadvantage: in the exoplanet field, their use (to date) has been limited almost exclusively to look for and study Jupiter-sized planets. Therefore, one of the key future technological goals will be to routinely reach a sub-arcsecond pointing stability and significantly increase the instrumental sensitivity to enable detecting and studying exoplanets much smaller than Jupiter.

The SmallSats currently under development and those that will be developed in the near future will fly at a time when the community will have access to an unprecedented number of observational facilities, many of them with an unprecedented size, such as the extremely large telescopes (ELTs) and JWST. One of the key goals for the whole community has to be finding the best ways for SmallSats to complement those large facilities. In the short term, the monitoring capability of SmallSats is an obvious unique niche, but we expect many creative scientists will find other tasks for SmallSats used in conjunction with much larger facilities.

Finally, in the near future, as SmallSat technology matures, the community will need to identify the best ways for using these facilities to address the big questions in exoplanet science, if at all possible. This is probably the major goal and challenge ahead of the exoplanet community interested in SmallSats.

## 10.4 Opportunities

The "SmallSat revolution" has already started: SmallSats offer a myriad of opportunities both scientific, technological, programmatic, and in respect to the growth of the exoplanet community.

One of the main scientific opportunities provided by SmallSats is that they uniquely enable us to monitor targets for long periods of time nearly without interruption, which is impractical or impossible from the ground (because of the day–night cycle) and also from larger (hence more expensive), shared-used space-based facilities. For these same reasons and because of the limited capabilities of the payloads, each SmallSat typically concentrates on a single science question, in order to get clear answers from the observations.

Thanks to the fast development of SmallSats, which is per se an opportunity, we can use these facilities to gain access to wavelengths that would be otherwise unreachable before the development and launch of the next major space mission(s). This is, for example, the case of the ultraviolet spectral range, which is currently observable for exoplanet science exclusively by HST, which has a limited lifetime ahead, while the next large mission with ultraviolet capability is not expected to happen before the late 2030s. This fact is one of the main drivers for the CUTE and SPARCS SmallSats.

On the technological and programmatic side, the fast development and low cost of SmallSats paves the way towards the formation of constellations of SmallSats, each capable of a different kind of observation and complementarity to each other, one of the characteristics of the BRITE constellation (Weiss et al. 2014). This capability may enable planning simultaneous measurements in multiple different wavelength bands, which otherwise may require winning observing time simultaneously on different observatories. Purely on the technological side, the fast development of SmallSats makes them ideal test facilities of new technology in space, thereby increasing the technology readiness level (TRL) of a given component enabling then its safe use on larger missions. The SPRITE CubeSat is one of those missions: it will study the structure and ionizing spectrum of nearby galaxies and test in space the next generation of coatings for far-ultraviolet astronomy to be employed for future larger missions (Fleming et al. 2019).

Without doubt the opportunity given by SmallSats is fascinating, even more so for the younger generation of scientists. Therefore, SmallSats give the opportunity to attract more students towards astrophysical research, exoplanets in particular, and to give new, fresh creativity, intellect, and energy to the community at large.

## 10.5 Challenges

The youth of SmallSat technology and the small size of these facilities bring numerous challenges, not only technological. Because of the limited volume and the risk of a short lifetime associated with low cost, SmallSats are typically not radiation hardened, therefore electronic components are often damaged by high-energy particles within a short period of time. However, the community has already realized that these are still precious, unique facilities that one would like to last for as long as possible, hence the need and challenge to find cheap, efficient, and miniaturized solutions for protecting the on-board electronics. This would also enable SmallSats to fly at higher altitudes, significantly increasing mission lifetimes. A further technological challenge the community will have to tackle to take full advantage of the opportunities given by SmallSats is the need of generating large collecting areas. This is possible, for example, by combining multiple SmallSats flying in formation or by implementing unfolding optics, e.g., as suggested by the SUPERSHARP project (Parry et al. 2018).

One of the hardest challenges that is faced when developing a SmallSat is finding the right balance between keeping costs low, development time short, and yet maintaining the required safety; that is, the right amount of testing and quality of the components in order to give a good probability that the mission will succeed. As more and more SmallSats for astrophysical research are developed finding this balance will become easier, but it will still require time until enough experience has been built up in the community. Help comes from the large number of new companies developing SmallSat components: this can decrease cost, but it also can increase the chance of encountering bad components. Again, this will require time until the community has identified the safe providers for SmallSat components.

The development of SmallSats will require the support of both funding and space agencies. First of all, only recently, some agencies became open to the idea of SmallSats, while others remain skeptical: as the first few SmallSats for exoplanet research (hopefully) succeed, that success will demonstrate to the agencies the power of these facilities. The support of the agencies will also be helpful for gaining launch slots, which are still available, but that are quickly taken up by SmallSats for commercial purposes. On the other hand, the relatively low cost of SmallSats will also enable non-agency-funded space missions.

## References

David, L., Lacour, S., Lapeyrere, V., Crouzier, A., & Nowak, M. 2018, in 42nd COSPAR Scientific Assembly, Vol. 42, E4.1-22-18

Egan, A., Fleming, B. T., & France, K. 2018, Proc. SPIE, 10699, 106990C

Fleming, B. T., France, K., Nell, N., et al. 2018, JATIS, 4, 014004

Fleming, B. T., France, K., Williams, J., et al. 2019, Proc. SPIE, 11118, 111180U

France, K., Loyd, R. O. P., & Youngblood, A. 2016, ApJ, 820, 89

Helvajian, H., & Janson, S. W. 2008, Small Satellites: Past, Present, and Future, (El Segundo, CA: Aerospace Press)
Knapp, M., Seager, S., Demory, B.-O., et al. 2020, AJ, 160, 23
Kubyshkina, D., Fossati, L., & Erkaev, N. V. 2018, A&A, 619, A151
Mason, J. P., Baumgart, M., Rogler, B., et al. 2017, JSSat, 6, 651
Mearns, R., & Trenti, M. 2018, arXiv:1808.06746
Engineering National Academies of Sciences and Medicine, 2016, Achieving Science with CubeSats: Thinking Inside the Box (Washington, DC: National Academies Press)
Parry, I., Queloz, D., Kennedy, G., et al. 2018, arXiv:1801.06111
Shkolnik, E. L., Ardila, D. R., Barman, T., et al. 2018, AGU Fall Meeting Abstracts, 2018, P24C-05
Shkolnik, E. L. 2018, NatAs, 2, 374
Sreejith, A. G., Fossati, L., Fleming, B. T., et al. 2019, JATIS, 5, 018004
Walker, G., Matthews, J., Kuschnig, R., et al. 2003, PASP, 115, 1023
Weiss, W. W., Rucinski, S. M., Moffat, A. F. J., et al. 2014, PASP, 126, 573
Winn, J. N., Matthews, J. M., Dawson, R. I., et al. 2011, ApJ, 737, L18
Roger, V. 2004, Icar, 170, 167

# Part III

## Exoplanet Theory and Characterization

# Chapter 11

# Orbital Dynamics and Architectures of Exoplanets

**Daniel Fabrycky**

## 11.1 Introduction

Over the past few decades, numerous programs have secured data to define planetary orbital architectures. Parallel orbital dynamics theory has sought to link these discoveries to the formation and evolution of planetary systems. Several waves of information have "come online" over the last 25 years. Orbital eccentricities spanning the whole range startled investigators and required explanation in the late 1990s, leading to scattering theories. Resonant period ratios and packed multi-planet systems were discovered in the early 2000s, and the dynamics of migration and stability took over many dynamists' research portfolios. Spin–orbit alignment found its place in the cabinet of architectural curios around 2010, prompting study of multi-body torquing interactions, but also sending theorists back to study the fundamentals of protoplanetary disks and tides. Mutual inclinations between planets became relevant while interpreting data from NASA's Kepler space telescope (Kepler) from 2011 to the present, which challenges theorists to revisit scattering for giant planets, but also to delve into hierarchical merging during the giant impact stage of terrestrial planet formation.

Questions that remain from the past 25 years, and may be addressable in the next decade, are:

- Are resonant chains a rare result of finely-tuned migration, or are they nearly ubiquitous transition phases between formation and the systems we observe?
- How should we interpret the regularities of planetary systems?
- To what extent are spin–orbit alignment properties set by the protoplanetary disk, by dynamical interactions between bodies, and by tidal interactions?
- The solar system has the potential to become unstable in the future; do the statistics of planetary architectures evolve over the lifetime of the star due to dynamical winnowing?

doi:10.1088/2514-3433/abfa8fch11  

We would also like to anticipate what the 2020s has in store. I suggest the next exciting frontier of orbital architectures is the history of systems throughout their stellar lifetimes. The techniques to assess stellar age are maturing, and opportunities to discover planets at young ages are opening—in a few cases it seems the formation event itself is being witnessed. Surely this latest wave of data will drive theory in new, exciting directions.

## 11.2 State of the Art

The state of the observational inputs were relatively recently reported in a review by Winn & Fabrycky (2015). The goal of this contribution is not to repeat such a wide-angle review, but to zoom in on particular questions that continue to gnaw at the community and highlight some of the results relevant to these topics.

### 11.2.1 Resonances

The orbital phases ($\lambda$) of satellites Io, Europa, and Ganymede around Jupiter have a resonant lock—their combination $\lambda_I - 3\lambda_E + 2\lambda_G$ holds steady at 180°. This three-body resonance, first understood by Laplace, is attributed to orbital expansion of the inner satellites due to the dissipation of tides they raise in Jupiter. The interaction of the pairs of bodies is seen more clearly in the Neptune–Pluto 2-body mean-motion resonance (MMR), in which the relative positions when Pluto comes to perihelion are governed by a 3:2 period ratio. The inferred orbital expansion in this case is Neptune's, due to planetesimal scattering.

A number of resonant chain systems have been noticed in various exoplanet data sets. See Figure 11.1, where the resonant chains with 3-body resonances (generalized Laplace resonances) are shown at the top, and systems of at least two back-to-back pairs near MMR are shown at the bottom. The first MMR among exoplanets was GJ 876 (Marcy et al. 2001); a third planet in the chain was discovered almost a decade later (Rivera et al. 2010). The directly-imaged system HR 8799 was suspected to be a resonant chain within weeks of its discovery (Fabrycky & Murray-Clay 2010), and that hypothesis has held strong with more data tracing the orbital trajectories (Wang et al. 2018) and more sophisticated analysis identifying the dynamical state of the system (Goździewski & Migaszewski 2018). In the Kepler era, Kepler-80 and Kepler-223 were noticed as having Laplace resonances in the first large catalog (Lissauer et al. 2011). Several others, including the famous TRAPPIST-1 (Luger et al. 2017), are now known to have 3-body resonances. One fascinating trend is that the chains consisting of sub-Neptunes consist of 3:2 and closer resonances, whereas the chains hosting giants have 2:1 resonances (Figure 11.1; for reference, the largest planets in Kepler 223 and Kepler 31 have radii slightly bigger than Neptune's). The interpretation is not yet certain: (1) planets of lower mass may simply form interior to the 2:1 from each other, and differential migration may tighten them to smaller period ratios from there, or (2) migration of embedded planets may be quick enough to overcome the torque that the 2:1

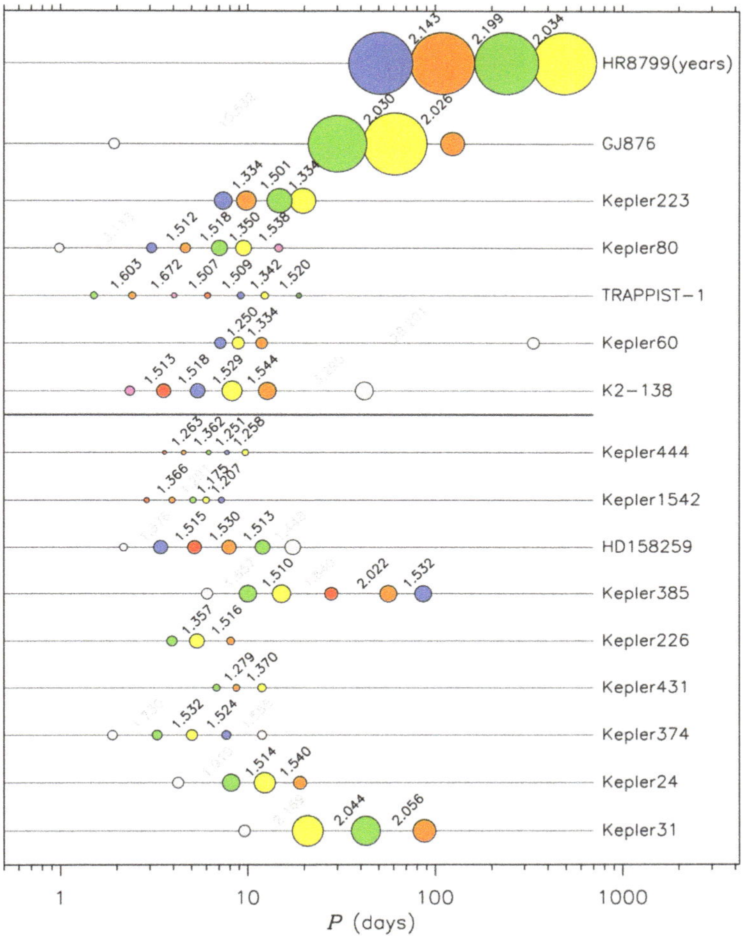


**Figure 11.1.** Gallery of resonant chain systems. Bright colors are for planets near resonances, selected by the criterion $0 < \frac{P_2}{P_1}\frac{j}{j+1} - 1 < 0.03$ with the period ratio written in black between and above each pair of planets; gray planets and gray numbers indicate planets and period ratios not deemed near-resonant. The systems above the heavy line exhibit 3-body Laplace resonances and have already been modeled as resonant chains in many journal articles. The selection of the systems below the heavy line was inspired by Jiang et al. (2020); who searched for resonant-rich systems—we require two adjacent pairs of planets fulfill the selection criterion above. These systems merit further dynamical investigation. Approximate relative size of the planets is plotted; these are probed by transits for most systems, but guessed for the directly-imaged system HR 8799 and the Doppler systems GJ 876 and HD 158259. The planets of HR 8799 are plotted at their orbital period in years, rather than days as for the rest of the systems.

resonance can provide, and the pair stabilizes at the 3:2 or narrower resonances (Quillen 2006).

*11.2.1.1 Transit Timing*

Resonant systems generate planet–planet interactions that are actually observable. In the Kepler era, transit timing variations became an extremely useful tool to confirm planet candidates. Observing these dynamical interactions has measured planetary masses and architectural properties like eccentricities and mutual inclinations. As a highlight, we now know the masses and orbital eccentricities of terrestrial planets in the habitable zone around one star, TRAPPIST-1, to few-percent precision (Agol et al. 2021).

Ground-based transit surveys could mainly find hot Jupiters; Kepler could find small planets but mainly around faint stars; the survey design of TESS has made Doppler a feasible method to confirm, and measure the masses of, small transiting planets. For those systems being discovered now, transit timing will likely play a *support* role, determining eccentricities to a precision of $\sim$0.01 (Hadden & Lithwick 2014), far better than the $\gtrsim$0.1 level that radial velocities can manage. Those measurements are important for theories to be discussed in Section 11.2.4.

### 11.2.2 Regularities in Systems

Most planetary systems, however, do not display resonances. Regularities in the Kepler discoveries show the way to learning more about how the more common systems form. Lissauer et al. (2011) and Ciardi et al. (2013) found a remarkable consistency between the radii of adjacent pairs, with a very slight preference for the outer planet being bigger within a system. Another regularity is orbital spacing, where a common logarithmic spacing is respected throughout a system (Weiss et al. 2018). This regular spacing signal is likely a true, astrophysical signature of the formation of planetary systems, because forward models (He et al. 2019) have not reproduced it based on selection effects alone (Gilbert & Fabrycky 2020).

It is the case that oligarchic growth is expected to yield comparably-sized bodies with a comparable spacing (Kokubo & Ida 1998). However, the final formation step of these bodies crossing on eccentric orbits could well break regularities. What is so remarkable about the radius regularity is that the *gas fraction* determines the radius of sub-Neptunes (Lopez & Fortney 2014), so the planets must have been embedded in the disk and accreted similar amounts of gas, and not lost it in a stochastic fashion during further giant impacts with each other. Fleshing out the full import of the regularities observed by Kepler will challenge formation theories for some time.

### 11.2.3 Spin–Orbit Alignment

Another architectural property indicative of formation is the tilt of a planet's orbit relative to its host star's spin. This property can be revealed by a spectroscopic signature during transit, the Rossiter effect. The review by Winn and Fabrycky (2015) covered most of the spin–orbit measurements to date. Several new trends

(Muñoz & Perets 2018) have been noted, and the Rossiter technique has been extended to smaller planets, for instance, Pi Men c (Hodžić et al. 2020).

A clever analysis of spot amplitudes, also useful for smaller planets, was pursued by Mazeh et al. (2015). They found that Sun-like stars have larger variability due to spots if they host transiting planets. The geometrical interpretation is that the star is equator-on, hence its near-equatorial spots are showing us their full face rather than being foreshortened; hence, spin–orbit alignment is statistically preferred. However, above an effective temperature of 6000 K, the spot amplitudes for stars with transiting systems reduces to amplitudes *lower* than stars without transiting planets. After accounting for selection effects, they concluded that hot stars being misaligned with their planets (Winn et al. 2010) was a feature that holds for small planets, too. I would point out that *if* the hotter stars have spots that are preferentially near the poles rather than the equator, this result could be interpreted as continued spin–orbit alignment. Spot location changing with temperature seems plausible, as the lack of convection zone for these hotter stars might cause such behavior.

Hot stars having high obliquities (i.e., spin–orbit misalignment) preferentially still needs to be understood. Otherwise, we cannot truly say we have a handle on why some planets have large spin–orbit misalignments. It is a true puzzle that is fundamental to planetary architectures, and it deserves more observational material in different parts of parameter space and perhaps a fresh perspective or theoretical approach.

### 11.2.4 Long-term Instability

All the foregoing properties are shaped by the formation events, but also by the long-term stability of systems. The modus operandi for many observers, after discovering a candidate planetary system, is to do a short numerical integration (short relative to the age of the system), which is easy to do with relatively user-friendly codes.[1] If their proposed system turns out to be violently unstable, it is unlikely to be the correct interpretation of the data.

However, this technique may miss out on instabilities that happen on many secular timescales. A triumph of the last few decades of solar system research is the realization that, not only is the terrestrial-planet system chaotic on a ~5 Myr Lyapunov timescale, this chaos has a chance of destabilizing Mercury's orbit in the future (Laskar 1994; Batygin & Laughlin 2008; Laskar & Gastineau 2009)!

One way of conceptualizing secular instabilities is to ask how much can planets change each other's orbits, not by changing semimajor axis and thus orbital energy, but by secular torques that change eccentricity magnitude and orientation, as well as inclination magnitude and its direction of tilt. The concept of angular momentum deficit (AMD) meets the challenge (Laskar 2000). This quantity is the amount of angular momentum one must extract from a planet to go from its presumably primordial coplanar and circular orbit, to its current orbit. Since total system angular momentum is conserved, planets can only trade AMD among themselves; the system as a whole cannot gain or lose it. Therefore as one planet becomes more

[1] REBOUND, downloadable at https://github.com/hannorein/rebound has the most support currently.

eccentric, other planets must become more circular. For certain arrangements of planets' semimajor axes, masses, and system AMD, it may be impossible for planets to cross orbits.

However, if a system has an AMD large enough to generate destructive crossings, then it may still be stable for long times. The key is that two or more secular resonances need to be active, and the system responding to each forcing simultaneously can drive it unstable (Lithwick & Wu 2011). Then the energy of the various eccentricity modes can flow freely through the system, sometimes yielding very high eccentricities for the planets that have lower masses or shorter-period orbits.

In the solar system, a very specific interaction, caused by Jupiter's precession due to mostly Saturn, and mediated by Venus, can cause Mercury to get excited to extremely high eccentricity (Batygin & Laughlin 2008; Boué et al. 2012). The final state is often a collision with other planets, and even a planetary collision with Earth, annihilating all life, is a possible outcome. So the inner system really is susceptible to instability because of the exact arrangement of the outer giant planets.

What was relatively recently appreciated, however, is that many inner exoplanetary systems—perhaps about 1/3—have giant planets exterior to them (Zhu & Wu 2018; Bryan et al. 2019). This number is about twice higher than it would be if there is no correlation between these types of planets. Turning it around, the data are consistent with the long-period gas giants *always* having small planets on interior orbits! That is great news for core accretion theorists (Zhou et al. 2005), who predicted such small planets, interpreting them as "failed cores" that did not quite reach the threshold to capture a massive gaseous envelope.

For the present discussion, the exterior giants hold a large reservoir of AMD that, if coupled to the inner system, can destabilize those small planets. For both the consistency of observational interpretation and understanding of how systems will survive for the star's full lifetime, more work on secular chaos is needed.

## 11.3 Opportunities

Several observational developments are opening opportunities to learn how systems evolve as a function of time.

Stellar modeling has had a radical makeover with the results of the Gaia satellite. Now we know the distance, to good precision, of almost all planet host stars, as well as their photometric measurements of brightness in a variety of bands. Combined, the size of the star, and hence evolutionary state, can be much more clearly defined. This technique can separate protostars that have not shrunk to the main sequence and old stars that have left it, from stars that are on the main sequence, the long-lived hydrogen-burning part of their lives.

Galactic orbits tend to start out on circular orbits near the plane of the Milky Way, where stars are born. Encounters with giant molecular clouds and spiral arms increase stellar orbits' eccentricity and inclination as a function of stellar age. As an individual example, one of the oldest planetary systems, Kepler-444 (Campante et al. 2015) has a large space velocity (Emily Gilbert, 2018 private communication). But due to their random-walk nature, space motions are not a reliable technique to determine the age of any one star; one could separate large samples of stars into classes and look for statistical dependence of the architecture. McTier & Kipping (2019) cautioned that spurious conclusions about how occurrence depends on Galactic space velocities can result if the comparison sample is not carefully tailored to take into account selection effects. When so doing, they found no dependence of planet occurrence on Galactic space velocity. Using the HARPS radial velocity sample, Bashi et al. (2020) reached a similar conclusion.

Gyrochronology (Sydney 2003) is a technique that has come into its own in the last ten years. After about 100 million years, main-sequence stars of the same subtype within FGK will have almost the same spin period, as traced by spot variability. They slow down in time, as magnetized winds carry away angular momentum. This phenomenon has been calibrated (Meibom et al. 2015; van Saders et al. 2019), and we can determine to within a factor of tens of percent how old stars are when they are between 0.1 and 1.0 Gyr. These are crucial times for dynamical instabilities to manifest.

Planets in disks observed by ALMA are an exciting way to determine the architectures of very young (even still-forming) systems, e.g., the PDS 70 system (Keppler et al. 2018). Even when planets are not directly imaged, the statistics of gaps can be compared with numerical simulations to determine the likely population of embedded planets (Zhang et al. 2018). Kinks in the velocity structure of the gas (Pinte et al. 2018, 2020) and kinematics of the entire gap (Teague et al. 2018) constitutes further dynamical evidence of those planets. By measuring the shape of those gaps and putting limits on shadowing phenomena (Matsakos & Königl 2017; Nealon et al. 2019), precise determinations of eccentricities and inclinations for these young planets should be possible.

## 11.4 Important Questions/Goals

Architectures may evolve on the gigayear timescale, which we may now be able to measure with these techniques. A brief introduction to three possible modes of evolution follows.

Due to tidal decay, are close-in massive planets consumed by their stars (Levrard et al. 2009; Matsakos & Königl 2015)? The first quite secure detection of planetary orbital decay is of WASP-12b, by Maciejewski et al. (2016). Jackson et al. (2009) used the stellar ages determined by observers to find hot Jupiters are around younger stars than average, concluding that they are consumed during the stars' main-sequence lifetime. Putting the conclusion of planetary consumption on stronger statistical footing, Hamer & Schlaufman (2019) used the Galactic velocity technique to reach the same conclusion. They also quantified the strength of tidal dissipation that accounts for that observation. Further confirmation or testing of this important evolutionary pathway is of great interest.

Do exoplanetary systems continue having eccentricity-generating instabilities during the main sequence? Kaib et al. (2013) noted that the very widest binaries have the most eccentric planets. They suggested that fly-bys during the Galactic wanderings of the star will drive the binaries to fluctuate eccentricity, and a periastron passage could disrupt the planetary system and result in very eccentric planets. If this is truly the mechanism, then it builds with time, and hence the very eccentric planets may preferentially populate old stars.

Do mean-motion resonances generate late-time instabilities? Barnes et al. (2015) found that resonant dynamics of planets could be chaotic, generating an instability on gigayear timescales. This result recalls that of Wisdom (1982), who found that the asteroids in 3:1 resonance with Jupiter could have very long residence times before acquiring enough eccentricity to cross Mars' orbit and eventually become Earth-crossing. A tentative detection of especial youth among the resonant planets (Koriski & Zucker 2011) was apparently recently overturned (Safsten et al. 2020). Meanwhile, the statistics of architectures of Kepler's multiple-planet systems may be explained if most systems migrated into chains of multiple planets in resonance, most of which disrupted (Izidoro et al. 2017). More investigations into the longevity of resonances, and especially any observational hints, is scientifically compelling.

## 11.5 Challenges

The main challenge to learning more about orbital architectures is that it may be a victim of its own success: it is hard to increase the numbers of planetary systems that are as easily characterized as the systems that have been found by Kepler. So far K2 and TESS have found only a fraction of the yield of the Kepler prime mission. However, they have found them around a much wider variety of stars (e.g., David et al. 2019). Therefore, probing the time dimension (i.e., across stellar age) is likely the most fruitful new dimension to explore in the study of orbital architectures and dynamics.

Summarizing all the great work going on in this field is a challenge in its own. An empirical look into this topic was published by Winn & Fabrycky (2015); and a theory-slanting review was recently published by Papaloizou (2020); which are both recommended for further reading.

This contribution has looked back at the history of exoplanets for key architectural observables that inform orbital dynamics and summarized the state of the art. It has glimpsed a new observable in the stellar age and proposed a few ways in which planetary architectures may be correlated with stellar age. May it serve as a challenge to both observers and theorists to deepen the understanding of planetary systems.

## Acknowledgments

I thank D. Seligman, G. Gilbert, and A. Hamann for input on a draft of this contribution.

## References

Agol, E., Dorn, C., Grimm, S. L., et al. 2021, PSJ, 2, 1
Barnes, R., Deitrick, R., Greenberg, R., et al. 2015, ApJ, 801, 101
Barnes, S. A. 2003, ApJ, 586, 464
Bashi, D., Zucker, S., Adibekyan, V., et al. 2020, A&A, 643, A106

Batygin, K., & Laughlin, G. 2008, ApJ, 683, 1207
Boué, G., Laskar, J., & Farago, F. 2012, A&A, 548, A43
Bryan, M. L., Knutson, H. A., Lee, E. J., et al. 2019, AJ, 157, 52
Campante, T. L., Barclay, T., Swift, J. J., et al. 2015, ApJ, 799, 170
Ciardi, D. R., Fabrycky, D. C., Ford, E. B., et al. 2013, ApJ, 763, 41
David, T. J., Petigura, E. A., Luger, R., et al. 2019, ApJL, 885, L12
Fabrycky, D. C., & Murray-Clay, R. A. 2010, ApJ, 710, 1408
Gilbert, G. J., & Fabrycky, D. C. 2020, AJ, 159, 281
Goździewski, K., & Migaszewski, C. 2018, ApJs, 238, 6
Hadden, S., & Lithwick, Y. 2014, ApJ, 787, 80
Hamer, J. H., & Schlaufman, K. C. 2019, AJ, 158, 190
He, M. Y., Ford, E. B., & Ragozzine, D. 2019, MNRAS, 490, 4575
Hodžić, V. K., Triaud, A. H. M. J., Cegla, H. M., Chaplin, W. J., & Davies, G. R. 2021, MNRAS, 502, 2893
Izidoro, A., Ogihara, M., Raymond, S. N., et al. 2017, MNRAS, 470, 1750
Jackson, B., Barnes, R., & Greenberg, R. 2009, ApJ, 698, 1357
Jiang, C.-F., Xie, J.-W., & Zhou, J.-L. 2020, AJ, 160, 180
Kaib, N. A., Raymond, S. N., & Duncan, M. 2013, Natur, 493, 381
Keppler, M., Benisty, M., Müller, A., et al. 2018, A&A, 617, A44
Kokubo, E., & Ida, S. 1998, Icar, 131, 171
Koriski, S., & Zucker, S. 2011, ApJL, 741, L23
Laskar, J. 1994, A&A, 287, L9
Laskar, J. 2000, PhRvL, 84, 3240
Laskar, J., & Gastineau, M. 2009, Natur, 459, 817
Levrard, B., Winisdoerffer, C., & Chabrier, G. 2009, ApJL, 692, L9
Lissauer, J. J., Ragozzine, D., & Fabrycky, D. C. 2011, ApJS, 197, 8
Lithwick, Y., & Wu, Y. 2011, ApJ, 739, 31
Lopez, E. D., & Fortney, J. J. 2014, ApJ, 792, 1
Luger, R., Sestovic, M., Kruse, E., et al. 2017, NatAs, 1, 0129
Maciejewski, G., Dimitrov, D., Fernández, M., et al. 2016, A&A, 588, L6
Marcy, G. W., Butler, R. P., Fischer, D., et al. 2001, ApJ, 556, 296
Matsakos, T., & Königl, A. 2015, ApJL, 809, L20
Matsakos, T., & Königl, A. 2017, AJ, 153, 60
Mazeh, T., Perets, H. B., McQuillan, A., & Goldstein, E. S. 2015, ApJ, 801, 3
McTier, M. A. S., & Kipping, D. M. 2019, MNRAS, 489, 2505
Meibom, S., Barnes, S. A., Platais, I., et al. 2015, Natur, 517, 589
Muñoz, D. J., & Perets, H. B. 2018, AJ, 156, 253
Nealon, R., Pinte, C., Alexander, R., Mentiplay, D., & Dipierro, G. 2019, MNRAS, 484, 4951
Papaloizou, J. C. B. 2020, The Orbital Architecture of Exoplanetary Systems, Oxford Research Encyclopedia of Planetary Science (Oxford: Oxford Univ. Press)
Pinte, C., Price, D. J., Ménard, F., et al. 2020, ApJL, 890, L9
Pinte, C., Price, D. J., Ménard, F., et al. 2018, ApJL, 860, L13
Quillen, A. C. 2006, MNRAS, 365, 1367
Rivera, E. J., Laughlin, G., Butler, R. P., et al. 2010, ApJ, 719, 890
Safsten, E. D., Dawson, R. I., & Wolfgang, A. 2020, AJ, 160, 214
Teague, R., Bae, J., Bergin, E. A., Birnstiel, T., & Foreman-Mackey, D. 2018, ApJL, 860, L12

van Saders, J. L., Pinsonneault, M. H., & Barbieri, M. 2019, ApJ, 872, 128
Wang, J. J., Graham, J. R., Dawson, R., et al. 2018, AJ, 156, 192
Weiss, L. M., Marcy, G. W., Petigura, E. A., et al. 2018, AJ, 155, 48
Winn, J. N., Fabrycky, D., Albrecht, S., & Johnson, J. A. 2010, ApJL, 718, L145
Winn, J. N., & Fabrycky, D. C. 2015, ARA&A, 53, 409
Wisdom, J. 1982, AJ, 87, 577
Zhang, S., Zhu, Z., Huang, J., et al. 2018, ApJL, 869, L47
Zhou, J. L., Aarseth, S. J., Lin, D. N. C., & Nagasawa, M. 2005, ApJL, 631, 631
Zhu, W., & Wu, Y. 2018, AJ, 156, 92

# Chapter 12

# Planet Formation

**Ravit Helled and Alessandro Morbidelli**

## 12.1 Introduction

Our Galaxy is full with planets. We now know that planets and planetary systems are diverse and come with different sizes, masses and compositions, as well as various orbital architectures. Although there has been great progress in understanding planet formation in the last couple of decades, both observationally and theoretically, several fundamental questions remain unsolved. This might not be surprising given the complexity of the process that includes various physical and chemical processes, and spans huge ranges of length-scales, masses, and timescales. In addition, planet formation cannot be directly observed but has to be inferred by gluing together different pieces of information into one consistent picture. Observations can provide information on the small solids (Andrews 2020) in young protoplanetary disks or information on mature planets, but the actual processes that determine how solids and gas accumulate to form planets remain obscure to measurements. The first planets embedded in protoplanetary disks are just being discovered (e.g., Keppler et al. 2018).

Planets are formed in protoplanetary disks around young stars (Williams & Cieza 2011; Testi et al. 2014; Alexander et al. 2014). Such disks are also diverse in terms of mass, size, and lifetime, which from the beginning may produce some diversity in the final planetary systems. It is the disk's composition and structure that determines the initial conditions for planetary formation and, once objects are formed in the disk, the interaction with the disk is of significant importance. Understanding the physical and chemical properties of protoplanetary disks is crucial since they determine the amount of different solid materials at given radial distance and time, and their lifetimes provide upper limits for the available timescale for planetary migration and giant planet formation.

## 12.2 State of the Art

*How do planets form?* remains a fundamental question in modern astrophysics. Studies on the origin of planets have been conducted throughout the last few

doi:10.1088/2514-3433/abfa8fch12

decades, although thoughts about the formation of the solar system started much longer ago, with the nebular hypothesis being introduced by Immanuel Kant (1755) and Pierre-Simon de Laplace (1796). A substantial milestone was the publication of Victor S. Safronov in 1969 (Safronov 1972) where a physical model for planet formation has been presented.

Until the mid-90s, before the first exoplanet around a Sun-like star had been discovered (Mayor & Queloz 1995), all planet formation models aimed to explain the properties of the solar system, Even today, the solar system remains a key benchmark for planet formation models, because we have a huge amount of information on its planets and its small bodies that allow testing the models in great detail.

The solar system, however, is a rather well-arranged planetary system with four terrestrial planets, two gas giants, and two ice giants,[1] in order of distance from the Sun. This leads to a mass function that has a maximum at Jupiter's location. The orbits of the planets have small inclinations relative to the plane orthogonal to the total angular momentum of the system, which itself is almost aligned with the Sun's equator. In addition, the orbits of all planets are almost circular and most of the planets rotate around their spin axis in the same direction they orbit the Sun. The farthest planet, Neptune is located at 30 au and the closest, Mercury, at ~0.4 au. Formation models were designed to explain these features as a generic and necessary outcome of the planet-formation process.

Our view of planetary systems completely changed with the discovery of exoplanets. New types of planets have been found—super-Earths and mini-Neptunes but also super-Jupiters; moreover, giant planets have been found on orbits very close to the host stars (hot Jupiters) but also extremely far from the star at tens of astronomical units, often on very eccentric orbits very inclined relative to the stellar equator; many packed systems of intermediate-mass planets have been discovered, confined within the orbit of Mercury, often in or near mean-motion resonances with each other. In summary, planetary systems can be dramatically different from our own. These observations challenge planet formation models tuned to explain the architecture of the solar system. Modern planet formation models need to explain not only the detailed structure of the solar system but also the wide diversity among planetary systems, as well as the trends observed in terms of stellar type, metallicity, etc.

The standard model for planet formation is known as "core accretion" (Stevenson 1982; Bodenheimer & Pollack 1986; Pollack et al. 1996; Alibert et al. 2005; Lissauer et al. 2009; D'Angelo et al. 2010; Wuchterl et al. 2000; D'Angelo & Lissauer 2018; Helled et al. 2014; Benz et al. 2014). The first stage of planetary formation is the build-up of a planetary "core" made of heavy-elements. The growth of the core could be a result of accretion of planetesimals (Pollack et al. 1996), pebbles (Lambrechts & Johansen 2012), or both (Alibert et al. 2018). This formation stage leads to the formation of solid protoplanets, often called *planetary embryos* if they are sub-terrestrial in mass or *planetary cores* if they are more massive. Their mass

[1] Note that the name "ice giants" for Uranus and Neptune is somewhat misleading since they may not be water-dominated and the materal in their deep interiors is not the solid phase (see Helled & Fortney 2020 and references therein).

depends on local conditions (planetesimal density, pebble flux, etc.). The growth of planetary embryos/cores is called *phase-1*. It should be noted that during phase-1, if the mass of the forming object exceeds ~2 $M_\oplus$ accretion of H–He can take place (Valletta & Helled 2019; Brouwers & Ormel 2020). Nevertheless, in this stage the composition of the object is clearly dominated by heavy elements, where $M_z \gg M_{H-He}$. If the growth of the core is rapid enough, and the gaseous disk is still present, the forming object can start to steadily accrete gas from the disk (*phase-2*). The nature and timescale of phase-2 depends on the accretion rates (planetesimals/pebbles), the properties of the solid (sizes, composition), and the thermal state of the protoplanet. During this stage, mini-Neptunes and Neptune-like planets can be formed, i.e., planets that are still heavy-element dominated but with bound envelopes of H–He. At the end of this phase the H–He mass can be comparable to the heavy-element mass, i.e., $M_z \sim M_{H-He}$. Once the heavy-element mass and the mass of the gaseous envelope become comparable (known as crossover mass), the growing planet begins to accrete H–He gas in an exponential fashion (*phase-3*, a.k.a. *runaway*). Planets that reach this formation stage rapidly become giant planets, i.e., planets dominated in mass by the gaseous envelopes where $M_z \ll M_{H-He}$, like Saturn and Jupiter. Runaway accretion of gas terminates either by dissipation of nebular gas or by other mechanisms that regulate the gas accretion as we discuss below. Once the giant planet reaches its final mass it contracts and

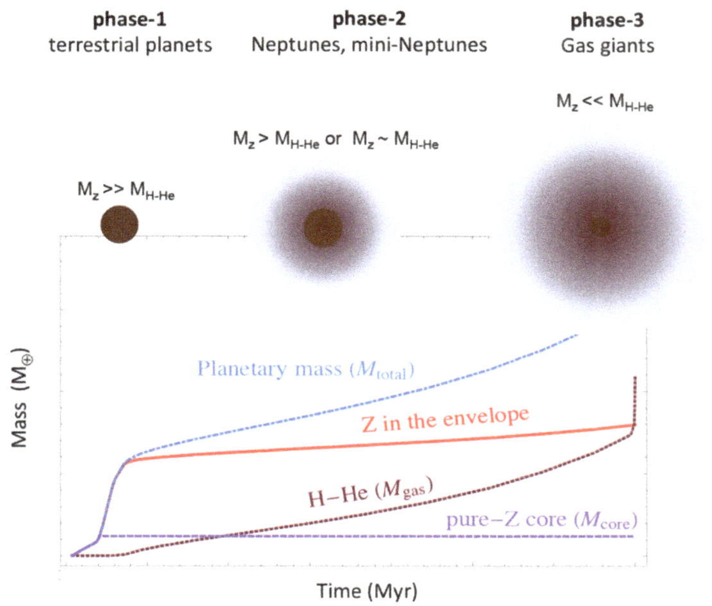


**Figure 12.1.** A sketch of the growth of a planet in the core accretion model. Shown is the planet's mass versus time up to the onset of *phase-3*, when runaway gas accretion begins. *Dashed purple line*: pure heavy-element actual core mass. *Dashed brown line*: gaseous (hydrogen and helium only) mass. *Solid red line*: heavy-element mass in the envelope. *Dashed blue line*: total planetary mass. Sketches of the forming object at each stage are shown above (not to scale). $M_z$ and $M_{H-He}$ correspond to the heavy-element and H–He mass, respectively.

cools down on a timescale of $10^9$ years (long-term planetary evolution). Figure 12.1 shows sketches of the three main phases of planet formation in the core accretion scenario and the expected formed objects at each stage.

## 12.3 Important Questions

Below we list some of the key open questions in planet formation theory.

*Where Do Planets Form?*

It was assumed for decades that the solar system planets formed where we observe them today. The discovery of hot Jupiters and later of smaller planets at very short periods suggested that planets can (significantly) change their orbital location. The most accepted explanation is orbital migration when the gaseous disk is still present. In fact planet migration in a disk was already predicted in 1980 (Goldreich & Tremaine 1980), but it attracted the attention of the community only after the discovery of hot Jupiters (Tanaka et al. 2002). Other scenarios that can explain the existence of hot Jupiters are mutual scattering of a system of giant planets (Beaugé & Nesvorný 2012) or *in-situ* formation (Batygin et al. 2016). Nevertheless, the existence of systems of medium-mass planets in resonance with each other (e.g., Trappist-1, Kepler-223) is a clear signature of migration. Also the heavy-element enrichment of warm Jupiters (Shibata et al. 2019) supports the scenario of planetary migration.

From the formation point of view, the ideal conditions for the formation of giant planets are near the ice line. However, when migration is included, giant planets forming at the snowline should all end up as hot Jupiters. This is in conflict with the observations suggesting that hot Jupiters are only $\sim$10% of that of cold Jupiters (i.e., Jovian-mass planets beyond 1 au from the central star), once observational biases are considered (Cumming et al. 2008; Mayor et al. 2011). With the currently expected migration rates, cold Jupiters should have started to form beyond $\sim$20 au (Bitsch et al. 2015) in order to migrate no closer to the star than their current orbits within the lifetime of the protoplanetary disk. However, it is still unclear where the planets that should have started to form at the snowline might have ended up. Therefore, it is fair to say that this topic is still being debated, and the link between planetary composition and formation location is being investigated intensively.

Interestingly, although the solar system is a well-arranged system, some of its observed properties can be explained if the planets moved around. The orbital structures of the populations of small bodies in the outer solar system require to have been sculpted during a phase of dynamical instability of the giant planets (see Nesvorný 2018 for a review). This dynamical instability is thought to have occurred after the disappearance of the gas from the protoplanetary disk and increased the mutual separations of the planets' orbits, also exciting somewhat their eccentricities and inclinations. During the gas disk phase, it has been proposed that Jupiter migrated down to 1.5–2 au before migrating outwards due to the effects of Saturn, as a means to explain the small mass of Mars and the depletion of the asteroid belt (Walsh et al. 2011). Even if the migration of Jupiter might not have been so extreme, Saturn is expected to have played a role in slowing down or reversing Jupiter's migration (Masset & Snellgrove 2001) and that all the outer planets entered in mean motion resonance with each other (Morbidelli et al. 2007). Unfortunately, there is no direct evidence that constrains the dynamical evolution of the young solar system. The uncertainties on planet migration challenge our ability to link the currently observed planetary orbits to their original formation locations and to assess the importance of dynamical interaction in young planetary systems.

*What is the Dominating Size of Solids that Build-up Planetary Embryos/Cores?* It is yet to be determined whether the formation of an embryo/core is dominated by the accretion of large (planetesimals) or small (pebbles) objects, or both. The growth of the core via planetesimals should stall once all planetesimals in the planet's vicinity have been accreted or scattered away (Levison et al. 2010) although when planets form in systems, scattering of planetesimals by one forming planet can brings planetesimals to the other forming planet which is forming nearby (Ida & Makino 1993). Giant planet formation models with planetesimal accretion suggest that the formation of giant planets takes a few Myr. However, there is a large uncertainty on this estimate since the planetary formation timescale can be decreased substantially if opacity reduction (Movshovitz et al. 2010) and/or envelope enrichment (Venturini et al. 2016; Valletta & Helled 2020) are considered. Unlike planetesimals, pebbles, being small objects, experience fast orbital decay due to gas drag and therefore continuously resupply solid material to the planet's vicinity; the core's accretional cross-section is also larger for pebbles than for planetesimals (Lambrechts & Johansen 2012). For pebble accretion the core formation timescale is inversely related to the flux of pebbles, which is unknown and depends on the dust growth rate and the disk's properties (e.g., size, density). Moreover, also the dynamical excitation of the growing planets affects the pebble accretion rate. In some cases formation via pebble accretion can be so efficient, that it overestimates the relatively low frequency of giant planets. Mechanisms that can delay/prevent substantial gas accretion are still being investigated.

A combination of the two scenarios, core formation via pebble accretion followed by accretion of planetesimals, can explain the growth history of Jupiter deduced from meteoritic constraints (i.e., the rapid growth to $\sim$20 $M_{\oplus}$ within 1 My, followed by a slow growth reaching $\sim$50 $M_{\oplus}$ not before 3 My; Kruijer et al. 2017). This was investigated in detail by Alibert et al. (2018) and Venturini & Helled (2020) where it was shown that these constraints can be fulfilled only if a heavy-element core is rapidly formed by pebble accretion, followed by a protracted phase of planetesimal accretion, at a rate of $\sim 10^{-6}$–$10^{-5}$ $M_{\oplus}$/yr, which delays the onset of runaway gas accretion for $\sim$3 Myr. In this scenario Jupiter's core mass is between 6 and 15 $M_{\oplus}$ with a heavy-element total mass of 20–40 $M_{\oplus}$. This formation scenario can also help explaining Jupiter's fuzzy core.

In the context of giant planet formation, whether dominated by pebble or planetesimal accretion, the forming core is expected to be composed of mostly heavy elements but can also have small amounts of H–He as we discuss above. Population synthesis models provide different predictions when assuming growth via pebbles or planetesimals (Brügger et al. 2020). Clearly a more detailed investigation of the topic is required. We encourage studies that combine both types of solids and investigate the transitions between the two and their dependence on the formation environment.

*How Do Planetesimals Form?*

Although it is still unknown whether the planetary growth is dominated by planetesimal accretion as discussed above, it is clear that planetesimals must form in protoplanetary disks. We see evidence for this in the solar system, where asteroids and Kuiper Belt objects are the leftover of planetesimal populations that were presumably much larger originally. In extrasolar systems, indirect evidence for massive planetesimal populations comes from the observations of debris disks. In these disks the dust must be continuously produced and mutual collisions among planetesimals is the favored mechanism for dust generation (Matthews et al. 2014). Nevertheless, the formation mechanism for planetesimals is still being actively investigated.

Very small grains can grow by collisional coagulation, sticking to each other by electrostatic forces. However, when the grains reach a size of about a millimeter, bouncing becomes the dominant outcome of collisions and further growth is inhibited (Zsom et al. 2010); this is known as "the bouncing barrier". The exact size of the bouncing barrier depends on the grains' compositions (see Güttler et al. 2010, and references therein for details). In addition, particles that are large enough to partially decouple from the motion of the gas experience rapid inward drift due to gas drag. This in turn leads to large relative velocities between particles with different sizes. As a consequence collisions become disruptive. Therefore collisions among dust particles are unlikely to lead to the formation of planetesimals, i.e., of objects of several kilometers in size. Moreover, even if somehow disruptive collisions could be avoided, the particles would rapidly drift into the host star. The drift speed is maximal for meter-size boulders, so this problem is known as the "meter-size barrier".

Great advances on planetesimal formation have been made in the last 15 years, as it was discovered that dust can form clumps due to a number of hydrodynamical effects (e.g., Youdin & Goodman 2005; Johansen et al. 2006, 2007; Cuzzi et al. 2008; Hartlep & Cuzzi 2020). The most promising and deeply studied of these effects is called the "streaming instability". Regardless of the actual dust-clumping mechanism, when a clumps becomes dense enough the dust remains bound by self-gravity against diffusion generated by turbulence and eventually settles to form a macroscopic object, with a characteristic size of $\sim$100 km. The problem is that efficient dust clumping can be triggered only if (i) the dust particles are large enough (decimeter-size, or Stokes' number of order unity) or (ii) the dust/gas ratio is a few times that of the average protosolar material (e.g., Johansen et al. 2009; Yang et al. 2017; Hartlep & Cuzzi 2020).

Therefore, planetesimal formation by this mechanism is possible only where there has been some preliminary pile-up of dust in the disk. Models of dust evolution show one location where sufficient pile-up should occur naturally: the snowline (e.g., Ida & Guillot 2016; Schoonenberg & Ormel 2017; Drażkowska & Alibert 2017). But, the planetesimals of the solar system are so different from each other in terms of chemical and isotopic compositions that they all could not have formed at the same location. However, see Lichtenberg et al. 2021, for a proposed explanation based on different accretion timescale and thermal evolution inside the parent bodies. In addition, there are still no models capable to predict the resulting radial mass distribution of planetesimals. Therefore, it is fair to say that we do not yet understand how planetesimals form throughout (i.e., not just at one or a few distinct locations) the solar system and exoplanetary systems.

*What is the Role of Gravitational Instability in Planet Formation?*

An alternative model for giant planets formation is "disk instability", where gaseous planets are formed by a local gravitational instability in the protoplanetary disk (see e.g., Boss 1997; Mayer et al. 2002; Durisen et al. 2007; Boley 2009). This model seems to be required to explain the formation of massive gaseous planets (above a few Jupiter masses), the existence of giant planets at very large radial distances such as in the HR8799 planetary system (e.g., Marois et al. 2010), the formation of planets in very short timescales (e.g., Lagrange et al. 2010), and the existence of giant planets around M stars (e.g., Morales et al. 2019).

The "disk instability" model is most efficient in massive disks and in cold regions (i.e., at large radial distances), as implied by the Toomre criterion (Toomre 1964) that is used to infer the conditions for disk fragmentation (see Helled et al. 2014 for review). However, the efficiency of the "disk instability" model, and the resulting planetary masses and their compositions, as well as the planetary orbital properties are still being investigated (Nayakshin 2017; Forgan et al. 2018; Müller et al. 2018).

Traditionally, the existence of a heavy-element core and non-stellar compositions of giant planets was used to discriminate between "core accretion" and "disk instability". However, it was shown that a core can form in both models and that the expected enrichment of giant planets is strongly affected by the formation environment and the early evolution of the planets (e.g., migration followed by planetesimal accretion, atmospheric loss) in both formation models. Therefore, the planetary composition and internal structure cannot be used in a simple manner to discriminate among these models.

A key observational constraint for giant planet formation models is the correlation between stellar metallicity and the occurrence rate of giant planets (Fischer & Valenti 2005). This correlation can easily be explained by the "core accretion" model as more metal-rich disks lead to more efficient core formation followed by gas accretion. For disk instability, it is yet to be determined whether disk fragmentation is more likely in metal-rich disks (see Helled et al. 2014, for details). Nevertheless, it should be noted that this strong metallicity correlation does not exist for massive giant planets that are several Jupiter-mass (e.g., Santos et al. 2017). This may imply that massive giant planets are formed by "disk instability" while Jupiter-mass planets are formed via "core accretion".

At the moment the main challenge of the "disk instability" model is to demonstrate that planetary-mass objects can form (and not accrete much during their early evolution) and survive in realistic disk conditions (Kratter & Lodato 2016). This is still work in progress and, because the results strongly depend on the numerical model, its resolution and assumptions, tendencies and correlations in the "disk instability" model are still unclear.

However, even if "disk instability" is not the main mechanism for giant planet formation, gravitational instabilities in young disks may represent the first trials of planet formation, affecting the disk's evolution and the conditions for effective planet formation at later times.

*What Are the Expected Compositions and Structures of Giant Planets?*

The compositions of giant planets can range from super-solar to solar and even sub-solar. The mass–radius relations of Jupiter and Saturn hint that they contain heavy elements in larger proportion than the protosolar gas; moreover the enrichment in heavy elements is different for the two planets. The diversity of planetary compositions of giant exoplanets can be probed for non-highly irradiated planets with measured masses and radii (Miller & Fortney 2011; Thorngren et al. 2019). Many of these planets have heavy-element enrichment that are much higher than the ones predicted from standard formation models. Some of these giant exoplanets possess around 100 M$_{\oplus}$ of heavy-elements. During the formation stages 1 and 2, the heavy-element mass is expected to be of the order of up to a few tens of $M_{\oplus}$, and accretion of heavy elements during runaway gas accretion should be very inefficient (Shibata & Ikoma 2019). Therefore, unless the solar nebula was extremely metal-rich (Venturini & Helled 2020) or giant impacts are invoked (Liu et al. 2015), it is not easy to explain these enrichments. For giant exoplanets an appealing explanation is planetesimal accretion during inward migration (Shibata et al. 2019). Since exoplanet observations are still biased toward close-in planets, the metallicities of giant planets on wider-orbit is yet to be determined. Such measurements would allow us to understand the dependence of the enrichment on the orbital distance.

The distribution of the heavy elements in giant planet interiors is still being investigated (Helled & Guillot 2018). While the assumption of a simple core+envelope structure is convenient from the modeling perspective, updated formation and structure models of giant planets show that the heavy elements are not uniformly mixed (Leconte & Chabrier 2012; Lozovsky et al. 2017). In addition, giant planets probably do not have distinct compact cores; they rather have an innermost region which is highly enriched with heavy-elements. This is because once the core mass becomes sufficiently large, the protoplanet can accrete H–He (phase-2) and the subsequently incoming solids dissolve in the atmosphere (Helled & Stevenson 2017; Valletta & Helled 2019; Brouwers & Ormel 2020). This leads to composition gradients in the deep interiors. Indeed, a fuzzy core seems to exist in Jupiter, as suggested by detailed structure models that fit its gravity field (Wahl et al. 2017; Debras & Chabrier 2019). In order to connect the formation and current-state structure it is important to model the long-term planetary evolution and investigate whether/how the heavy-element distribution changes with time (Vazan et al. 2018; Müller et al. 2020).

*What Controls the Final Mass of Giant Planets?*

Giant planets come at different masses and it is still not understood what mechanism stops the gas accretion. Originally, it was assumed that the planet stops accreting gas when the gaseous disk disappears. Giant planets open gaps in the disk, but gap opening is not an effective way to shut-off gas accretion, because the gas penetrates into the gap from the surface of the disk (Morbidelli et al. 2014). With or without the gap, the flux of gas to the planet's orbit is essentially the same as the accretion rate of the star from the disk. With a typical value of $10^{-8}M_{\odot}$/yr, a planet like Jupiter can grow in $10^5$ yr (Lubow et al. 1999; D'Angelo et al. 2003; Tanigawa & Ikoma 2007; Lissauer et al. 2009; Tanigawa & Tanaka 2016; Lambrechts et al. 2019; Tanaka et al. 2020). Thus, there is the need to understand how most giant planets limited their growth to a mass comparable to Jupiter's or just a few times larger. In the solar system, we still need to understand how Jupiter and Saturn reached their terminal masses.

*What is the Role of Giant Impacts in Planetary Formation?*

In the solar system it seems that giant impacts played a key role in shaping the planetary properties. Giant impacts are favorable explanations for the formation of the Earth's Moon (Canup & Asphaug 2001), the high iron-to-rock ratio in Mercury (Chau et al. 2018), Uranus' obliquity and regular satellites (Kegerreis et al. 2018; Reinhardt et al. 2020; Ida et al. 2020), and even Jupiter's fuzzy core (Liu et al. 2019), Mars hemisphere dichotomy (Marinova et al. 2008) and its moons (Canup & Salmon 2018), and the Pluto–Charon system (Canup 2005). This suggests that too many protoplanets tend to form in the disk, so that the system is initially unstable. The impacts reduce the number of massive objects and increase the masses of the survivors, until a stable configuration is obtained.

Therefore it is likely that giant impacts also played an important role in exoplanetary systems (Inamdar & Schlichting 2015; Bonomo et al. 2019). The fact that the planetary properties can be changed due to giant impacts suggest that one has to be careful when linking the currently observed properties with the planet's origin.

*How Long Does it Take for Planets to Form?*

Giant planets are H–He dominated and therefore must have completed their formation within the lifetime of the protoplanetary disk (because they accreted the gas from the disk). Since the average age of protoplanetary disks is ~3 Myr (Mamajek 2009), giant planets are expected to form within this timescale. However, in terms of formation models a couple of Myr make a great difference in the required disk's conditions (which determine the accretion rates), the formation efficiency, the early evolution, as well as the predicted masses and compositions of giant planets in various formation models (Pollack et al. 1996; Alibert et al. 2005; Movshovitz et al. 2010; Helled et al. 2014; Helled & Bodenheimer 2014; Alibert et al. 2018; Paardekooper & Johansen 2018; Mordasini 2018; Bitsch et al. 2019; Venturini & Helled 2020).

The terrestrial planets might have formed on a longer timescale than the disk's lifetime. The Moon-forming event happened several 10 My after gas photo-evaporation, according to radioactive chronometers (Kleine et al. 2009). Mars, instead, ended its accretion within 4-5 My (Dauphas & Pourmand 2011). However, it is not known what was the mass of the Earth when the gas disappeared. In the classic model of Wetherill (1992), the Earth should have had a mass comparable to Mars, but modern models, accounting also for pebble accretion, argue for a much more substantial mass (Schiller et al. 2018; Popovas et al. 2018; Lammer et al. 2020, so that planet migration should have played a role in terrestrial planet formation (Johansen et al. 2021; Brož et al 2021). The Earth must have been sufficiently massive to retain the material delivered by the cometary shower associated to the giant planet instability, whose amount is constrained by comparing the elemental and isotopic compositions of noble gases in the Earths atmosphere with those of comet 67P/C-G (Marty et al. 2016, 2017).

*How Do Intermediate-Mass/Size Planets Form?*

Exoplanet data suggest that planets with intermediate masses (a few to a few tens of $M_\oplus$) and radii (1.5–4 $R_\oplus$) are extremely common in the Galaxy. These planets are often referred to as super-Earths and mini-Neptunes. The first, represent planets with relatively high densities (refractory materials), some kind of large-scale Earth, while the latter represent planets with non-negligible amounts of H–He atmospheres, small or comparable version of Uranus/Neptune. It is not easy to learn about the formation of such planets without having tighter constraints on their bulk compositions and their original formation location. With accurate measurements of the planetary mass and radius, the average planetary density can be determined, and can be used to (non-uniquely) infer the planetary bulk composition.

The formation of intermediate-mass/size planets remains a challenge to planet formation models (Helled & Bodenheimer 2014). The formation of mini-Neptunes and super-Earths is expected to occur at larger radial distances (via pebble/planetesimal accretion) followed by inward migration (Bitsch et al. 2015; Venturini & Helled 2017). If intermediate-mass planets form at large radial distances and migrate inward, they are expected to contain a large amount of volatile elements, water in particular, because their formation started at the snowline. However, the analysis of the M–R relationships for super-Earths that are up to $\sim$3 $M_\oplus$ and radii smaller than 2 $R_\oplus$ and have no atmosphere indicate that these planets are made of refractory elements (Owen & Wu 2017; Jin & Mordasini 2018; Lozovsky et al. 2018). This could imply that these planets have formed *in-situ* from the accumulation of refractory dust at the inner edge of the protoplanetary disk (Chatterjee & Tan 2014; Flock et al. 2019).

Future measurements of atmospheric compositions could assist in this direction as well as advances in theory. Because intermediate-mass/size planets lie between terrestrial and giant planets in a mass–radius diagram (Otegi et al. 2020), knowledge of both groups is required and we suggest that this is a topic that should be investigated in detail in future research, as the return is expected to be enormous.

*Have We Discovered Extrasolar Terrestrial Planets Yet?*

Because of observational biases, planets as small as the terrestrial planets of the solar system have been discovered only around low-mass stars. Around solar-mass stars, so far only super-Earths have been detected, where some of them are expected to have Earth-like compositions. Due to observational biases, in all cases the discovered planets are significantly closer to the parent star than our own terrestrial planets, although some of those orbiting low-mass stars are also in the so-called habitable zone. However, in terms of formation process, it is unclear whether these planets had formation paths similar to the terrestrial planets of the solar system (Lambrechts et al. 2019). If we identify “terrestrial planets” by their formation process the question in the title of this section has all its relevance. We are sill missing a complete understanding of what determines the characteristics of terrestrial planets, with Venus and Earth being a clear example. Finally, it also highlights the need to identify which of the specific aspects of Earth’s formation have been fundamental to determine the properties that make our planet a truly habitable world and to assess the likelihood that the same happened on other planets.

## 12.4 Opportunities

The near and far future seem bright. The ongoing and future observations of planets and planetary systems, as well as the theoretical developments, are expected to significantly improve our understanding of planet formation. Some of the key opportunities are listed below:

- The ongoing TESS (NASA) and CHEOPS (ESA; Broeg et al. 2013) missions as well as the availability of new spectrographs for improved radial velocity measurements (e.g., ESPRESSO, ESO), and coronagraphs for direct imaging (e.g., SPHERE+, ESO). All these instruments will broaden the census of extrasolar planets and extend it to a region of the mass-distance diagram not yet sampled.
- The PLATO ESA mission (Rauer et al. 2014) will provide an understanding of the architecture of planetary systems, probing a larger range of radial distances, with well constrained ages of the systems. This can be used to further constrain planet formation and evolution models.
- ALMA (Ansdell et al. 2016) keeps providing data on protoplanetary disks, the birth environments of planets. This is then used to constrain the formation timescale and the expected properties of protoplanetary disks and their diversity.
- JWST (Greene et al. 2016) and ARIEL (Tinetti et al. 2018) are expected to provide key information on the atmospheric composition of gaseous-rich planets. This, together with an accurate determination of the planetary mean density can further constrain the planetary bulk composition, as well as their formation and evolution histories.
- Direct imagining surveys can be used to better understand giant planet formation and put constraints on the entropies and luminosities of young giant planets.
- Solar system missions: Within the solar system, various missions are dedicated to better understand planets and small bodies. Particularly relevant will be the missions to Uranus and/or Neptune (Fletcher et al. 2020) and the upcoming missions to Venus. The properties of these planets are poorly known, but it is now clear that they represent a category of planets that are numerous around other stars.

## 12.5 Challenges

As some questions are being solved, new questions arise; models reach the next level of complexity where more detailed processes must be included; different pieces of the puzzle must be fit into a unified model. We reach a point where the details matter, which makes models more complex but also allow achieving a deeper understanding of the topic.

We envision future progress in the following directions:

- **Combined solar-system and exoplanetary science:** While solar system science focuses on explaining the detailed properties of the objects in our specific planetary system, exoplanetary science provides a broader view of planetary systems allowing to build statistics and identify trends and correlations. However, each exoplanetary system is much less observationally constrained than our own. These two fields are clearly complementary and we urge to bridge the two in future research.

- **Combining dynamics with thermodynamics:** At the moment planet formation models can be broadly divided in two categories: some are based on a careful treatment of the thermodynamics and other physical processes, but assume that the planets form *in-situ* and independently from the other objects of the system; others are based on dynamical simulations that model properly the interaction of the growing planet with the disk and the other objects, but the physical structure of the planet is modeled in an over-simplified way. Clearly, a unified formation model is required in the future.
- **Linking current-state structure and origin:** In order to link the internal structure of a planet with its origin, a good understanding of its evolution is required. This is because the three aspects of formation, evolution and internal structure are linked: the formation process determines the primordial planetary composition, internal structure and thermal state. These determine the heat transport mechanism, the potential re-distribution of the materials, and the planetary long-term evolution which could also include evaporation and outgassing. The planetary evolution then determines the current-state internal structure. As a result, in order to use the current-state planetary properties with formation, modeling the planetary evolution properly is required.

The challenges in planet formation theory should not be viewed as obstacles but as exciting opportunities to advance the field and reach a new level of understanding of how planets form. After all, we live on a planet which is part of a planetary system, and only with a better knowledge of how planets form can we understand our uniqueness.

## References

Alexander, R., Pascucci, I., Andrews, S., Armitage, P., & Cieza, L. 2014, in Protostars and Planets VI, ed. H. Beuther, R. S. Klessen, C. P. Dullemond, & T. Henning (Tucson, AZ: Univ. Arizona Press), 475

Alibert, Y., Mordasini, C., Benz, W., & Winisdoerffer, C. 2005, A&A, 434, 343

Alibert, Y., Venturini, J., Helled, R., et al. 2018, NatAs, 2, 873

Andrews, S. M. 2020, ARA&A, 58, 483

Ansdell, M., Williams, J. P., van der Marel, N., et al. 2016, ApJ, 828, 46

Batygin, K., Bodenheimer, P. H., & Laughlin, G. P. 2016, ApJ, 829, 114

Beaugé, C., & Nesvorný, D. 2012, ApJ, 751, 119

Benz, W., Ida, S., Alibert, Y., Lin, D., & Mordasini, C. 2014, in Protostars and Planets VI, ed. H. Beuther, R. S. Klessen, C. P. Dullemond, & T. Henning (Tucson, AZ: Univ. Arizona Press), 691

Bitsch, B., Izidoro, A., Johansen, A., et al. 2019, A&A, 623, A88

Bitsch, B., Lambrechts, M., & Johansen, A. 2015, A&A, 582, A112

Bodenheimer, P., & Pollack, J. B. 1986, Icar, 67, 391
Boley, A. C. 2009, ApJL, 695, L53
Bonomo, A. S., Zeng, L., Damasso, M., et al. 2019, NatAs, 3, 416
Boss, A. P. 1997, Sci., 276, 1836
Broeg, C., Fortier, A., Ehrenreich, D., et al. 2013, EPJWC, 47, 03005
Brouwers,, & Ormel, 2020, A&A, 634, A15
Brož, M., Chrenko, O., Nesvorný, D., & Dauphas, N. 2021, NatAs,
Brügger, N., Burn, R., Coleman, G., Alibert, Y., & Benz, W. 2020, A&A, 640, A21
Canup, R., & Salmon, J. 2018, SciA, 4, eaar6887
Canup, R. M. 2005, Sci., 307, 546
Canup, R. M., & Asphaug, E. 2001, Natur, 412, 708
Chatterjee, S., & Tan, J. C. 2014, ApJ, 780, 53
Chau, A., Reinhardt, C., Helled, R., & Stadel, J. 2018, ApJ, 865, 35
Cumming, A., Butler, R. P., Marcy, G. W., et al. 2008, PASP, 120, 531
Cuzzi, J. N., Hogan, R. C., & Shariff, K. 2008, ApJ, 687, 1432
D'Angelo, G., Durisen, R. H., & Lissauer, J. J. 2010, in Exoplanets, ed. S. Seager (Tucson, AZ: Univ. Arizona Press), 319
D'Angelo, G., Kley, W., & Henning, T. 2003, ApJ, 586, 540
D'Angelo, G., & Lissauer, J. J. 2018, in Handbook of Exoplanets (Berlin: Springer), 140
Dauphas, N., & Pourmand, A. 2011, Natur, 473, 489
Debras, F., & Chabrier, G. 2019, ApJ, 872, 100
Drażkowska, J., & Alibert, Y. 2017, A&A, 608, A92
Durisen, R. H., Boss, A. P., Mayer, L., et al. 2007, in Protostars and Planets V, ed. B. Reipurth, D. Jewitt, & K. Keil (Tucson, AZ: Univ. Arizona Press), 607
Fischer, D. A., & Valenti, J. 2005, ApJ, 622, 1102
Fletcher, L. N., Helled, R., Roussos, E., et al. 2020, P&SS, 191, 105030
Flock, M., Turner, N. J., Mulders, G. D., et al. 2019, A&A, 630, A147
Forgan, D. H., Hall, C., Meru, F., & Rice, W. K. M. 2018, MNRAS, 474, 5036
Goldreich, P., & Tremaine, S. 1980, ApJ, 241, 425
Greene, T. P., Line, M. R., Montero, C., et al. 2016, ApJ, 817, 17
Güttler, C., Blum, J., Zsom, A., Ormel, C. W., & Dullemond, C. P. 2010, A&A, 513, A56
Hartlep, T., & Cuzzi, J. N. 2020, ApJ, 892, 120
Helled, R., & Fortney, J. J. 2020, RSPTA, 378, 20190474
Helled, R., & Stevenson, D. J. 2017, ApJ, 840, L4
Helled, R., Bodenheimer, P., Podolak, M., et al. 2014, in Protostars and Planets VI, ed. H. Beuther, R. S. Klessen, C. P. Dullemond, & T. Henning (Tucson, AZ: Univ. Arizona Press), 643
Helled, R., & Bodenheimer, P. 2014, ApJ, 789, 69
Helled, R., & Guillot, T. 2018, in Handbook of Exoplanets (Berlin: Springer), 44
Ida, S., & Guillot, T. 2016, A&A, 596, L3
Ida, S., & Makino, J. 1993, Icar, 106, 210
Ida, S., Ueta, S., Sasaki, T., & Ishizawa, Y. 2020, NatAs, 4, 880
Inamdar, N. K., & Schlichting, H. E. 2015, MNRAS, 448, 1751
Jin, S., & Mordasini, C. 2018, ApJ, 853, 163
Johansen, A., Henning, T., & Klahr, H. 2006, ApJ, 643, 1219
Johansen, A., Ronnet, T., & Bizzarro, M. 2021, SciA, 7, eabc0444
Johansen, A., Oishi, J. S., Mac Low, M.-M., et al. 2007, Natur, 448, 1022

Johansen, A., Youdin, A., & Mac Low, M.-M. 2009, ApJL, 704, L75
Kegerreis, J. A., Teodoro, L. F. A., Eke, V. R., et al. 2018, ApJ, 861, 52
Keppler, M., Benisty, M., Müller, A., et al. 2018, A&A, 617, A44
Kleine, T., Touboul, M., Bourdon, B., et al. 2009, GeCoA, 73, 5150
Kratter, K., & Lodato, G. 2016, ARA&A, 54, 271
Kruijer, T. S., Burkhardt, C., Budde, G., & Kleine, T. 2017, PNAS, 114, 6712
Lagrange, A. M., Bonnefoy, M., Chauvin, G., et al. 2010, Sci., 329, 57
Lambrechts, M., Lega, E., Nelson, R. P., Crida, A., & Morbidelli, A. 2019, A&A, 630, A82
Lambrechts, M., & Johansen, A. 2012, A&A, 544, A32
Lambrechts, M., Morbidelli, A., Jacobson, S. A., et al. 2019, A&A, 627, A83
Lammer, H., Leitzinger, M., Scherf, M., et al. 2020, Icar, 339, 113551
Leconte, J., & Chabrier, G. 2012, A&A, 540, A20
Lichtenberg, T., Drażkowska, J., Schönbächler, M., Golabek, G. J., & Hands, T. O. 2021, Science, 371, 365
Levison, H. F., Thommes, E., & Duncan, M. J. 2010, AJ, 139, 1297
Lissauer, J. J., Hubickyj, O., D'Angelo, G., & Bodenheimer, P. 2009, Icar, 199, 338
Liu, S.-F., Agnor, C. B., Lin, D. N. C., & Li, S.-L. 2015, MNRAS, 446, 1685
Liu, S.-F., Hori, Y., Müller, S., et al. 2019, Natur, 572, 355
Lozovsky, M., Helled, R., Dorn, C., & Venturini, J. 2018, ApJ, 866, 49
Lozovsky, M., Helled, R., Rosenberg, E. D., & Bodenheimer, P. 2017, ApJ, 836, 227
Lubow, S. H., Seibert, M., & Artymowicz, P. 1999, ApJ, 526, 1001
Mamajek, E. E. 2009, in AIP Conf. Proc. 1158, Exoplanets and Disks: Their Formation and Diversity, ed. T. Usuda, M. Tamura, & M. Ishii (Melville, NY: AIP), 3
Marinova, M. M., Aharonson, O., & Asphaug, E. 2008, Natur, 453, 1216
Marois, C., Zuckerman, B., Konopacky, Q. M., Macintosh, B., & Barman, T. 2010, Natur, 468, 1080
Marty, B., Altwegg, K., Balsiger, H., et al. 2017, Sci, 356, 1069
Marty, B., Avice, G., Sano, Y., et al. 2016, E&PSL, 441, 91
Masset, F., & Snellgrove, M. 2001, MNRAS, 320, L55
Matthews, B. C., Krivov, A. V., Wyatt, M. C., Bryden, G., & Eiroa, C. 2014, in Protostars and Planets VI, ed. H. Beuther, R. S. Klessen, C. P. Dullemond, & T. Henning (Tucson, AZ: Univ. Arizona Press), 521
Mayer, L., Quinn, T., Wadsley, J., & Stadel, J. 2002, Sci., 298, 1756
Mayor, M., Marmier, M., Lovis, C., et al. 2011, arXiv:1109.2497
Mayor, M., & Queloz, D. 1995, Natur, 378, 355
Miller, N., & Fortney, J. J. 2011, ApJL, 736, L29
Morales, J. C., Mustill, A. J., Ribas, I., et al. 2019, Sci, 365, 1441
Morbidelli, A., Szulágyi, J., Crida, A., Lega, E., Bitsch, B., Tanigawa, T., & Kanagawa, K. 2014, Icar, 232, 266
Morbidelli, A., Tsiganis, K., Crida, A., Levison, H. F., & Gomes, R. 2007, AJ, 134, 1790
Mordasini, C. 2018, in Handbook of Exoplanets (Berlin: Springer), 143
Movshovitz, N., Bodenheimer, P., Podolak, M., & Lissauer, J. J. 2010, Icar, 209, 616
Müller, S., Helled, R., & Cumming, A. 2020, A&A, 638, A121
Müller, S., Helled, R., & Mayer, L. 2018, ApJ, 854, 112
Nayakshin, S. 2017, PASA, 34, e002
Nesvorný, D. 2018, ARA&A, 56, 137

Otegi, J. F., Bouchy, F., & Helled, R. 2020, A&A, 634, A43
Owen, J. E., & Wu, Y. 2017, ApJ, 847, 29
Paardekooper, S.-J., & Johansen, A. 2018, SSRv, 214, 38
Pollack, J. B., Hubickyj, O., Bodenheimer, P., et al. 1996, Icar, 124, 62
Popovas, A., Nordlund, Å., Ramsey, J. P., & Ormel, C. W. 2018, MNRAS, 479, 5136
Rauer, H., Catala, C., Aerts, C., et al. 2014, ExA, 38, 249
Reinhardt, C., Chau, A., Stadel, J., & Helled, R. 2020, MNRAS, 492, 5336
Safronov, V. S. 1972, Evolution of the Protoplanetary Cloud and Formation of the Earth and Planets, Translated from Russian (Jerusalem: Israel Program for Scientific Translations, Keter Publishing House)
Santos, N. C., Adibekyan, V., Figueira, P., et al. 2017, A&A, 603, A30
Schiller, M., Bizzarro, M., & Fernandes, V. A. 2018, Natur, 555, 507
Schoonenberg, D., & Ormel, C. W. 2017, A&A, 602, A21
Shibata, S., Helled, R., & Ikoma, M. 2019, arXiv:1911.02292
Shibata, S., & Ikoma, M. 2019, MNRAS, 487, 4510
Stevenson, D. J. 1982, P&SS, 30, 755
Tanaka, H., Murase, K., & Tanigawa, T. 2020, ApJ, 891, 143
Tanaka, H., Takeuchi, T., & Ward, W. R. 2002, ApJ, 565, 1257
Tanigawa, T., & Ikoma, M. 2007, ApJ, 667, 557
Tanigawa, T., & Tanaka, H. 2016, ApJ, 823, 48
Testi, L., Birnstiel, T., Ricci, L., et al. 2014, in Protostars and Planets VI, ed. H. Beuther, R. S. Klessen, C. P. Dullemond, & T. Henning (Tucson, AZ: Univ Arizona Press), 339
Thorngren, D. P., Marley, M. S., & Fortney, J. J. 2019, RNAAS, 3, 128
Tinetti, G., Drossart, P., Eccleston, P., et al. 2018, ExA, 46, 135
Toomre, A. 1964, ApJ, 139, 1217
Valletta, C., & Helled, R. 2019, ApJ, 871, 127
Valletta, C., & Helled, R. 2020, ApJ, 900, 133
Vazan, A., Helled, R., & Guillot, T. 2018, A&A, 610, L14
Venturini, J., Alibert, Y., & Benz, W. 2016, A&A, 596, A90
Venturini, J., & Helled, R. 2017, ApJ, 848, 95
Venturini, J., & Helled, R. 2020, A&A, 634, A31
Wahl, S. M., Hubbard, W. B., Militzer, B., et al. 2017, GeoRL, 44, 4649
Walsh, K. J., Morbidelli, A., Raymond, S. N., O'Brien, D. P., & Mandell, A. M. 2011, Natur, 475, 206
Wetherill, G. W. 1992, Icar, 100, 307
Williams, J. P., & Cieza, L. A. 2011, ARA&A, 49, 67
Wuchterl, G., Guillot, T., & Lissauer, J. J. 2000, in Protostars and Planets IV, ed. V. Mannings, A. P. Boss, & S. S. Russell (Tucson, AZ: Univ. Arizona Press), 1081
Yang, C. C., Johansen, A., & Carrera, D. 2017, A&A, 606, A80
Youdin, A. N., & Goodman, J. 2005, ApJ, 620, 459
Zsom, A., Ormel, C. W., Güttler, C., Blum, J., & Dullemond, C. P. 2010, A&A, 513, A57

AAS | IOP Astronomy

ExoFrontiers
Big questions in exoplanetary science
Nikku Madhusudhan

# Chapter 13

# Planetary Migration

**John C. B. Papaloizou**

## 13.1 Introduction

Studies of planet migration derived from disk planet interactions began before the discovery of exoplanets (see Papaloizou et al. 2007; Baruteau et al. 2014; for reviews). The potential importance of migration for determining orbital architectures being realized, the field received greater attention soon after the initial discoveries of exoplanets. Early studies based on very simple disk models indicated very fast migration times for low mass planets that raised questions about its relevance. However, more recent studies, made possible with improving resources, that considered improved physics and disk models revealed processes that could halt or reverse this migration. That in turn led to a focus on special regions in the disk where migration could be halted. In this way the migration of low mass planets could be reconciled with formation theories. In the case of giant planets which have a nonlinear interaction with the disk, the migration should be slower and coupled to the evolution of the disk which is where attention needs to be focused.

This review is primarily concerned with processes where migration is connected with the presence of the protoplanetary disk as described above. However, it should be noted that migration may be induced by disk-free gravitational interactions among planets or with binary companions. This will be discussed relatively briefly.

## 13.2 State of the Art Opportunities and Challenges

The classification of the main types of migration applicable to single planets: type I, type II, and type III, which apply for different planet masses and disk physics, has been established. In addition systems of planets have been considered in order to investigate the origin and sustainability of commensurabilities. However, the disk models normally considered are primitive in comparison to those that are becoming available which involve magnetic fields, detailed vertical structure and dust. Access to steadily improving computer resources and numerical tools should enable studies

doi:10.1088/2514-3433/abfa8fch13  

of disk planet interactions in their context. We may then obtain a more complete picture of how and where migration stalls for low mass planets and the speed of migration of giant planets. After brief descriptions of the main types of migration in Sections 13.3–13.3.5 we give an outline of some of the issues that may be addressed with improved resources in future in Sections 13.4–13.4.4.

## 13.3 Disc Planet Interaction and the Main Types of Migration

A planet of mass, $M_p$, orbiting within a protoplanetary disk interacts gravitationally with nearby disk gas (Baruteau et al. 2014). We shall assume the planet, moving in a circular Keplerian orbit with angular velocity, $\Omega_p$, at radius, $r_p$, is moving hypersonically with Mach number, $\mathcal{M}$, in the range 10–50. The disk aspect ratio, or ratio of vertical semi-thickness to radius is $\sim 1/\mathcal{M}$. Consider nearby material that is streaming past the planet supersonically on circular orbits with the minimum distance from the planet exceeding $2r_p/(3\mathcal{M})$. Assuming that pressure effects can be neglected we can view this material as undergoing a scattering as it passes by (see e.g., Lin & Papaloizou 1979, 1993; Papaloizou & Terquem 2006).

### 13.3.1 Scattering Calculation

Consider a ring of disk material orbiting interior to the planet at radius $r_p - b$, where $b$ is small. As a fluid element of the ring undergoes its closest approach, its trajectory is deflected through an angle $\delta$ given by

$$\tan(\delta/2) = GM_p/(bv^2), \tag{13.1}$$

where the relative speed $v = 3\Omega_p b/2$. For small $\delta$ the specific angular momentum transferred to the planet is

$$\Delta j = r_p v(1 - \cos\delta) \sim r_p v\delta^2/2 = 2r_p v\left(GM_p/(bv^2)\right)^2. \tag{13.2}$$

Considering the interior disk to consist of an ensemble of rings the rate of angular momentum transfer to the planet is

$$dJ/dt = \int_{b_0}^{\infty} \Sigma\left(3\Omega_p b/2\right)\Delta j db = (8/27)\Sigma r_p^4 q^2 \Omega_p^2 \left(r_p/b_0\right)^3, \tag{13.3}$$

where $\Sigma$ is the dick surface density, $b_0$, is the smallest allowed $b$ which may correspond to an inner disk edge and $q = M_p/M_*$. A corresponding calculation and expression applies to material orbiting exterior to the planet with the transfer of angular momentum being from the planet to the disk which rotates more slowly.

For an embedded low mass planet the cut off distance $b_0$ should be the distance at which the flow becomes supersonic, thus $b_0 = 2r_p/(3\mathcal{M})$. For large mass planets that make a gap in the disk, $b_0$ should be the distance to the gap edge. The net torque on the planet then depends on the difference between contributions from the inner and outer disk.

### 13.3.2 The Effect of Coorbital Material

Coorbital material no longer streams past the planet but undergoes horseshoe turns, reversing direction as it approaches the planet e.g., as in going from $A$ to $A'$ in Figure 13.1 (middle panel). As such a turn moves disk material from exterior to interior to the planet, angular momentum is transferred to it. After moving around the horseshoe material that was at $A'$ undergoes a close turn from $C$ to $C'$. Consequently angular momentum is removed from the planet. It then moves round the horseshoe back to $A$. For a non-zero net torque there must be a difference between the two types of closely approaching material. This has to be brought about by rapid enough transport of angular momentum and possibly heat brought about by material outside the coorbital region. If this does not occur the torque saturates as the two types of approach cancel each other out. The unsaturated torque on low mass planets is found to depend on the gradients of the inverse specific vorticity, $\Sigma/\Omega_p$, and the entropy if the material is not barotropic or efficiently cooled. A well-known example is the planet trap (Masset et al. 2006) where the surface density, and hence $\Sigma/\Omega_p$ decreases rapidly inwards. Then angular momentum transport associated with the horseshoe turn $A \rightarrow A'$ will exceed that from the turn $C \rightarrow C'$ in Figure 13.1 resulting in net transport to the planet that can halt inward migration. Note that coorbital dynamics is also affected by entropy gradients (Paardekooper & Mellema 2006), magnetic fields (e.g., Terquem 2003; Guilet et al. 2013) and heat transport between the planet and coorbital material (Masset 2017).

### 13.3.3 Type I Migration

This affects planets with masses up to those characteristic of ice giants embedded in a protoplanetary disk. To estimate the migration rate we use Equation (13.3) with, $b_0 = 2r_p/(3\mathcal{M})$. The difference between outer and inner torques introduces a factor $\sim 5/\mathcal{M}$, which measures the departure from strict symmetry, with the torque from the

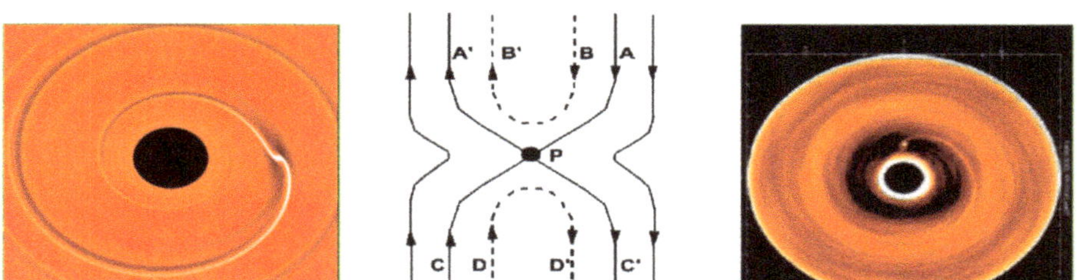


**Figure 13.1.** Left panel: a five Earth mass planet orbits in the type I migration regime while embedded in a laminar disk with a surface density profile similar to that of the minimum mass solar nebula (MMSN). The form of the surface density is indicated (light high dark low). Note the outer wake is stronger leading to a negative torque on the planet. Middle panel: a schematic illustration of the part of the horseshoe region in the neighborhood of a low mass planet. The direction of flow along streamlines is indicated. Note that the streamlines that leave the plot at the bottom wrap around to become the streamlines entering the plot from the top. Right panel: a five Jupiter mass orbiting in a prominent gap in a disk with MHD turbulence is illustrated in the right panel. Note that the direction of orbital rotation is anticlockwise.

outer disk being larger (see Figure 13.1) Thus the total torque acting on the planet is estimated as

$$dJ/dt \sim -(8/27)\Sigma r_p^4 q^2 \Omega_p^2 (r_p/b_0)^3 (5/\mathcal{M}) = -5\Sigma r_p^4 q^2 \Omega_p^2 \mathcal{M}^2, \quad (13.4)$$

leading to the inward migration rate

$$\tau_{\rm mig} \sim 10^5 {\rm yr}\left(M_*(4\pi\Sigma r_p^2)^{-1}(500)^{-1}\right)\left(2\pi\Omega_p^{-1}(12{\rm yr})^{-1}\right)\left(3\times 10^{-5} q^{-1}\right)\left(400\mathcal{M}^{-2}\right). \quad (13.5)$$

The effect of coorbital material should be included but this is ineffective in locally isothermal disks in which $\Sigma/\Omega_p$ is constant such the MMSN. The estimate (13.5) then indicates a rapid inward migration timescale $\sim 10^5$ yr for a 10 $M_\oplus$ planet at 5 astronomical units (au). This is much shorter than protoplanetary disk lifetimes $\sim 10^7$ yr, leading some to arbitrarily drop planetary migration from consideration (e.g., Hansen & Murray 2012). However, this neglects coorbital torques in more general disk models that could reverse or halt this migration (e.g., see discussion of a planet trap in Section 13.3.2). Thus instead of migrating inwards protoplanets stall at preferred locations where mass growth may occur.

### 13.3.4 Type II Migration

This applies to protoplanets massive enough to make a gap in the disk. Typically, their mass should exceed that of Saturn (for further discussion see the review by Papaloizou & Terquem 2006). The surface density profile adjusts so that angular momentum exchange with the planet, which repels disk material, is balanced by internal transport through the disk at gap edges. Denoting the angular momentum flow through the disk by, $\mathcal{F}$, making use of Equation (13.3) this balance requires that at an edge

$$\mathcal{F} = (8/27)\Sigma r_p^4 q^2 \Omega_p^2 (r_p/b_0)^3. \quad (13.6)$$

The cut off distance is the semi-width of the gap. We estimate a minimum value, $b_0 = (3/2)r_H = (3/2)(q/3)^{1/3} r_p$, where $r_H$ is the Hill radius. For any realizable gap with larger $b_0$ we have

$$\mathcal{F} \leqslant 0.26 \Sigma r_p^4 q \Omega_p^2. \quad (13.7)$$

For a disk in which angular momentum transport results from an effective kinematic viscosity, $\nu$, assumed to result from a form of turbulence, $\mathcal{F} = 3\pi\nu\Sigma r_p^2 \Omega_p$. From (13.7) we then get

$$36\nu/(r_p^2 \Omega_p) \leqslant q. \quad (13.8)$$

This is marginally satisfied for a Saturn mass planet for a typical value of $\nu$. For a *local* gap to be produced we also require the planets gravity to dominate pressure effects within the Hill sphere leading to the requirement that $r_H/r_p = (q/3)^{1/3} > (1/\mathcal{M})$ (see e.g., Lin & Papaloizou 1993). A five Jupiter mass orbiting in a prominent gap in a turbulent disk is illustrated in Figure 13.1. When there is a deep gap, even though there may be mass flow across it, torque balance across the gap results in radial migration on the same timescale as the general local evolution of the disk as indicated by Lin & Papaloizou 1986 (for more recent discussion of this aspect see Robert et al. 2018 and

Scardoni et al. 2020). This is typically $2 \times 10^5$ yr in standard models resembling the MMSN with $h = (\mathcal{M})^{-1} = 0.05$ at 5 au (see e.g., Nelson et al. 2000). However, this time increases with the local orbital period.

### 13.3.5 Type III Migration

This very rapid migration can occur for planets that form partial gaps in sufficiently massive disks (Masset & Papaloizou 2003; Peplinski et al. 2008). It involves a feedback process operating through coorbital torques that greatly amplifies the effect of an already existing possibly weak torque, $-\mathcal{T}$.

Suppose the planet migrates inwards at a rate $\dot{a}$ (a corresponding parallel discussion applies for outward migration). Matter flows through the horseshoe region via the separatrix at a rate $\dot{m} = 2\pi\Sigma r_p|\dot{a}|$. This produces a torque acting on the planet of magnitude $|\dot{J}| = 2\pi\Sigma|\dot{a}|r_p^2\Omega_p x_s$, where $x_s$ is the half width of the horseshoe region. As the planet moves it drags coorbital material with it. Noting that the mass in this region is $4\pi\Sigma_s r_p x_s$, where $\Sigma_s$ is the surface density in the gap. This is added to the planet mass when considering torque balance which gives

$$\begin{aligned} &\frac{1}{2}(M_p + 4\pi\Sigma_s r_p x_s) r_p \Omega_p |\dot{a}| = |\dot{J}| - \mathcal{T} = 2\pi\Sigma|\dot{a}|r_p^2\Omega_p x_s - \mathcal{T}. \\ &\text{Hence} \quad \frac{1}{2}(M_p - \delta m) r_p \Omega_p |\dot{a}| = -\mathcal{T}, \end{aligned} \tag{13.9}$$

where, $\delta m = 4\pi\Sigma_s r_p x_s(\Sigma - \Sigma_s)$, is the coorbital mass deficit. When the coorbital mass deficit approaches the planet mass migration can increase rapidly, the direction determined by the seeding torque $-\mathcal{T}$. In practice the migration rate can increase until the time to migrate through the horseshoe region, $2x_s/|\dot{a}|$, becomes comparable to the horseshoe libration period $8\pi r_p/(3\Omega_p x_s)$. This leads to a characteristic fast migration time

$$r_p/|\dot{a}| = (2/3)(r_p/x_s)^2 P_{\text{orb}} = (8/3^{7/3})q^{-2/3}P_{\text{orb}}, \tag{13.10}$$

where we adopt $x_s = 3r_H/2$ and $P_{\text{orb}}$ is the orbital period.

### 13.3.6 High Eccentricity Migration

After the protoplanetary disk has dispersed secular gravitational interactions among the planets in a multi-planetary system may lead to a planet on an orbit with very high eccentricity, which has close approaches to the central star leading to tidal interaction that circularizes the orbit which conserves its angular momentum (e.g., Wu & Lithwick 2011; Naoz et al. 2013; Papaloizou 2020). The semimajor axis may thus undergo a large decrease leading to an orbit with a period of a few days. High orbital inclinations relative to the initial orbital plane may also be generated. In principle the Kozai–Lidov mechanism could attain a similar result through the action of a binary companion in a highly inclined orbit. However, it has been argued that there are not enough companions of the required type for this to be viable except in a small number of cases (see Dawson & Johnson 2018; for a review).

## 13.4 Important Questions/Goals for Future Studies

### 13.4.1 Planet Migration in Disks with MHD Winds and Ordered Magnetic Fields

Hot Jupiters have masses >0.3 $M_J$ and orbital periods <10 days. They can be seen as a group in the top left of the left panel of Figure 13.2. About one in ten giant planet systems contain a hot Jupiter (see e.g., Papaloizou 2020, and references therein). Disk migration has been proposed as a mechanism for moving them from beyond the ice line to their present close in orbits. For a review of the applicability of this mechanism as well as others involving gravitational interactions among planets or with binary companions leading to high eccentricity migration after the disk has dispersed (see Dawson & Johnson 2018). Note that the left panel of Figure 13.2 indicates a separation between giant planets with orbital periods >~100 days that may have formed beyond an ice line with later modest migration and the hot Jupiters that underwent a more extreme form. Type II migration is the most likely mechanism applying to giant planets (see e.g., Lin & Papaloizou 1986; Baruteau et al. 2014) with hot Jupiters potentially undergoing more extreme type III migration (see e.g., Masset & Papaloizou 2003; Papaloizou et al. 2007).

As the timescale for type II migration at a few au is significantly less than the disk lifetime (e.g., Nelson et al. 2000) a slowing down mechanism is needed. Advantage may be taken of the scaling of the migration rate with viscosity when there are regions of the disk where the viscosity is expected to be small. This requires consideration of recent models that include magnetic fields and non-ideal magneto-hydrodynamics (MHD; e.g., Bai 2017; Riols et al. 2020). Three dimensional (3D) simulations show the disk inner region to be mainly laminar with large scale toroidal fields and wind-driven accretion. This is in contrast to earlier simulations that

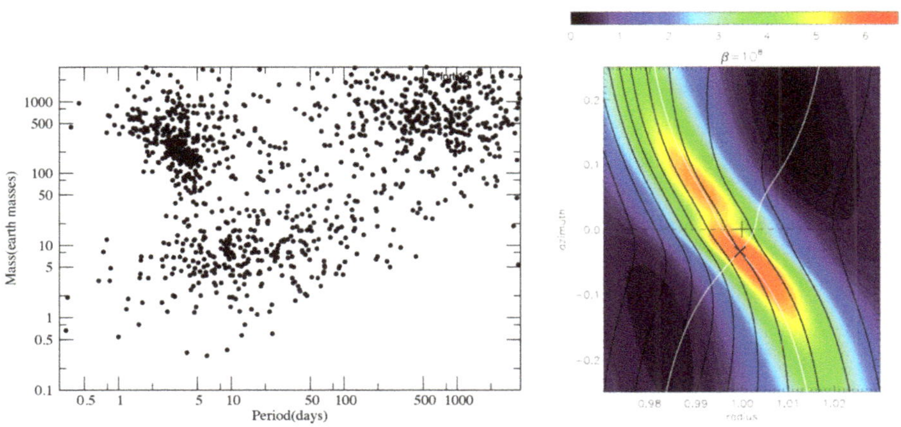


**Figure 13.2.** Left panel: the dependence of the planet mass, $M_p$, on orbital period, the data was taken from the NASA exoplanet archive: https://exoplanetarchive.ipac.caltech.edu/index.html (as of January 2019). Right panel: magnetic energy contours in the coorbital region of a ~7 $M_\oplus$ planet, indicated by a + sign, where there was an initial toroidal field. The gas flow around the horseshoe turns produces regions of field concentration. The black lines denote field lines and the. white lines horseshoe separatrices (reproduced from Guilet et al. 2013).

studied gap formation assuming ideal MHD (e.g., the simulation in Nelson & Papaloizou 2003; illustrated in the right panel of Figure 13.1).

In the presence of MHD winds with negligible explicit viscosity, gap formation and type II migration are still expected but with the angular momentum loss rate from the outer disk required to supply the wind used in Equation (13.7). No such balance applies to the inner disk if it is disconnected from the outer disk. However, magnetic linkage could affect this, enabling angular momentum transfer between the inner and outer disk independent of the planet. This may cause its migration to differ from that found for a viscous disk, potentially slowing it down. Simulations also need to be consider the role of the vertical shear instability (e.g., Stoll & Kley 2015) and incorporate processes leading to photo-evaporation and disk dispersal in order to investigate final planetary configurations (e.g., Alexander & Pascucci 2012).

Studies of type III migration also need to be done with recent disk models in 3D. Here there is significant material in the coorbital region, some of which may be associated with vortex production that can interrupt the migration (Lin & Papaloizou 2010) as well as accrete onto the planet. It has been assumed that the planet does not accrete significantly (e.g., Peplinski et al. 2008), though significant accretion might inhibit the mechanism. It is possible that the planet fills its critical Hill sphere or Roche lobe and is unable to accept mass on a rapid type III migration timescale. We note that the thermal timescale of a planet filling its Roche lobe may be estimated as $t_{\rm KH} \sim GM_p^2/(4\pi(2r_H/3)^3\sigma T_e^4)$, where $T_e$ is the effective temperature and $\sigma$ is Stefan's constant. This may be written

$$t_{\rm KH} \sim 81\pi M_p/(8P_{\rm orb}^2\sigma T_e^4) \sim 1255(M_p/M_\eta)(P_{\rm orb}/(1{\rm yr}))^{-3}(T_e/(300{\rm K}))^{-4}P_{\rm orb}, \quad (13.11)$$

which may plausibly exceed the characteristic time for type III migration given by (13.10) leading to thermally restricted accretion for a Saturn mass planet at ~1 au. The issues highlighted here need to be investigated in more detail.

### 13.4.2 Tidal Dissipation and High Eccentricity Migration

Orbital evolution produced as a result of tidal interaction with the central star may be important for giant planets brought close in by disk migration or high eccentricity migration occurring after the disk has dispersed. In the latter case, as the energy that has to be dissipated may greatly exceed the binding energy of the planet (Ivanov & Papaloizou 2004), the interaction may be strongly nonlinear, potentially leading to disruption (Guillochon et al. 2011). In this context the issue of how much energy is dissipated in the planet and whether it can be efficiently radiated away is important for determining possible final orbital parameters of hot Jupiters. Further studies are needed.

### 13.4.3 Coorbital Torques, Saturation and Halting Type I Migration

The role of coorbital torques in halting type I migration has been mentioned in Section 13.3.3. As the action of coorbital torques or horseshoe drag may determine disk locations where planets can grow it is important that they be investigated with the most accurate disk models that incorporate magnetic fields where relevant. These can prevent the torques from saturating when there is an effective viscosity (see Section 13.3.2) as shown for a disk with ideal MHD and a simple equation of state by Baruteau et al. 2011. However, what happens in the non-ideal case with ordered fields and winds is unclear.

The effect of a large scale toroidal magnetic field has been considered in two dimensional (2D) model disks by Terquem (2003) for ideal MHD and Guilet et al. (2013) when there is enough magnetic diffusivity to enable disk gas to diffuse across field lines as horseshoe turns are executed (see the right panel of Figure 13.2). In the former inviscid case, horseshoe turns are prevented through the operation of Alfvèn resonances and the migration can reverse. In the latter case field concentrations can produce surface density changes that modify the torques potentially reversing migration (see Figure 13.2). These simulations adopted a laminar viscosity with Reynolds' number of $1.2 \times 10^5$ and Prandtl number of unity in most cases. This needs to be looked at in 3D models with non-ideal MHD without applied viscosity. One might anticipate instabilities leading to vortex production for sufficiently large planet masses (e.g., Li et al. 2009).

Another issue associated with horseshoe turns in 3D models is the role of vertical stratification (see McNally et al. 2020). They find that buoyancy oscillations lead to a speeding up inward migration. This needs to be considered with active MHD and thermal diffusion. Additional effects arise from heat transport connecting the planet and disk material (Masset 2017). Asymmetries in the horseshoe region (note the displacement of the planet from the central stagnation point in the right panel of Figure 13.2) and an offset between the circular motion of the planet and disk material lead to asymmetries in the heat transport and density structure that can produce migration stalling torques. The effects of accretion of mass and angular momentum and a magnetic connection should be studied in 3D.

### 13.4.4 Multiplanet Systems, Commensurabilities and Conversion from Inward to Outward Migration, Convergent to Divergent Migration and the Grand Tack

Systems of planets often do not behave as if the components were independent. Convergent migration of planet pairs leads to mean motion resonances yet their incidence is modest, being about 30% of giant planet pairs with period ratios <3.5 and 20% of all pairs with period ratio 3 are close to a commensurability (see e.g., Papaloizou 2020 for discussion). There are rare systems of three or more planets forming resonant chains. If convergent migration of components is common during or post formation some disruptive mechanism is required to disrupt them either before or after the disk dispersal (e.g., Izidoro et al. 2017). Up to now studies of disk planet interactions in these contexts have mostly adopted simple disk models and been in 2D.

The joint migration of Jupiter and Saturn was first studied by Masset & Snellgrove (2001). They found that the system switched from inward to outward migration after an initial rapid inward migration of Saturn caused the system to lock into a

3:2 commensurability. This scenario has been adopted in the Grand Tack scenario proposed to operate during the formation of the solar system (see e.g., Pierens et al. 2014). The process which occurs when an inwardly migrating giant planet is caught up by a lighter one has been proposed as a mechanism that can push planets out to large distances (see Crida et al. 2009). These simulations involve type II migration as well as fast possibly type III migration of an outer lighter giant planet. Thus the issues highlighted in Section 13.4.1 apply. Consideration of realistic disk models with allowance for the mass growth fed by material in coorbital regions is needed to see if the outer planet always has the lowest mass as required in these scenarios.

*13.4.4.1 Formation and Subsequent Evolution of Commensurabilities*

Giant planets with periods >100 days may not have migrated through large distances (see left panel of Figure 13.2). The absence of super Earths with bare cores composed of volatiles and the relative rarity of commensurabilities in two planet systems indicates the same applies to them (see e.g., Papaloizou & Terquem 2019; Papaloizou 2020). But the existence of near commensurabilities supports a picture of at least restricted migration in some cases. The formation of low order commensurabilities through the convergent migration of two planets is well documented (see e.g., Baruteau et al. 2014). However, initially convergent migration can be reversed through wake planet interactions when the density response produced by one planet reacts back on the other. This has been seen when giant planets interact with incoming super Earths (Podlewska-Gaca et al. 2012) and for pairs of Neptune mass planets (Baruteau & Papaloizou 2013). It results in a departure from commensurability and may help reduce the number of expected strict commensurabilities. These studies should be extended to incorporate the final stages of the evolution of the protoplanetary disk. The location and evolution of structural features constituting migration traps and interaction with planetesimals and dust will be important.

**Summary**

Studies of disk planet interactions began with simple models before the discovery of exoplanets about forty years ago. While studies of migration resulting from disk free processes such as gravitational interactions leading to high eccentricity migration followed by tidal interaction with the central star began shortly after the first discoveries of exoplanets. Advances in our understanding of protoplanetary disks coupled with improving numerical tools and computational resources offer the prospect of more realistic modeling of planet formation and early evolution. This will lead to improved comparisons with observations of structured protoplanetary disks (see e.g., Francis & van der Marel 2020) as well as improved predictions for the form of planetary systems.

## References

Alexander, R., & Pascucci, I. 2012, MNRAS, 422, 3415
Bai, X.-N. 2017, ApJ, 845, 75

Baruteau, C., Fromang, S., Nelson, R. P., & Masset, F. 2011, A&A, 533, A84
Baruteau, C., & Papaloizou, J. 2013, MNRAS, 430, 1764
Baruteau, C., Crida, A., Paardekooper, S.-J., et al. 2014, in Protostars and Planets VI, ed. H. Beuther, R. S. Klessen, C. P. Dullemond, & T. Henning (Tucson, AZ: Univ. Arizona Press), 667
Crida, A., Masset, F., & Morbidelli, A. 2009, ApJL, 705, L148
Dawson, R. I., & Johnson, J. A. 2018, ARA&A, 56, 175
Francis, L., & van der Marel, N. 2020, ApJ, 892, 111
Guilet, J., Baruteau, C., & Papaloizou, J. C. B. 2013, MNRAS, 430, 1764
Guillochon, J., Ramirez-Ruiz, E., & Lin, D. 2011, ApJ, 732, 74
Hansen, B. M. S., & Murray, N. 2012, ApJ, 751, 158
Ivanov, P. B., & Papaloizou, J. C. B. 2004, MNRAS, 347, 437
Izidoro, A., Ogihara, M., Raymond, S. N., et al. 2017, MNRAS, 470, 1750
Li, H., Lubow, S. H., Li, S., & Lin, D. N. C. 2009, ApJ, 690, L52
Lin, D. N. C., & Papaloizou, J. C. B. 1979, MNRAS, 186, 799
Lin, D. N. C., & Papaloizou, J. C. B. 1986, ApJ, 309, 846
Lin, D. N. C., & Papaloizou, J. C. B. 1993, in Protostars and Planets III (Tucson, AZ: Univ. Arizona Press), 749
Lin, M.-K., & Papaloizou, J. C. B. 2010, MNRAS, 405, 1473
Masset, F. S. 2017, MNRAS, 472, 4204
Masset, F. S., Morbidelli, A., Crida, A., & Ferreira, J. 2006, ApJ, 642, 478
Masset, F., & Papaloizou, J. C. B. 2003, ApJ, 588, 494
Masset, F., & Snellgrove, M. 2001, MNRAS, 320, L55
McNally, C. P., Nelson, R. P., Paardekooper, S.-J., Benitez-Llambay, P., & Gressel, O. 2020, MNRAS, 493, 4382
Naoz, S., Farr, W. M., Lithwick, Y., Rasio, F. A., & Teyssandier, J. 2013, MNRAS, 431, 2155
Nelson, R. P. 2005, A&A, 443, 1067
Nelson, R. P., Papaloizou, J. C. B., Masset, F., & Kley, W. 2000, MNRAS, 318, 18
Nelson, R. P., & Papaloizou, J. C. B. 2003, MNRAS, 339, 993
Papaloizou, J. 2020, Oxford Research Encyclopaedia of Planetary Science (Oxford: Oxford University Press)
Papaloizou, J. C. B., Nelson, R. P., Kley, W., Masset, F. S., & Artymowicz, P. 2007, in Protostars and Planets V (Tucson, AZ: Univ. Arizona Press), 655
Papaloizou, J. C. B., & Terquem, C. 2006, RPPh, 69, 119
Papaloizou, J. C. B., & Terquem, C. 2019, MNRAS, 482, 530
Paardekooper, S. J., & Mellema, G. 2006, A&A, 459, L17
Peplinski, A., Artymowicz, P., & Mellema, G. 2008, MNRAS, 386, 179
Pierens, A., Raymond, S. N., Nesvorny, D., & Morbidelli, A. 2014, ApJL, 795, L11
Podlewska-Gaca, E., Papaloizou, J. C. B., & Szuszkiewicz, E. 2012, MNRAS, 421, 1736
Rein, H. 2012, MNRAS, 427, L21
Riols, A., Lesur, G., & Menard, F. 2020, A&A, 639, A95
Robert, C. M. T., Crida, A., Lega, E., et al. 2018, A&A, 617, A98
Scardoni, C. E., Rosotti, G. P., Lodato, G., & Clarke, C. J. 2020, MNRAS, 492, 1319
Stoll, M. H. R., & Kley, W. 2015, A&A, 572, A77
Terquem, C. E. J. M. L. J. 2003, MNRAS, 341, 1157
Wu, Y., & Lithwick, Y. 2011, ApJ, 735, 109

# Chapter 14

## Astrochemistry and Planet Formation

**Ewine F. van Dishoeck and Edwin A. Bergin**

### 14.1 Introduction

Stars and planets are formed deep inside dense clouds. When these clouds collapse, the surrounding gas and dust become part of the infalling envelopes and rotating disks, thus providing the basic material from which new planetary systems are made. Today we are finding that much of the chemical composition of the planet-building material is likely set in the cold pre- and protostellar stages and preserved en route to planet and comet construction sites. With new observational techniques, astronomers can now zoom into these planet-forming disks and study them on scales comparable with the orbit of Saturn in our own solar system (Andrews 2020).

Astrochemistry, also known as molecular astrophysics, is the study of the formation, destruction and excitation of molecules in astronomical environments. By observing many lines of the same species, molecules are excellent diagnostics of the physical conditions and their high resolution spectral profiles provide unique information on the kinematics in the regions where they reside. More than 200 different molecules have now been detected in interstellar space (McGuire 2018). The main questions in astrochemistry therefore include: how, when and where are these molecules produced? What do they tell us about temperatures, densities, gas masses, ionization rates, radiation fields, and dynamics of the star-forming clouds and planet-forming disks? From these quantities what can be learned about the physics of star and planetary birth? How are they cycled through the various phases of stellar evolution, from birth to death? How far does chemical complexity go? And, most far-reaching, can they form the building blocks for life elsewhere in the universe? While organics are a central focus of the latter question, water, in the form of ice, stands out as a reservoir of abundant elemental oxygen throughout all phases, hinting that life's solvent started its journey in space (van Dishoeck et al. 2014).

This brief overview summarizes some of the main results and questions regarding the chemistry along this journey from cloud to disk. References are limited primarily to review papers of the many hundreds of papers in this rapidly growing field;

doi:10.1088/2514-3433/abfa8fch14  

citations to individual papers are mostly to relevant results in the last few years and are highly incomplete. Readers are encouraged to cite original papers.

## 14.2 State of the Art

Recent overviews of the field of astrochemistry have been given by Tielens (2013) and Yamamoto (2017) and refs cited. Fourteen challenges have been formulated by van Dishoeck (2018).

### 14.2.1 How to Observe Molecules

Molecules can be observed through their electronic, vibrational or rotational transitions at optical/UV, infrared and millimeter wavelengths, respectively. Several of the most powerful new telescopes deployed over the past decades have been particularly well suited to observe interstellar molecules. The bulk of the molecules are detected through their rotational transitions at millimeter wavelengths, and various single-dish and interferometers at millimeter wavelengths with increasingly sensitive broad-band detectors have become available culminating in the Atacama Large Millimeter/submillimeter Array (ALMA). ALMA has the combined sensitivity and spatial resolution to zoom in on planet-forming disks. Millimeter observations require the molecule to have a permanent dipole moment to be detectable.

Infrared observations are highly complementary. Space missions like the Infrared Space Observatory, Spitzer Space Telescope and the Herschel Space Observatory at mid- and far-infrared wavelengths were particularly well suited to study $H_2O$ and molecules like $CO_2$ that are abundant in our Earth's atmosphere. Moreover, solid state species (silicates, ices), complex molecules (PAHs, fullerenes) and small molecules without a dipole moment like $CH_4$, $C_2H_2$, and $CO_2$ can be uniquely observed through their vibrational transitions. Warm $H_2$ can also be detected through mid-infrared lines, but cold $H_2$ in dark clouds is invisible and can only be traced indirectly through other molecules, most notably through far-infrared lines of HD. These space-based data have been complemented by studies of selected molecules using ground-based 8 m optical/infrared telescopes equipped with high resolution spectrometers. The James Webb Space Telescope (JWST), to be launched late 2021, will be the next big jump in mid-infrared capabilities.

### 14.2.2 Astrochemical Models

Chemistry in the interstellar medium, where densities are typically $10^3$–$10^6$ cm$^{-3}$, cannot be described by the laws of thermodynamics, except in the densest stellar and planetary atmospheres well shielded from radiation, with densities $>10^{13}$ cm$^{-3}$. In interstellar space, the chemistry is controlled by two-body reactions and abundances can be obtained through kinetics involving large networks of reactions. These networks contain gas-phase reactions (ion–molecule, neutral–neutral) as well as gas–grain and grain–surface chemistry, with UV photoprocesses playing a role in both gaseous and solid phases. Thanks to many decades of laboratory experiments and quantum chemical calculations by physicists and chemists, deep insight into these processes has been obtained and their reaction rates under different conditions have been quantified.

Compilations of the rate coefficients together with codes that solve the coupled differential equations of these networks include the UMIST 2013 database,[1] the KIDA database,[2] and the UCLCHEM code.[3] Programs tailored to exoplanetary atmospheres include VULCAN[4] and LEVI (Hobbs et al. 2019).

To run an astrochemical model, a prescription of the temperature and density structure of the source, the incident UV radiation field, the cosmic ray ionization rate, as well as the abundances of the (volatile) elements that are not locked up in refractory solids but that can cycle between the gas and ice phases are needed. Given the uncertainties in each of these parameters as well as in individual rate coefficients, agreement between observations and models at the factor of a few level is generally considered to be good.

## 14.3 Important Questions + Goals: Chemistry from Clouds to Planets

The various stages in the formation of a new planetary system are illustrated in Figure 14.1. The first three stages typically take a few million years, with a clear drop in gas and dust content after ~3 Myr. A key recent finding is that planet formation must already start early, in the embedded phase of star formation a few $\times 10^5$ yr after cloud collapse, in order to account for the observed solid mass in exoplanet cores (Tychoniec et al. 2020). This is consistent with the solar system record as Fe-rich meteorites sample the cores of large bodies that formed, in many instances, within the first few 100,000 years (Kruijer et al. 2017). Thus, much of the chemistry of planet-forming material is already set in the cold pre- and protostellar stages. On the other hand, there is clear evidence that molecules and dust were vaporized in the inner young solar nebula and then re-condensed, based on elemental abundance patterns and minerals found in meteorites (Connelly et al. 2012). Key questions therefore include

- What determines the chemical composition of planet-building material: inheritance or full reset? At what disk radius is the cross over (for a given type of star)? Can the composition of the young still-forming protostellar disks be characterized enough to reveal answers to these questions?
- If inherited from the collapsing cloud, to what extent is the gas and ice composition modified en route from cloud to disk (e.g., by accretion shocks), and within the protoplanetary disk midplane?
- Ice rich pebbles drift inwards during the main phase of planetary assembly. What is the fate of pebbles drifting from the outer tens to hundreds of au into the inner disk (<10 au)? How does this impact planet formation and its chemistry?

[1] http://udfa.ajmarkwick.net/index.php?mode=downloads
[2] kida.astrophy.u-bordeaux.fr
[3] https://uclchem.github.io/
[4] https://github.com/exoclime/VULCAN

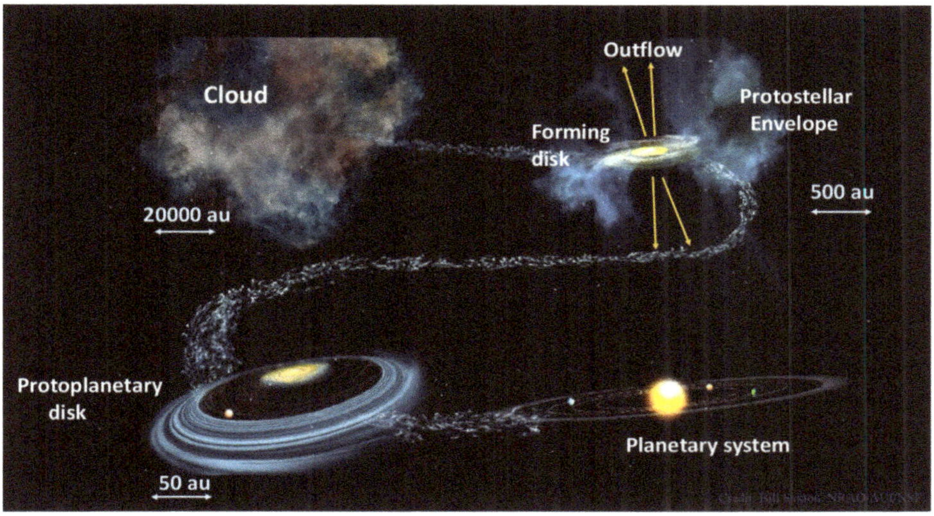


**Figure 14.1.** Stages in the formation of a new planetary system from a dark interstellar cloud. Much of the chemistry of planet-forming material is likely set in the cold pre- and protostellar stages. Also, there is strong evidence that the first steps of planet(esimal) formation take place early, in the embedded protostellar phase of star formation (typically few $\times 10^5$ yr after collapse). Artist impression from Bill Saxton (NRAO/AUI/NSF), annotated by the authors.

- How quickly are ices locked up in large planetesimals, and what are the C/O or C/N ratios with radius in the midplane? How are these affected by pebble drift and dust traps? How do the snowlines move with time? How important is vertical mixing?
- Are the majority of the heavy volatile elements in a giant planetary atmosphere accreted as ice or as gas? How are they modified in the atmosphere itself and does a dilute core mix with the envelope? What is the importance of LTE versus kinetic chemistry in the outer layers?
- What can the composition of exoplanetary atmospheres ultimately tell us about their formation location, given the above questions?
- What can we learn about terrestrial planet formation and the delivery of needed volatile elements from astronomical observations?

### 14.3.1 Cold Dark Clouds

Chemistry starts in cold dark clouds prior to star formation with have typical temperatures of 10 K and densities of $10^4$–$10^5$ cm$^{-3}$. They exhibit a variety of chemical characteristics, the most important of which are ice formation and heavy deuterium fractionation (Bergin & Tafalla 2007; Caselli & Ceccarelli 2012; Ceccarelli et al. 2014). Most of these chemical signatures are transferred to the protostellar stage where they are observed in the cold outer parts of the collapsing envelope (Figure 14.1).

Infrared observations of ice features toward reddened background stars show that the formation of water ice starts once the cloud extinction $A_V$ is a few mag (Boogert et al. 2015; Öberg et al. 2011). The formation of water ice on the surfaces of grains is now well characterized in the laboratory and models, and tested against observations of cold water gas and ice (van Dishoeck et al. 2013, 2014). $CH_4$, $NH_3$, and some $CO_2$ ice are also made in this early phase, with the amount of ice rapidly increasing as the cores become more centrally concentrated. All of these molecules are made on the grain surfaces, from reactions of atomic O, C, and N with atomic hydrogen; they are not accreted from the gas.

At densities around $10^5$ $cm^{-3}$, the timescales for freeze-out become shorter than the lifetime of the cloud core, and CO (the dominant form of volatile carbon at these high densities) rapidly depletes from the gas onto the grains. This CO-rich ice can subsequently react with atomic H to form $H_2CO$ and $CH_3OH$, a process that has been demonstrated to proceed rapidly at low temperatures in the laboratory. Various other routes can transform CO to $CO_2$ and more complex organic ices, even at very low temperatures.

A clear signature of cold interstellar chemistry are high abundances of deuterated molecules such as DCN, with DCN/HCN ratios at least three orders of magnitude higher than the overall [D]/[H] ratio of $2 \times 10^{-5}$. Even doubly- and triply-deuterated molecules such as $D_2CO$ and $ND_3$ have been detected. This huge fractionation has its origin in two factors. First, the zero-point vibrational energy of deuterated molecules is lower than that of their normal counterparts because of their higher reduced mass. This makes their production reactions exothermic. In cold cores, most of the fractionation is initiated by the $H_3^+ + HD \leftrightarrow H_2D^+ + H_2$ reaction which is exothermic by about 230 K. $H_2D^+$ then transfers a deuteron to CO or $N_2$ or another species. Since CO is the main destroyer of both $H_3^+$ and $H_2D^+$, their abundances are even further enhanced when CO is removed from the gas. Similarly, $CH_2D^+$ is enhanced at low temperatures and can enhance deuterium in organic molecules. The second, related effect is that the gaseous atomic D/H ratio is enhanced, which implies that relatively more D than H arrives on the grain to react with CO to make deuterated versions of formaldehyde, methanol and more complex organic molecules, and powering deuteration of water ice.

Key references summarizing this stage are Caselli & Ceccarelli (2012), and Boogert et al. (2015).

### 14.3.2 Protostellar Envelopes and Young Disks

Once a protostar has formed in the center, its luminosity sets up a temperature gradient in the gas and dust. Temperatures increase from 10 K in the outer envelope to a few hundred K in the innermost region close to the protostar. This can result in numerous chemical changes: radicals become mobile in and on icy surfaces and recombine to form even more complex molecules, whereas increased UV and x-rays from the protostar-disk accretion boundary trigger further chemistry in the ice and gas. When dust temperatures become high enough for ices to sublimate, molecules do so presumably in a sequence according to their binding energies, with volatile species like CO and $N_2$ sublimating first.

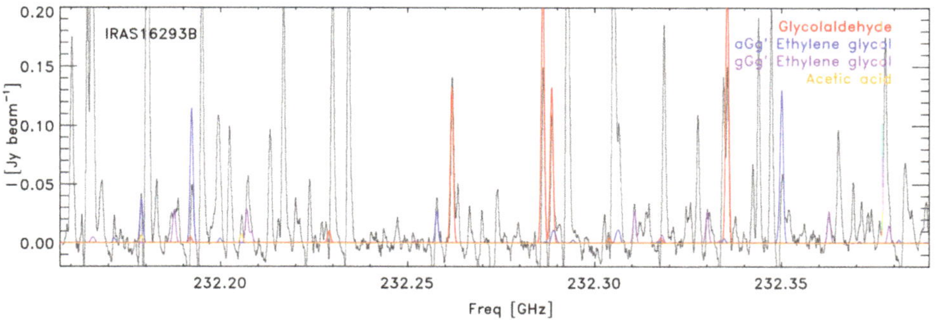


**Figure 14.2.** Blow-up of a small part of the ALMA spectral survey toward the young solar-mass protostar IRAS16293B on scales of 50 au, illustrating the wealth of lines (roughly 1 line every 3 km $s^{-1}$) and the identification of complex organic molecules like glycolaldehyde and ethylene glycol. Credit: Jørgensen et al. (2016), reproduced with permission © ESO.

Once dust temperatures of >100 K are reached, even the strongly-bound water and methanol ice sublimate, together with any minor molecules trapped in them, resulting in particularly rich gas-phase millimeter spectra (Figure 14.2). This inner 100 K zone is called the "hot core." Here high-temperature gas-phase reactions between sublimated molecules can result in "second generation" complex organic molecules (Balucani et al. 2015).

Much of the extensive chemical complexity seen in hot cores is thought to be largely assembled in ices. A major question is whether this rich chemistry is incorporated in the forming disks (Figure 14.1), and if so in what form. Further, the full extent of chemical complexity occurring on grain surfaces is still uncertain. It is clear that some molecules of "pre-biotic" significance are found in ices, such as simple sugars and peptide bonds (Figure 14.2), but these are still small compared with molecules found in biological systems as well as the macromolecular material found in meteorites and comets (Section 3.5).

The solar system record suggests that there must have been a mixture of material where the chemistry is reset with pre-existing (or less processed) material "inherited" from the interstellar medium (Section 3.5). In this regard, it is unclear whether the material entering the disk experiences a strong accretion shock, and if so, whether this shock completely resets the chemistry or whether strongly-bound ices survive. At face value observations find an interesting transition in the disk seen in hydrocarbon emission and SO, but in general the overall temperature of these systems appears to be rather cool (<50 K).

Key references for this section include Herbst & van Dishoeck (2009), Ceccarelli et al. (2014), and Jørgensen et al. (2020).

### 14.3.3 Planet-forming Disks

ALMA has been transformational for studies of planet-forming disks, and will likely remain so for the coming decade(s) (Andrews 2020). ALMA and high contrast infrared imaging of disks are providing striking pictures of the first steps of planet formation: rings, gaps, cavities, asymmetric structures, and/or spiral arms. Although there are many possible interpretations of these structures, they do indicate that growth of dust grains to pebbles and planetesimals is taking place, starting even in the embedded phase.

Disks remain challenging to study both observationally and theoretically, however. Observationally, molecular lines are very weak since disks are small (typically less than 1″ on the sky) and their mass is only 1% of that of the collapsing cloud. Theoretically, they are a challenge since they cover a huge range of densities and temperatures in at least two dimensions, from >1000 K in the inner disk and upper layers, to 10 K in the outer midplane, and from densities of $>10^{13}$ $cm^{-3}$ in the inner midplane down to $10^5$ $cm^{-3}$ in the upper outer layers. UV radiation fields from the central star impinging on the surface layers can be as high as $10^5$ times the interstellar radiation field at 10 au from the star, thereby ionizing atoms and dissociating molecules. Thus, different types of chemistry are important in different parts of the disks (Henning & Semenov 2013; Bergin & Cleeves 2018; Öberg & Bergin 2021; Figure 14.3).

Moreover, gas and dust are largely decoupled (except for the smallest grains): dust grains grow to pebble size (few cm), settle to the midplane and drift in radially (Figure 14.3). If they encounter a pressure bump, dust traps can form where particles can grow to even larger, planetesimal sizes. Gas/dust ratios can therefore differ significantly from 100. While ALMA is particularly well suited to study the gas and mm-sized dust in the outer disk (>10 au for gas, few au for dust), JWST and other mid-infrared facilities probe the important inner 10 au planet-forming zones of disks.

The decreasing temperature in the radial direction sets up a range of snowlines, i.e., radii where molecules freeze-out onto the grains, defined as the half-gas, half-ice point. Because of the vertical temperature gradient, the 2D snow surfaces are actually curved (Figure 14.3). Snowlines are thought to play a significant role in planet formation, since ice coating of grains enhances the solid mass, potentially promotes coagulation of grains to larger particles and/or require a higher collision velocity for destruction (Blum 2018; Pinilla & Youdin 2017), effects which are

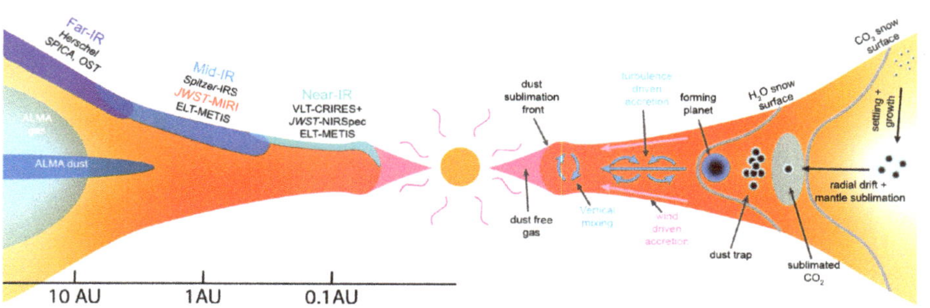


**Figure 14.3.** Overview of the structure of planet-forming disks. The left-hand side shows the different parts of the inner disk probed by various instruments. The right-hand part illustrates some important physical and chemical processes, including gas accretion and mixing, as well as grain growth, settling and radial drift of icy pebbles, which sublimate at snowlines. Figure by A. Bosman, PhD thesis, Leiden University.

particularly prominent just outside the snowline. Thus it is thought that water ice coated grains grow to larger sizes and provide larger pebbles for planet formation which has implications on growth via pebble accretion (Morbidelli et al. 2015). They also control the bulk elemental composition of the icy planetesimals and gas from which exoplanetary atmospheres are built.

**Cold outer disks.** Most molecules detected in disks are simple species containing only a few atoms. Only two complex molecules, $CH_3CN$ and $CH_3OH$, have been detected so far, and only barely. Even the brightest disks do not show rich line forests such as found for hot cores. This lack of gas-phase lines is because the bulk of the disks are cold with molecules frozen out. In fact, the CO snow surface is resolved in highly flared systems—thus the comet-forming zone can essentially be seen by eye in some ALMA images. The most important snowline, that of water, is generally out of reach because it typically occurs at a few au for T Tauri stars, too small for ALMA to pick up. Only systems undergoing luminosity outbursts such as V883 Ori or young disks with enhanced accretion luminosity, for which the water snowlines have moved out to tens of au, offer the opportunity for direct detection and imaging of complex chemistry (Lee et al. 2019).

The overall picture of the outer disk is therefore that much of the volatile oxygen and carbon is locked up in ices in large bodies. Since the dominant $H_2O$ and $CO_2$ ices have more oxygen than carbon, the ice is oxygen rich, with an overall C/O ratio that is lower than the interstellar or solar abundance (Figure 14.4; Öberg et al. 2011). In contrast, the gas is carbon rich even though it is overall depleted in carbon and oxygen. Once C/O >1, the chemistry completely changes and small hydrocarbon molecules become abundant. This can indeed explain the strong observed $C_2H$ and c-$C_3H_2$ emission in some disks, although models suggest that midplane chemistry tends to evolve toward gaseous C/O significantly less than 1 depending on the level of ionization (Eistrup et al. 2018). JWST will be able to observationally constrain the ice composition in planet-forming zones, but only for a handful of near edge-on disks and then only in the intermediate layers.

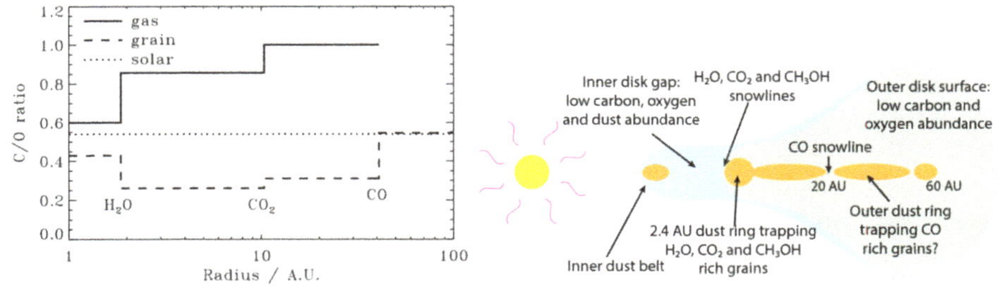


**Figure 14.4.** Left: midplane C/O ratios in gas and ice as function of disk radius, with major steps occurring at the $H_2O$, $CO_2$ and CO snowlines. The ice is oxygen rich, whereas the gas has high C/O but is overall depleted in carbon and oxygen. Reproduced from Öberg et al. (2011). © 2011. The American Astronomical Society. All rights reserved. Right: illustration how abundances in the inner disk can be affected by dust traps locking up volatiles in the outer disk beyond their snowlines, as in this case for the TW Hya disk. Credit: Bosman & Banzatti (2019), reproduced with permission © ESO.

**Warm inner disk.** The warm gas in the upper layers of the inner disk (<10 au) emits strongly at mid-infrared wavelengths. Indeed, a dense forest of lines due to simple molecules has been detected by Spitzer and ground-based infrared telescopes, providing a glimpse of the chemistry in that region (Pontoppidan et al. 2014). Here the high temperatures can completely "reset" the chemistry. Key abundant molecules without a dipole moment such as $CO_2$ and $C_2H_2$ are observed, together with $H_2O$, OH, and HCN, allowing tests of major C, O, and N reservoirs and high-temperature chemistry. The degree to which these data can probe the disk midplane chemistry and ice sublimation, relevant for planet formation models, is still unclear however, since they can only be observed if effective vertical mixing takes place. On the other hand, the recent observational evidence for meridional flows suggests that giant planets accrete a significant fraction of their gas from the upper disk layers (Teague et al. 2019). JWST will be poised to provide much deeper searches for $CH_4$, $NH_3$ and minor species, and detect isotopologues to constrain line optical depth, and follow the inner disk chemistry (at least in the surface layers) from the youngest embedded disks to the debris disk stage, and across the stellar type range.

Key references for this stage include Henning & Semenov (2013), Pontoppidan et al. (2014), Andrews (2020), and Öberg & Bergin (2021).

### 14.3.4 Disk Evolution and Planet(esimal) Formation

Disk structure and chemistry is not static but evolving, due to grain growth and inward drift of dust particles (Figure 14.3; Pinilla & Youdin 2017). At the same time, mass and angular momentum transport in disks, both inward and outward, are still poorly understood. How does this affect the chemistry? Recent ALMA data indicate either surprisingly low gas/dust ratios or a large fraction (up to a factor of 100) of volatile carbon, carried by CO, missing from the gas phase. The latter case, that CO is missing from layers with temperatures >20 K (the CO sublimation temperature), has support in terms of consistency with accretion rates, the mass needed to make planetary systems, and, in a handful of instances, detections of HD to constrain the gas mass (Bergin & Williams 2018). Thus, at face value, it appears that planet formation removes volatiles from the gas.

The first clue actually came from Herschel–HIFI observations which revealed surprisingly weak gaseous $H_2O$ lines in disks. Since cold $H_2O$ vapor in the outer disks results from UV photodesorption of water ice coating small grains in intermediate layers, the observed weak emission indicates that this water ice is spatially confined to only a part of the disk: radially, vertically or both. The most plausible interpretation is that most of the water is locked up in large icy bodies (at least pebble size) in the midplane (Du et al. 2017). A second clue comes from models of dust coagulation and growth which suggest that water and CO on the surface can be caught up in the grain evolution and locked within icy pebbles and planetesimals (Krijt et al. 2020). In the case of CO it may be combined with an active disk chemistry transforming CO into $CH_3OH$, $CO_2$, or hydrocarbons.

Do these heavy elements return to the gas when the pebbles drift inward and cross the snowlines (Figures 14.3, 14.4)? If so, one would expect inner disk abundances inside the water snowline to be similar to, or even higher than, the stellar abundances. This is not always the case: there is significant evidence that dust traps located beyond the major snowlines in the outer disk can lock up large fractions of carbon and oxygen preventing these pebbles to drift inward and resulting in inner disks that are depleted in these elements (Figure 14.4, right; Kama et al. 2015; McClure et al. 2020).

So how do we then determine the disk chemical composition just prior to planet formation? Much of the chemistry of planet formation is unfortunately hidden from our view, including the bulk C- and O-containing species. One option is to observe warmer, younger disks where less of the material is frozen out and fewer planetesimals have formed, such as the outbursting case described above. Observational studies show that only the youngest, deeply embedded sources have high enough temperatures to reveal the full chemical richness (Section 3.2; van't Hoff et al. 2020). The forming disks around Class 0 sources, like those for IRAS16293-2422 (Figure 14.2), may therefore well provide the most detailed information of chemistry on solar-system scales, even if the material has not yet settled into a Keplerian structure. That chemistry, in turn, may have been set already to a large degree in the dense cold core just prior to and during collapse, arguing for an "inheritance" rather than a "reset" scenario, at least in the outer disk.

Key references for this section include Pinilla & Youdin (2017), Du et al. (2017), Bergin & Williams (2018), and Krijt et al. (2020).

### 14.3.5 Clues from Our Solar System: Comets and Meteorites

The solar system record shows evidence of both a full reset of the chemistry as well as inheritance from the interstellar medium. The pattern of elemental abundances in the Earth's lithosphere and in meteorites suggests that this material formed within a hot (>1500 K) atomic gas in which all original interstellar material was vaporized, followed by a sequence of mineral condensation as the gas cooled (McDonough & Sun 1995). This cannot be the case for all solar system materials, however, as the deuterium enrichment in Earth's water, and indeed all solar system water, implies a cold (<30 K) source and cometary abundances appear to follow that of the interstellar medium (Altwegg et al. 2019).

Comets and other icy bodies in our own solar system provide the best clues to the chemistry in the cold part of the natal solar nebula disk and the level of change since their formation 4.5 billion years ago. Comparisons between cometary and interstellar abundances, including those from the Rosetta mission to comet 67P/C-G, have shown some tantalizing similarities (Mumma & Charnley 2011; Drozdovskaya et al. 2019). The organic complexity in comets goes further than has been detected so far in young disks, with even 4- or 5-carbon organic molecules, alkanes, aromatics and even the simplest amino acid, glycine, detected (Altwegg et al. 2019). More bright comets like Hale–Bopp that can be probed with ALMA and modern infrared instruments will provide further insight into their origin, including through determining their D/H ratio in water.

Primitive carbonaceous meteorites also have a rich organic composition, but much of that inventory is either macromolecular or altered via aqueous reactions in liquid water (Alexander et al. 2007). Thus, in terms of the origins of biology and the importance of molecules produced via cold chemistry in the interstellar medium, a key question is the amount of material that was supplied to the Earth that originated in the asteroid belt or beyond. The fate of this material on the Earth itself is, of course, a central issue (see Section 3.6).

Key references for this section include McDonough & Sun (1995), Altwegg et al. (2019), and Alexander et al. (2007).

### 14.3.6 Exoplanets and Their Atmospheres

**Detection of young exoplanets with molecules.** One new frontier is the use of ALMA to infer the presence of hidden planets, first using dust emission (Andrews 2020) followed by constraints from CO emission showing deep gas cavities and gaps (van der Marel et al. 2016). Even more recently, the presence of hidden planets based on kinematical information has been gleaned from ALMA CO emission maps (Teague et al. 2018; Pinte et al. 2018; and other work by their teams). Figure 14.5 provides an example of the overall emission (dust and CO) along with the exquisite measured velocity field. This topic is very much in its infancy but has promise to potentially isolate the locations of hidden planets, set unique constraints on the $H_2$ gas pressure, and reveal the presence of ordered gas motions.

A central aspect of this work is the close collaboration with dynamical theorists. Indeed, the wealth of dust substructures in disks points to a larger number of giant exoplanets than found by direct imaging at 10–100 au distance, suggesting a so-far unseen population of sub-Jupiter mass planets that needs independent confirmation (Zhang et al. 2018; Lodato et al. 2019). The kinematical work is pure molecular astrophysics, but has interesting astrochemical links. Different molecular tracers

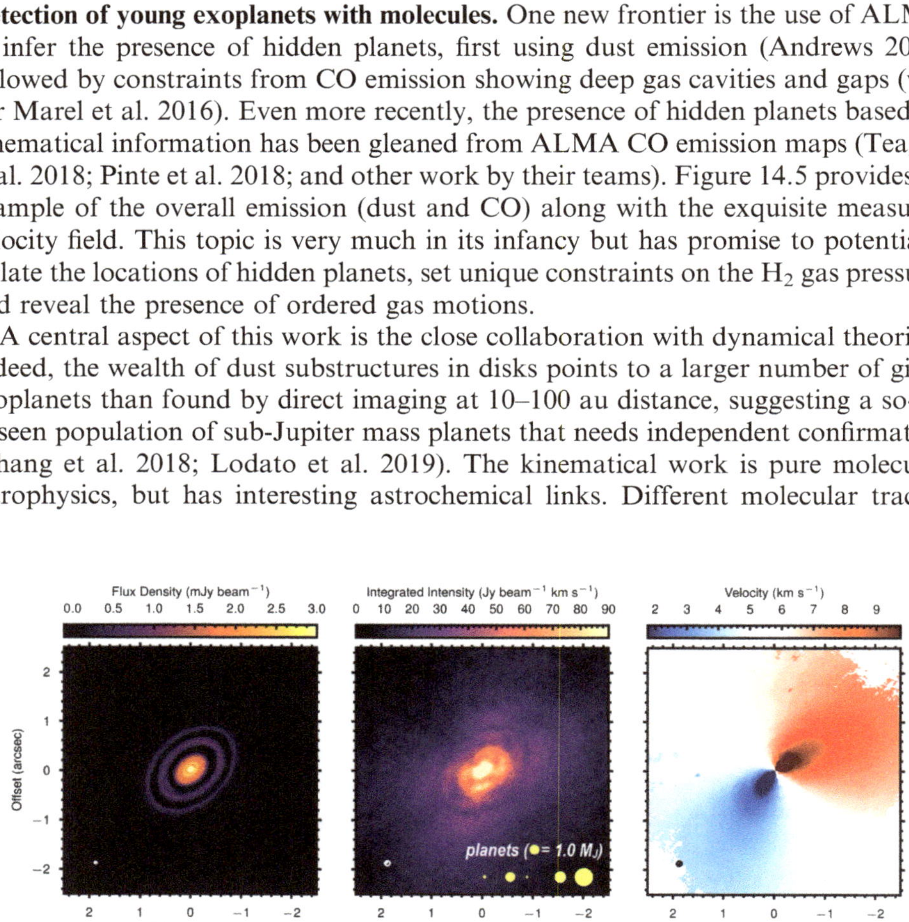


**Figure 14.5.** High spatial resolution ALMA image of the 1.3 mm dust emission from the HD 163296 disk ($d$ = 101 pc) showing several dust rings (left) and $^{12}$CO $J$ = 2–1 emission (middle) published by Isella et al. (2018). The right-hand panel is the $H_2$ gas velocity field in $^{12}$CO 2–1, reproduced from Teague et al. (2019).  The rough (inclination not removed) locations and masses of planets estimated to be present are shown at the bottom of the middle panel as yellow circles. The circles on the left side of the figure (white/black) represent the angular resolution of the data.

probe different layers and one might use this information to study the overall physics and chemistry of material directly related to forming planets.

**Exoplanet atmospheres.** The chemical composition and origin of exoplanet atmospheres, from super-Earths to mini-Neptunes and Jovian planets, is clearly a new frontier for astrochemistry (Madhusudhan 2019) and discussed in other chapters in this book. One of the ultimate goals is to link the planetary atmosphere composition with its formation history in the natal protoplanetary disk. On the one hand, the changing C/O ratio with disk radius could provide such a probe of the formation location (Figure 14.4). However, the route from disk gas+dust to a mature planet is long and involves many steps, each of them with significant uncertainties (e.g., Mordasini et al. 2016; Cridland et al. 2019). For example, are the heavy elements (i.e., other than H and He) in a giant planet atmosphere accreted mostly from the gas or delivered by icy pebbles? How does the migration history affect the outcome? How about dust traps? It is also clear that individual molecule abundances will be fully reset in giant planet atmospheres, preserving only the overall C/O, C/N, O/H etc. abundance ratios. And even those could be affected if part of the atmosphere material is cycled to the planetary core.

**Rocky exoplanets, lessons from Earth.** The atmospheres of rocky terrestrial planets may have an even more complicated history. Well outside the water snowline, the planets are built up largely from planetesimals that are roughly half rock and half ice. When these planets move inward, water becomes liquid, resulting in ocean planets or water worlds. Inside the snowline, the planets are usually thought to be very dry. Computing the atmospheric composition of terrestrial exoplanets is significantly more complex than that of giant exoplanets (Kaltenegger 2017). In this light, a central conundrum lies in the Earth's carbon content, which is orders of magnitude below the carbon content seen in cometary material and the Sun. Overall, this hints at loss of the main carriers of carbon (and also nitrogen) in Earth's planetary building blocks in the inner few au of the presolar nebula, and in disks more generally. It also requires consideration of many additional processes that are operative during each stage of the formative process of terrestrial worlds (e.g., Bergin et al. 2015; Gail & Trieloff 2017), followed by billions of years of evolution on a living active planet (Foley 2019).

A secondary atmosphere can however be formed through impacts of (icy) planetesimals in the late formation stages, including by comets such as 67P/C-G, which can deliver water and organics to the young planet. Given the architecture of the solar system it is not clear if significant mass from comet-forming zones can be brought to Earth (van Dishoeck et al. 2014). However, the giant planets could have migrated and noble gas analyses of 67P are consistent with some cometary supply to the young Earth including most of the organic material, even if not the water (Marty et al. 2017). The survival of these organics depends on planetesimal size and impact speed, and whether the volatile material can perhaps be shielded by a protective layer on the parent body. If they do, there would indeed be a direct link between interstellar molecules and the building blocks for life on new planets.

Key references for this section include Disk Dynamics Collaboration et al. (2020), Mordasini et al. (2016), Madhusudhan (2019), and Bergin et al. (2015).

## 14.4 Opportunities

There are several frontiers in front of us. On short timescales ALMA remains a pioneering instrument and JWST is on the horizon. The combination of these two observatories will be quite powerful as we describe below.

- ALMA will be able to survey hundreds of young disks in the embedded phase for their chemical composition and thus search for similarities and differences in the chemistry prior to planet formation across stellar mass and age range. ALMA will also reveal whether substructures in young disks are common or not, which could point to locations where planets are typically formed.
- ALMA has inferred the presence of volatile depletion in disk systems beyond the CO snowline, and in a few instances interior to the CO snowline. JWST will be able to probe the inner disk composition (within a few au, including O and C budgets) in the upper disk atmosphere as well as that of ices in the intermediate layer for a few near edge-on disks. Will ALMA and JWST tell the same story?
- In a handful of systems the stellar abundances, and that of accreted/ejected material, provide a crucial link to explore the composition of gas in the inner disk that reaches the star. Linking these studies to ALMA/JWST analyses may provide unique information on the composition of hidden material.
- Models of grain growth to pebble size and their radial drift are becoming increasingly sophisticated and are being coupled with chemistry. Since pebble accretion is believed to be a primary mode of planetary growth, these models (combined with observations) are a needed step.
- ALMA has just begun to explore the kinematical landscape of planet formation. A clear question is whether any planets inferred to be present by molecular kinematics, or emission structures, are confirmed by ground-based or space-based direct detection. Within this landscape the link between chemistry and 3D kinematics presents an inviting field to explore the composition of gas that is being supplied from the surface to the planet-forming zone.
- High spectral resolution infrared spectrometers on future extremely large telescopes, such as ELT/METIS, will be able to spatially resolve the chemistry and kinematics of warm gas down to a few au and also image, in tandem with ALMA, gas and dust in circumplanetary disks. VLTI/Gravity (+) provides new opportunities to obtain spatially resolved spectra of young exoplanets and their disks on even smaller scales.

## References

Alexander, C. M. O. 'D., Fogel, M., Yabuta, H., & Cody, G. D. 2007, GeCoA, 71, 4380

Altwegg, K., Balsiger, H., & Fuselier, S. A. 2019, ARA&A, 57, 113

Andrews, S. M. 2020, ARA&A, 58, 483

Balucani, N., Ceccarelli, C., & Taquet, V. 2015, MNRAS, 449, L16

Bergin, E. A., Blake, G. A., Ciesla, F., Hirschmann, M. M., & Li, J. 2015, PNAS, 112, 8965

Bergin, E. A., & Tafalla, M. 2007, ARA&A, 45, 339

Bergin, E. A., & Cleeves, L. I. 2018, in Handbook of Exoplanets, ed. H. J. Deeg, & J. A. Belmonte (Berlin: Springer), 137
Bergin, E. A., & Williams, J. P. 2018, arXiv:1807.09631
Blum, J. 2018, SSRv, 214, 52
Boogert, A. C. A., Gerakines, P. A., & Whittet, D. C. B. 2015, ARA&A, 53, 541
Bosman, A. D, & Banzatti, A. 2019, A&A, 632, L10
Caselli, P., & Ceccarelli, C. 2012, A&ARv, 20, 56
Ceccarelli, C., Caselli, P., Bockelée-Morvan, D., et al. 2014, in Protostars and Planets VI, ed. H. Beuther, R. Klessen, K. Dullemond, & T. Henning (Tucson, AZ: Univ. Arizona Press), 859
Connelly, J. N, Bizzarro, M., Krot, A. N., et al. 2012, Sci, 338, 651
Cridland, A. J., Eistrup, C., & van Dishoeck, E. F. 2019, A&A, 627, A127
Disk Dynamics Collaboration, Armitage, P. J., Bae, J., et al. 2020, arXiv:2009.04345
Drozdovskaya, M. N., van Dishoeck, E. F., Rubin, M., Jørgensen, J. K., & Altwegg, K. 2019, MNRAS, 490, 50
Du, F., Bergin, E. A., Hogerheijde, M., et al. 2017, ApJ, 842, 98
Eistrup, C., Walsh, C., & van Dishoeck, E. F. 2018, A&A, 613, A14
Foley, B. J. 2019, ApJ, 875, 72
Gail, H.-N., & Trieloff, M. 2017, A&A, 606, A16
Henning, T., & Semenov, D. 2013, ChRv, 113, 9016
Herbst, E., & van Dishoeck, E. F. 2009, ARA&A, 47, 427
Hobbs, R., Shorttle, O., Madhusudhan, N., & Rimmer, P. 2019, MNRAS, 487, 2242
Isella, A., Huang, J., Andrews, S. M., et al. 2018, ApJL, 869, L49
Jørgensen, J. K., van der Wiel, M. H. D., Coutens, A., et al. 2016, A&A, 595, A117
Jørgensen, J. K., Belloche, A., & Garrod, R. T. 2020, ARA&A, 58, 727
Kaltenegger, L. 2017, ARA&A, 55, 433
Kama, M., Folsom, C. P., & Pinilla, P. 2015, A&A, 582, L10
Krijt, S. D., Bosman, A. D., & Zhang, K. 2020, ApJ, 899, 134
Kruijer, T. S., Burkhardt, C., Budde, G., & Kleine, T. 2017, PNAS, 114, 6712
Lee, J.-E., Lee, S., Baek, G., et al. 2019, NatAs, 3, 314
Lodato, G., Dipierro, G., Ragusa, E., et al. 2019, MNRAS, 486, 453
Madhusudhan, N. 2019, ARA&A, 57, 617
Marty, B., Altwegg, K., Balsiger, H., et al. 2017, Sci, 356, 1069
McClure, M. K., Dominik, C., & Kama, M. 2020, A&A, 642, L15
McDonough, W. F., & Sun, S. S. 1995, ChGeo, 120, 223
McGuire, B. A. 2018, ApJS, 239, 17
Morbidelli, A., Lambrechts, M., Jacobson, S., & Bitsch, B. 2015, Icar, 258, 418
Mordasini, C., van Boekel, R., Mollière, P., Henning, T., & Benneke, B. 2016, ApJ, 832, 41
Mumma, M. J., & Charnley, S. B. 2011, ARA&A, 49, 471
Öberg, K. I., Boogert, A. C. A., Pontoppidan, K. M., et al. 2011, ApJ, 740, 109
Öberg, K. I., & Bergin, E. A. 2021, PhR, 893, 1
Öberg, K. I., Murray-Clay, R., & Bergin, E. A. 2011, ApJL, 743, L16
Pinilla, P., & Youdin, A. 2017, in Formation, Evolution, and Dynamics of Young Solar Systems, Astrophysics and Space Science Library Vol. 445 (Berlin: Springer), 91
Pinte, C., Price, D. J., & Ménard, F. 2018, ApJL, 860, L13
Pontoppidan, K. M., Salyk, C., Bergin, E. A., et al. 2014, in Protostars & Planets VI, ed. H. Beuther, R. S. Klessen, C. P. Dullemond, & T. Henning (Tucson, AZ: Univ. Arizona Press), 363

Teague, R., Bae, J., & Bergin, E. A. 2019, Natur, 574, 378
Teague, R., Henning, T., Guilloteau, S., et al. 2018, ApJ, 864, 133
Tielens, A. G. G. M. 2013, RvMP, 85, 1021
Tychoniec, Ł., Manara, C. F., & Rosotti, G. P. 2020, A&A, 640, A19
van der Marel, N., van Dishoeck, E. F., Bruderer, S., et al. 2016, A&A, 585, A58
van Dishoeck, E. F., Bergin, E. A., Lis, D. C., & Lunine, J. I. 2014, in Protostars and Planets VI ed. H. Beuther, R. S. Klessen, C. P. Dullemond, & T. Henning (Tucson, AZ: Univ. Arizona Press), 835
van Dishoeck, E. F., Herbst, E., & Neufeld, D. A. 2013, ChRv, 113, 9043
van Dishoeck, E. F. 2018, in IAU Symp. 332, Astrochemistry VII: Through the Cosmos from Galaxies to Planets, ed. M. Cunningham, T. J. Millar, & Y. Aikawa (Cambridge: Cambridge Univ. Press), 3
van't Hoff, M. L. R., van Dishoeck, E. F., Jørgensen, J. K., & Calcutt, H. 2020, A&A, 633, A7
Yamamoto, S. 2017, Introduction to Astrochemistry: Chemical Evolution from Interstellar Clouds to Star and Planet Formation (Berlin: Springer)
Zhang, S., Zhu, Z., Huang, J., et al. 2018, ApJL, 869, L47

AAS | IOP Astronomy

ExoFrontiers
Big questions in exoplanetary science
Nikku Madhusudhan

# Chapter 15

# Debris Disks

**Mark C. Wyatt**

## 15.1 Introduction

*Debris disk* is a catch-all term that can be used to refer to any component of a planetary system that is not an actual planet. In the solar system this refers to the asteroids and comets in the Asteroid and Kuiper Belts as well as the dust and gas derived from them, such as the zodiacal cloud. Studying the structure of extrasolar debris disks provides unique constraints on the underlying planetary system and on the processes of planet formation and protoplanetary disk evolution. Debris disks also have important implications for processes affecting the planets, such as impact events that may strip an atmosphere or deliver volatiles. Their presence or absence also has consequences both positive and negative for the detectability of exoplanets in the system. Here I describe five of the most important ways in which debris disks will contribute to our understanding of extrasolar planetary systems in the coming years.

## 15.2 Placing the Solar System in Context

Debris disks were first discovered over 30 years ago from the far-infrared emission of circumstellar dust heated by the star (Aumann et al. 1984). This continues to be the main method of discovery and several hundred such cold debris disks are known, with temperatures that place the dust at 10s of au from their stars (i.e., at distances comparable with our own Kuiper Belt). Most recently Spitzer and Herschel surveys have quantified the incidence of detectable cold emission as ~20% across the spectral type range A–K (Eiroa et al. 2013; Thureau et al. 2014; Sibthorpe et al. 2018) and has shown that such disks decay in brightness as the stars age (Rieke et al. 2005; Wyatt 2008) in a manner that can be explained by mass loss due to collisional erosion within a population of planetesimal belts (Wyatt et al. 2007; Löhne et al. 2008; Gáspár et al. 2013). However, we are not yet able to detect true analogs to the solar system's debris disk (Booth et al. 2009; Vitense et al. 2012). Thus we cannot answer how typical the solar system is in terms of its Kuiper Belt and zodiacal dust

doi:10.1088/2514-3433/abfa8fch15

levels (i.e., where it sits among the ~80% of stars without detectable debris). This is important to place our best example of a habitable planetary system into context, especially if debris played a role in its habitability and in the evolution of life (e.g., Alvarez et al. 1980; Patel et al. 2015; see also Section 15.4).

While Herschel had the sensitivity to achieve detections at a few times solar system levels (Eiroa et al. 2013), such photometric surveys are fundamentally limited in that a Kuiper Belt analog emits less than 1% of the stellar flux at all wavelengths, whereas neither the stellar flux nor the instrumental calibration are known that accurately (e.g., Liseau et al. 2013). Essentially this has limited our current knowledge to the brightest 20% of extrasolar Kuiper Belts and the brightest ~1% of exozodiacal clouds (i.e., analogs to the cometary dust seen in the inner solar system; Kennedy & Wyatt 2013). To detect true Kuiper Belt analogs, these would have to be resolved from the stellar emission, which requires high resolution imaging capabilities with well characterized instrumental systematics to reach the high contrast of the disk relative to the star, as well as techniques to mitigate background confusion at some wavelengths (e.g., in the far-IR where there is a high density of extragalactic sources; Nguyen et al. 2010).

Recent improvements in nulling mid-IR interferometric techniques have resulted in detections of exozodiacal emission with LBTI at 30 times solar system levels (Ertel et al. 2020). However, this still means that it is only the brightest ~20% of exozodi that have been detected, and inferences about lower dust levels have to be gleaned by extrapolation. Yet, such low levels are important not only for their astrophysical context (i.e., how the solar system's cometary dust levels compare with other stars), but also because this dust is a serious hindrance in our ability to detect emission from exo-Earths and its presence or absence must be factored in to the design of telescopes such as HabEx, LUVOIR and LIFE (Roberge et al. 2012; Quanz et al. 2019). For now such extrapolations suggest that the median exozodi level could be as low as 3 times solar system levels (Ertel et al. 2020), which would be good news for exo-Earth detection. However, this remains to be confirmed with more sensitive observations and we can only confidently say this median is <27 times solar system levels.

Fortunately there is a range of instrumentation coming on-line that will help progress toward the detection of true analogs to the solar system's debris disk, and so our understanding of the solar system's place within the wider population. The LBTI techniques and instrumentation can be improved, and JWST provides high resolution and sensitivity mid-IR imaging that is imminent. Shortly thereafter, even higher resolution capabilities will become available with ELT and coronagraphic imaging with WFIRST. Further in the future far-IR observatories such as Spica and the Origins Space Telescope, if approved, would have the capability of detecting true Kuiper Belt analogs toward the nearest stars (Roelfsema et al. 2018). In the meantime, studies of the brightest 20% of debris disks also provide a perspective on how the solar system may have looked early in its history, before the system-wide instability that depleted the Kuiper Belt (Booth et al. 2009; see also Section 15.3), or shortly after giant impact events like that which formed the Moon (Jackson & Wyatt 2012; see also Section 15.4).

## 15.3 Constraints on Planetary System Architecture and Evolutionary History

The proximity of nearby stars (10s of pc) and the large size of most debris disks (typical radii in the range 30–150 au) means that, for the brightest disks their emission can be spatially resolved providing a map of the dust distribution in the system. Indeed the first debris disk was imaged around the star $\beta$ Pictoris soon after its discovery (Smith & Terrile 1984), though it was not until 14 years later that the next five debris disk images were taken (Holland et al. 1998; Greaves et al. 1998; Jayawardhana et al. 1998). By now, well over 100 debris disks have been spatially resolved, at a range of wavelengths from the optical, at which starlight scattered by the dust is seen, and where the stellar emission has to be suppressed by coronagraphic techniques, to the sub-mm, at which thermal emission from mm- to cm-sized dust is seen that is thought to trace the parent planetesimals (for a review, see Hughes et al. 2018).

In general the structures observed in debris disks are well described as planetesimal belts. However, these *belts* range in width from narrow rings ($\Delta r/r \approx 0.1$) to broad disks ($\Delta r/r > 1$). Some of those which are broad have radial sub-structure such as gaps which may be carved by planets, (e.g., Marino et al. 2018, 2019, 2020), reminiscent of similar structures seen in protoplanetary disks (Andrews et al. 2018), albeit over a narrower range of radii. The inner edges of many disks are found to be sharp, which has been inferred to be evidence for sculpting by a planet at the inner edge (Chiang et al. 2009). However, there are other possible interpretations, such as the radial profile being set by where planetesimals were able to form within protoplanetary disks, such as at snowlines (e.g., Schoonenberg & Ormel 2017). Indeed there is emerging evidence that the radii of debris belts may correlate with snowline locations (Matrà et al. 2018).

The vertical structure of disks that are not face-on can also now be resolved, with scale heights of order a few % being inferred and linked to stirring processes by bodies embedded within or near to the disks (Daley et al. 2019; Matrà et al. 2019). Non-axisymmetric structures in the disks also provide evidence for planet-sized bodies, the most unambiguous of which is the warp in the $\beta$ Pictoris disk, since that may be linked to the known planet in the system (Lagrange et al. 2010; Apai et al. 2015). Other asymmetric structures, such as eccentric disks, clumps and spirals (Wyatt 2005; Kalas et al. 2013; Dent et al. 2014), have also been interpreted as evidence for planets. However, more examples of systems where known planets can be linked to observed disk structures would boost confidence in what can be inferred from disk structures about underlying planetary systems. Nevertheless, we can be confident that even small planets (e.g., as small as Neptune or even Earth) would leave an imprint in the structure of a debris disk (e.g., Wyatt et al. 1999; Pearce & Wyatt 2014; Shannon et al. 2016), moreover one that would be readily detectable with current imaging capabilities. Thus high resolution observations of disk structure have the potential to pinpoint the location of planets that would otherwise be extremely hard to detect using any other techniques (e.g., Neptune analogs), and also can inform about their evolutionary history (e.g., providing evidence of past migration; Wyatt 2003).

We are truly in an era of high resolution disk imaging, with ALMA continuing to image new disks that can be followed up at even higher resolution. JWST looks set to provide further high resolution images both of exo-Kuiper Belts and of any (mid-planetary system) dust structures that may lie inside these. Longer term, coronagraphic imaging with WFIRST and ELT will provide further advances. These disk images provide our best window into outer planetary systems and the mechanisms involved in their formation and evolution. Theoretical modeling techniques will also continue to be developed to find ways to extract more concrete evidence for the presence of planets from the images, e.g., using kinematic structures in the gas (Dent et al. 2014; Teague et al. 2018), and this field will further benefit when direct imaging capabilities allow the detection of planets toward more than a few % of stars.

## 15.4 Planetary Bombardment

As noted in Section 15.2, there remains uncertainty about the level of cometary activity in the inner few au of nearby stars, as traced by their level of exozodiacal dust. However, it is already clear that for 20% of stars, the level of dust in their inner regions is orders of magnitude higher than in the solar system. These exozodi detections correlate with the presence of an outer belt (Mennesson et al. 2014; Ertel et al. 2020), but it is debated whether this dust is dragged inwards from the outer belt (e.g., Rigley & Wyatt 2020) or if the material arrives in the inner regions on larger planetesimals through scattering among a planetary system before being released by sublimation or disintegration (e.g., Di Sisto et al. 2009; Marboeuf et al. 2016). It will soon be possible to test such models by using JWST (and later WFIRST and ELT) to search for dust at intermediate ~10 au separations. However, it is already evident that dust levels in some systems are too high to be explained by drag processes and so planetesimal delivery is required (Defrère et al. 2015; Rigley & Wyatt 2020).

If exozodis are replenished by planetesimals this opens up the possibility that some of these planetesimals may collide with planets in the system, an interaction with implications for conditions on the planets' surfaces. For example, planetesimal bombardment can both deplete a pre-existing atmosphere and deliver complex organic molecules and water, as well as volatiles that can be released to supply a secondary atmosphere (Melosh & Vickery 1989; Chyba 1990). The level of planetesimal bombardment in a system depends on a number of factors, most of which are completely unknown, such as the architecture of the outer planetary system. While this means there is potential for constraining that planetary system from observations of exozodiacal dust and its relation to any outer belt (i.e., providing complementary information to that discussed in Section 15.3; e.g., Faramaz et al. 2017; Wyatt et al. 2017; Marino et al. 2018), there are a number of theoretical hurdles. Notably, there is still not a complete model that links the level of zodiacal dust to its origin in comets (Nesvorný et al. 2010). Clearly there is scope for improved models that include all of the processes at play in debris disks, including dynamical interactions, collisions, radiation forces, gas release, gas drag, etc. (e.g., Kral et al. 2013; Pearce et al. 2020).

While the exozodiacal dust observations show that it is plausible that planetesimal bombardment is ongoing in a significant fraction of systems, it should be acknowledged that bombardment does not need to be ongoing to have (or have had) a significant effect on a planet. High levels of warm dust are more common around stars in the first few 100 Myr of their lives (e.g., Siegler et al. 2007; Kennedy & Wyatt 2013), which could be indicative of an early bombardment phase as planetesimal belts are being cleared. Indeed, the solar system's terrestrial planets are thought to have experienced high bombardment levels early in their history (Morbidelli et al. 2018). In general what matters for a planet is its integrated bombardment, the timing being less critical, and so it will be important to understand what these young bright warm disks tell us about typical bombardment histories.

This evidence for bombardment is opening up new lines of research, since it is now possible to consider for example how this would affect a planet's propensity for habitability. It has been shown that there is a shoreline in planet mass versus semimajor axis space that divides planets that would be expected to grow secondary atmospheres under bombardment, and those that would be expected to have their atmospheres eroded (Wyatt et al. 2020). This process would be competing with the effects of evaporation due to irradiation by the star, but also with secondary outgassing as well as geological processes. Such interactions have been considered for solar system bodies where the bombardment history is better constrained, but it is becoming relevant to consider such processes in an exoplanetary context, given the growing body of constraints on exoplanetary atmospheres and geology.

Whereas systems with bright hot dust were discussed above in the context of external bombardment onto the planets, it has also been interpreted as evidence for ongoing planet formation processes (Wyatt & Jackson 2016). Specifically, this could be the dust created in the final giant impact stage of terrestrial planet growth. There is evidence for this in one system from the silica composition of its dust inferred from the infrared spectrum (Lisse et al. 2009). The rapid variability in some systems (Meng et al. 2014) is also explained in this context (Jackson et al. 2014; Su et al. 2019). If so, observations of these extreme debris disks tell us something about the timing and duration of the giant impact phase, and potentially about the frequency of terrestrial planet formation (e.g., Jackson & Wyatt 2012). For example, for Sun-like stars this dust is seen within a few 0.1 au of the star, which could be because this is where planets tend to form for such stars. This is perhaps unsurprising given the prevalence of close-in super-Earth systems seen in transit surveys, but this could mean that solar system-like configurations are rare. The coming years will see more examples of such extreme disks, as well as studies of their variability. We can also expect advances in the models for their interpretation which need to include a wider range of physical effects (e.g., due to the optical depth of the dust). There are also puzzles that for now defy easy explanation, such as the rapid dispersal of one disk (Melis et al. 2012), and the high fraction of stars with significant quantities of $\sim$1000K dust (e.g., Ertel et al. 2014). Solutions to these puzzles may come in the form of theoretical insights (e.g., Rieke & Gáspár 2016; Pearce et al. 2020) or from further observational characterization of the phenomena.

## 15.5 Planetesimal Composition

A fundamental property of the circumstellar material is its composition. This could provide information on where and when the planetesimals replenishing debris disk dust formed, as well as how those bodies may have been subsequently altered by physical and geological processes.

The spectrum of the dust emission can in some cases be used to infer what it is made of. Spectral features in the mid-IR associated with silicates are often used in this way. However, most debris disks are too cold to emit in the mid-IR. Such spectral features are also sensitive to the size of the dust, being present only if the dust is smaller than the mid-IR wavelengths that characterize the feature. Nevertheless, this has been used to show that the bright warm dust in a few systems is silicate-rich and bears a similarity to dust compositions found in the solar system (Lisse et al. 2012), while in one case the abundance of silica implied formation in a recent giant impact (Lisse et al. 2009). Herschel opened up a new possibility for the colder disks which was to identify forsterite features at 69 $\mu$m, allowing the fraction of Fe and Mg to be determined (de Vries et al. 2012). However, this was only achieved for two disks. Future far-IR observatories like Spica could make such measurements routine. Scattered light observations also have the potential to probe the dust composition, e.g., through its color or phase function (e.g., Debes et al. 2008), although such inferences are again prone to sensitivity to the size of the dust and are limited by how well we can model dust optical properties. Nevertheless, those optical properties can be characterized using techniques like the discrete dipole approximation (e.g., Arnold et al. 2019), and empirical comparisons with solar system objects are also possible (Hedman & Stark 2015). The current set of observations with instruments like GPI and SPHERE have led to a recent increase in the number of scattered light detections (Olofsson et al. 2020; Esposito et al. 2020), and these will gain momentum with JWST.

The volatile content of the planetesimals is harder to probe through spectral features, though there are mid- and far-IR features due to ice that could be detected by JWST and Spica (Kim et al. 2019). However, ALMA is providing a growing body of observations of gas in debris disks that is thought to have a secondary origin, i.e., to have been released from volatile-rich planetesimals, rather than being a remnant of the protoplanetary disk (Dent et al. 2014; Marino et al. 2016, 2020). Most commonly detected is CO, but there were a few detections of atomic carbon and oxygen with Herschel (e.g., Cataldi et al. 2014), and neutral carbon detections with ALMA are becoming more prevalent (Higuchi et al. 2017; Cataldi et al. 2018; Kral et al. 2019; Cataldi et al. 2020). The paradigm within which these observations are interpreted is that of release of volatiles during the collisional cascade process, with those molecules photodissociating relatively quickly to produce an atomic gas disk that spreads viscously (Kral et al. 2016, 2017). The relative abundances of the observed gas and dust constrain their abundances on planetesimals which looks to be similar to solar system comets (Marino et al. 2016; Matrà et al. 2017). This paradigm is not without its challenges, for example there remains debate about whether the gas is produced stochastically rather than in steady state (Cataldi et al. 2018, 2020). We can expect significant observational progress on this in the coming years, first with ALMA and SOFIA, but in the future with Spica. These observational advances will have to go hand-in-hand with improvements in the models to accurately couple the hydrodynamical evolution of the gas with its thermodynamics and chemistry.

Perhaps the most direct measurement of planetesimal composition comes from observations of white dwarf atmospheres (e.g., Farihi 2016). These are frequently seen to be polluted with metals that have short sinking times, and so require constant replenishment that is inferred to be from planetesimals that are tidally disrupted before forming a disk that accretes onto the star (Jura & Young 2014; Hollands et al. 2018). Multiple species have been observed in the atmospheres of many stars allowing the bulk composition of the pollutants to be measured, which can then be interpreted in terms of the temperature at which the planetesimal formed and the differentiation processes that it underwent since formation (Bonsor et al. 2020). While such inferences are possible, there are still many unknowns; e.g., there is no direct evidence for the location of the parent planetesimal belt (Farihi et al. 2014), the mechanism which caused the planetesimal to arrive at the star is not known, as are the details of the tidal disruption and disk accretion process (Veras 2016). With large numbers of white dwarfs being discovered in the era of large all-sky surveys like Gaia, SDSS, ZTF and LSST (Fusillo et al. 2019), and follow-up spectroscopy being performed to characterize any pollution, we can anticipate new observational results in the coming years, as well as significant theoretical progress on the various steps needed for the interpretation.

## 15.6 Protoplanetary Disk Dispersal

The recent far-IR surveys of Spitzer and Herschel have reinforced the paradigm in which stars are born with a planetesimal belt that subsequently erodes through collisional processing that passes mass from large planetesimals down to dust that is eventually expelled by radiation forces (Wyatt et al. 2007; Marino et al. 2020). This model fits the statistics for the fraction of stars seen to have detectable emission as a function of age in different wavebands. However, as already noted this necessarily cannot explain the 80% of stars without detectable disks. Further analysis of the current detections is also leading to new insights, such as a correlation of planetesimal belt radius with stellar mass (Matrà et al. 2018). The model also avoids the question of how and when the planetesimal belts formed, rather presupposing that these are present, with a collisional cascade already set up, as soon as the protoplanetary disk disperses. However, with the discovery of gas in debris disks, and the knowledge that dust in protoplanetary disks is not primordial but subject to a continual cycle of growth and destruction, it is becoming clear that the distinction between the two types of disk is not well defined (Wyatt et al. 2015). This is particularly noticeable when considering the youngest systems, <20 Myr or so, which means that observations of such young stars can inform on the processes of protoplanetary disk dispersal and of the birth of a debris disk.

Studies of young stars generally involve first identifying associations of stars that appear co-eval and with similar distances and space motions. However, the relevant kinematic information has until now not been available for the nearest star-forming regions that lie at ~150 pc. While it has nevertheless been relatively unambiguous to identify protoplanetary disks in this way (i.e., the class II stars), this is harder for the young stars without such an obviously bright disk (i.e., the class III stars), with background stars often mistakenly falling in this category (Manara et al. 2018). Gaia is changing that, and the coming years will see improved membership of the different nearby associations from which to study their disk populations (Cánovas et al. 2019; Luhman 2020). At the same time, ALMA observations of such regions have until now been relatively shallow, i.e., only sensitive to protoplanetary disks (e.g., Ansdell et al. 2016; Williams et al. 2019). However, deeper ALMA observations can probe for dust emission at levels of the brightest debris disks in the nearest star-forming regions, and can also search for remnant primordial gas that may have yet to disperse after the dust has been depleted (e.g., Owen & Kollmeier 2019). Thus it is timely now to perform a census of the youngest debris disks as well as to search for protoplanetary disks in the final stages of dissipation (e.g., Lovell et al. 2021). Far-IR surveys are still more sensitive than those in the sub-mm, and for those we will have to wait for Spica, which would be able to set limits on the presence of dust that are comparable with those for nearby stars (except where imaging has been able to probe fainter dust, see Section 15.2).

Significant advances have also been made in our understanding of protoplanetary disks, both in their observational characterization and in theoretical advances in modeling the processes within them (for a recent review, see Andrews 2020). These will continue along with the advances in debris disks described in this chapter, and together it should be possible to infer the sequence of events that lead to the depletion of a protoplanetary disk. Observations of the gas will be particularly telling, since this component dominates the mass of a protoplanetary disk, and its dynamical influence is responsible for much of the dust morphology in such disks. Observations of the dispersal of that component in a disk wind would constrain that process (Haworth & Owen 2020), as well as its impact on dust and planetesimals in the disk (Carrera et al. 2017; Owen & Kollmeier 2019), and planet formation more generally.

## 15.7 Opportunities and Challenges

Debris disks represent an opportunity, by providing unique information about exoplanetary systems. Observationally one challenge is to detect debris disks as faint as those in the solar system, particularly given how relatively bright the host star is. Theoretically the challenge is to extract reliable information in the face of so many unknowns about the architecture, history and detailed physics of the underlying planetary system.

## References

Alvarez, L. W., Alvarez, W., Asaro, F., & Michel, H. V. 1980, Sci, 208, 1095
Andrews, S. M. 2020, ARA&A, 58, 483
Andrews, S. M., Huang, J., & Pérez, L. M. 2018, ApJL, 869, L41

Ansdell, M., Williams, J. P., van der Marel, N., et al. 2016, ApJ, 828, 46
Apai, D., Schneider, G., Grady, C. A., et al. 2015, ApJ, 800, 136
Arnold, J. A., Weinberger, A. J., Videen, G., & Zubko, E. S. 2019, AJ, 157, 157
Aumann, H. H., Gillett, F. C., Beichman, C. A., et al. 1984, ApJL, 278, L23
Bonsor, A., Carter, P. J., Hollands, M., et al. 2020, MNRAS, 492, 2683
Booth, M., Wyatt, M. C., Morbidelli, A., Moro-Martín, A., & Levison, H. F. 2009, MNRAS, 399, 385
Cánovas, H., Cantero, C., Cieza, L., et al. 2019, A&A, 626, A80
Carrera, D., Gorti, U., Johansen, A., & Davies, M. B. 2017, ApJ, 839, 16
Cataldi, G., Brandeker, A., Olofsson, G., et al. 2014, A&A, 563, A66
Cataldi, G., Brandeker, A., Wu, Y., et al. 2018, ApJ, 861, 72
Cataldi, G., Wu, Y., Brandeker, A., et al. 2020, ApJ, 892, 99
Chiang, E., Kite, E., Kalas, P., Graham, J. R., & Clampin, M. 2009, ApJ, 693, 734
Chyba, C. F. 1990, Natur, 343, 129
Daley, C., Hughes, A. M., Carter, E. S., et al. 2019, ApJ, 875, 87
de Vries, B. L., Acke, B., Blommaert, J A D L, et al. 2012, Natur, 490, 74
Debes, J. H., Weinberger, A. J., & Schneider, G. 2008, ApJL, 673, L191
Defrère, D., Hinz, P. M., Skemer, A. J., et al. 2015, ApJ, 799, 42
Dent, W R F, Wyatt, M. C., Roberge, A., et al. 2014, Sci, 343, 1490
Di Sisto, R. P., Fernández, J. A., & Brunini, A. 2009, Icar, 203, 140
Eiroa, C., Marshall, J. P., Mora, A., et al. 2013, A&A, 555, A11
Ertel, S., Absil, O., Defrère, D., et al. 2014, A&A, 570, A128
Ertel, S., Defrère, D., Hinz, P., et al. 2020, AJ, 159, 177
Esposito, T. M., Kalas, P., Fitzgerald, M. P., et al. 2020, AJ, 160, 24
Faramaz, V., Ertel, S., Booth, M., Cuadra, J., & Simmonds, C. 2017, MNRAS, 465, 2352
Farihi, J. 2016, NewAR, 71, 9
Farihi, J., Wyatt, M. C., Greaves, J. S., Bonsor, A., Sibthorpe, B., & Panić, O. 2014, MNRAS, 444, 1821
Gáspár, A., Rieke, G. H., & Balog, Z. 2013, ApJ, 768, 25
Fusillo, N. P. G., Tremblay, P.-E., Gänsicke, B. T., et al. 2019, MNRAS, 482, 4570
Greaves, J. S., Holland, W. S., Moriarty-Schieven, G., et al. 1998, ApJL, 506, L133
Haworth, T. J., & Owen, J. E. 2020, MNRAS, 492, 5030
Hedman, M. M., & Stark, C. C. 2015, ApJ, 811, 67
Higuchi, A. E., Sato, A., Tsukagoshi, T., et al. 2017, ApJL, 839, L14
Holland, W. S., Greaves, J. S., Zuckerman, B., et al. 1998, Natur, 392, 788
Hollands, M. A., Gänsicke, B. T., & Koester, D. 2018, MNRAS, 477, 93
Hughes, A. M., Duchêne, G., & Matthews, B. C. 2018, ARA&A, 56, 541
Jackson, A. P., & Wyatt, M. C. 2012, MNRAS, 425, 657
Jackson, A. P., Wyatt, M. C., Bonsor, A., & Veras, D. 2014, MNRAS, 440, 3757
Jayawardhana, R., Fisher, S., Hartmann, L., et al. 1998, ApJL, 503, L79
Jura, M., & Young, E. D. 2014, AREPS, 42, 45
Kalas, P., Graham, J. R., Fitzgerald, M. P., & Clampin, M. 2013, ApJ, 775, 56
Kennedy, G. M., & Wyatt, M. C. 2013, MNRAS, 433, 2334
Kim, M., Wolf, S., Potapov, A., Mutschke, H., & Jäger, C. 2019, A&A, 629, A141
Kral, Q., Thébault, P., & Charnoz, S. 2013, A&A, 558, A121

Kral, Q., Wyatt, M., Carswell, R. F., Pringle, J. E., Matrà, L., & Juhász, A. 2016, MNRAS, 461, 845
Kral, Q., Marino, S., Wyatt, M. C., Kama, M., & Matrà, L. 2019, MNRAS, 489, 3670
Kral, Q., Matrà, L., Wyatt, M. C., & Kennedy, G. M. 2017, MNRAS, 469, 521
Lagrange, A. M., et al. 2010, Sci, 329, 57
Liseau, R., Montesinos, B., Olofsson, G., et al. 2013, A&A, 549, L7
Lisse, C. M., Chen, C. H., Wyatt, M. C., et al. 2009, ApJ, 701, 2019
Lisse, C. M., Wyatt, M. C., Chen, C. H., et al. 2012, ApJ, 747, 93
Löhne, T., Krivov, A. V., & Rodmann, J. 2008, ApJ, 673, 1123
Lovell, J. B., Wyatt, M. C., Ansdell, M., et al. 2021, MNRAS, 500, 4878
Luhman, K. L. 2020, AJ, 160, 186
Manara, C. F., Prusti, T., Comeron, F., et al. 2018, A&A, 615, L1
Marboeuf, U., Bonsor, A., & Augereau, J. C. 2016, P&SS, 133, 47
Marino, S., Carpenter, J., Wyatt, M. C., et al. 2018, MNRAS, 479, 5423
Marino, S., Flock, M., Henning, Th, et al. 2020, MNRAS, 492, 4409
Marino, S., Matrà, L., Stark, C., et al. 2016, MNRAS, 460, 2933
Marino, S., Yelverton, B., Booth, M., et al. 2019, MNRAS, 484, 1257
Marino, S., Zurlo, A., Faramaz, V., et al. 2020, MNRAS, 498, 1319
Marino, S., Bonsor, A., Wyatt, M. C., & Kral, Q. 2018, MNRAS, 479, 1651
Matrà, L., Dent, W R F, Wyatt, M. C., et al. 2017, MNRAS, 464, 1415
Matrà, L., Marino, S., Kennedy, G. M., et al. 2018, ApJ, 859, 72
Matrà, L., Wyatt, M. C., Wilner, D. J., et al. 2019, AJ, 157, 135
Melis, C., Zuckerman, B., Rhee, J. H., et al. 2012, Natur, 487, 74
Melosh, H. J., & Vickery, A. M. 1989, Natur, 338, 487
Meng, H. Y. A., Su, K. Y. L., Rieke, G. H., et al. 2014, Sci, 345, 1032
Mennesson, B., Millan-Gabet, R., Serabyn, E., et al. 2014, ApJ, 797, 119
Morbidelli, A., Nesvorny, D., Laurenz, V., et al. 2018, Icar, 305, 262
Nesvorný, D., Jenniskens, P., Levison, H. F., et al. 2010, ApJ, 713, 816
Nguyen, H. T., Schulz, B., Levenson, L., et al. 2010, A&A, 518, L5
Olofsson, J., Milli, J., Bayo, A., Henning, T., & Engler, N. 2020, A&A, 640, A12
Owen, J. E., & Kollmeier, J. A. 2019, MNRAS, 487, 3702
Patel, B. H., Percivalle, C., Ritson, D. J., Duffy, C. D., & Sutherland, J. D. 2015, NatCh, 7, 301
Pearce, T. D., Krivov, A. V., & Booth, M. 2020, MNRAS, 498, 2798
Pearce, T. D., & Wyatt, M. C. 2014, MNRAS, 443, 2541
Quanz, S. P., Absil, O., Angerhausen, D., et al. 2019, arXiv:1908.01316
Rieke, G. H., Gáspár, A., & Ballering, N. P. 2016, ApJ, 816, 50
Rieke, G. H., Su, K Y L, Stansberry, J. A., et al. 2005, ApJ, 620, 1010
Rigley, J. K., & Wyatt, M. C. 2020, MNRAS, 497, 1143
Roberge, A., Chen, C. H., Millan-Gabet, R., et al. 2012, PASP, 124, 799
Roelfsema, P. R., Shibai, H., Armus, L., et al. 2018, PASA, 35, e030
Schoonenberg, D., & Ormel, C. W. 2017, A&A, 602, A21
Shannon, A., Bonsor, A., Kral, Q., & Matthews, E. 2016, MNRAS, 462, L116
Sibthorpe, B., Kennedy, G. M., Wyatt, M. C., et al. 2018, MNRAS, 475, 3046
Siegler, N., Muzerolle, J., Young, E. T., et al. 2007, ApJ, 654, 580
Smith, B. A., & Terrile, R. J. 1984, Sci, 226, 1421
Su, K. Y. L., Jackson, A. P., & Gáspár, A. 2019, AJ, 157, 202

Teague, R., Bae, J., Bergin, E. A., Birnstiel, T., & Foreman-Mackey, D. 2018, ApJL, 860, L12
Thureau, N. D., Greaves, J. S., Matthews, B. C., et al. 2014, MNRAS, 445, 2558
Veras, D. 2016, RSOS, 3, 150571
Vitense, C., Krivov, A. V., Kobayashi, H., & Löhne, T. 2012, A&A, 540, A30
Williams, J. P., Cieza, L., Hales, A., et al. 2019, ApJL, 875, L9
Wyatt, M. C. 2003, ApJ, 598, 1321
Wyatt, M. C. 2005, A&A, 440, 937
Wyatt, M. C. 2008, ARA&A, 46, 339
Wyatt, M. C., Bonsor, A., Jackson, A. P., Marino, S., & Shannon, A. 2017, MNRAS, 464, 3385
Wyatt, M. C., Dermott, S. F., Telesco, C. M., et al. 1999, ApJ, 527, 918
Wyatt, M. C., Kral, Q., & Sinclair, C. A. 2020, MNRAS, 491, 782
Wyatt, M. C., Panić, O., Kennedy, G. M., & Matrà, L. 2015, Ap&SS, 357, 103
Wyatt, M. C., Smith, R., Su, K Y L, Rieke, G. H., Greaves, J. S., Beichman, C. A., & Bryden, G. 2007, ApJ, 663, 365
Wyatt, M. C., & Jackson, A. P. 2016, SSRv, 205, 231

AAS | IOP Astronomy

ExoFrontiers
Big questions in exoplanetary science
**Nikku Madhusudhan**

# Chapter 16

## Exoplanetary Interiors

**Nadine Nettelmann and Diana Valencia**

### 16.1 Introduction

With the turn of the century, two discoveries opened new pathways in planetary science. Those were the first mass-estimate of an exoplanet around a Sun-like star, 51 Peg b (Mayor & Queloz 1995) through the radial velocity (RV) method and the first radius measurement of an exoplanet, HD209458b (Charbonneau et al. 2000) through transit light curve (TLC) analysis. These initial discoveries pointed to the challenges of understanding the atmosphere, interior, and evolution of exoplanets including the possibility of mass loss of planets on close orbits that are exposed to strong irradiation (Guillot et al. 1996). They raised the question of heating and inflation mechanisms (Guillot 2002), and finally, of the nature of these objects in terms of composition compared to the known planets in the solar system. The field of exoplanet interior modeling was born.

However the interior of a planet is not directly accessible to observations. The most important observational parameters to infer internal structure refer to their orbits (semimajor axis a, eccentricity $e$), the star (effective temperature $T_\star$ and radius $R_\star$ governing the equilibrium temperature, age $\tau_\star$), gravity (planet mass $M_p$, radius $R_p$, tidal response Love number $k_2$), shape (tidal response $h_2$), as well as atmospheric spectra and phase curves. Models must also rely on assumptions. Common assumptions include hydrostatic equilibrium and, for the case for close-in planets, a 1:1 spin–orbit rotational resonance. Planet formation models can provide additional constraints.

The first discovered and characterized exoplanets were gaseous planets owing to observational biases, but technological advancements have made discovering smaller planets a routine endeavor, albeit still challenging. In fact, we now know that super-Earths are the most common class of planets on orbital distances out to 500 days (Hsu et al. 2019), and while the questions regarding interiors are similar to that of gaseous planets, there are also marked differences. For example, the temperature structure has a negligible effect on the radius of a rocky planet compared to one that has an envelope. We outline big science questions pertaining to each class of planets:

- **What is the amount of heavy elements in a planet and do all planets possess an iron-rock core?** Low-mass exoplanets are plagued with compositional degeneracies arising from trade-offs between the different building blocks: iron cores, silicate mantles, water content, and hydrogen+helium gas layers. This means that interior models constrained by $M_p$ and $R_p$ allow for large spread of possible compositions. Famous examples are the warm Neptune GJ436b and the sub-Neptune GJ1214b. Their mass and radius could be explained by a light (few % $M_p$), dry H/He atmosphere atop a massive rock-iron core but as well by a 90% (GJ436b) to 100% (GJ1214b) water interior (Nettelmann et al. 2010, 2011; Valencia et al. 2013; Morley et al. 2017). This motivates to search for additional constraints. For GJ436b, formation models were invoked to reduce the degeneracy (Figueira et al. 2009), atmospheric spectra suggest a high-$Z$ atmosphere (Morley et al. 2017) and thus water-rich interior with just a small core. GJ1214b's transmission spectrum is flat due to clouds/hazes (Kreidberg et al. 2014), its internal structure remains hidden. An alternative, promising, and novel approach for exoplanets can be measuring their tidal response in form of the Love numbers $h_2$ and $k_2$ (Kramm et al. 2011; Padovan et al. 2018).
- **How much and through what mechanisms are the interiors of planets heated or delayed from cooling?** Many strongly irradiated gaseous planets require an additional heat source to explain their large radii. Recently, statistical analysis of a large sample of inflated planets in the mass range 0.5–10 $M_J$ has allowed to empirically quantify the amount of extra heating (Thorngren & Fortney 2018). But some individual mysteries exist. An example is the hot Saturn WASP-39b, where even interior models that incorporate heating fall short of explaining the high atmospheric metallicity inferred from transmission spectroscopy (Wakeford et al. 2018; Poser et al. 2019). In that regard, exoplanetary interiors are reaching the level of challenges we encounter in the solar system: state-of-art albeit simple Jupiter models that employ most advanced H/He-equations of state fail to explain observed super-solar atmospheric heavy element abundances (Stevenson 2020).
- **What is the origin of the observed populations in the radius–period diagram?** In the solar system, the outer planets are large, gaseous, and cold while the inner ones are small, rocky, and warm. Planets closer in, at shorter orbital distances, receive stronger irradiation. There is observational indication that these classes, gas-rich versus gas-poor, are separated by a gap in radius–irradiation space also in extrasolar systems, albeit for different reasons. Occurrence rates of Kepler planets around FGK stars revealed a peak at compact size ($<1.5\ R_E$) and very high irradiation ($>100\ S_E$—irradiation received by Earth), and a second peak at large radius ($>2\ R_E$) and low irradiation values ($<60\ S_E$; Fulton et al. 2017). The negative slope of the radius–irradiation valley is consistent with compact transiting planets being the result of atmospheric evaporation of gas-born planets (Owen & Wu 2013;

Lopez & Fortney 2013). However, by extending the analysis to include M dwarf stars, the slope of the valley was found (Cloutier & Menou 2020) to reverse sign making it inconsistent with the mass loss scenario and more compatible with a gas-poor formation environment (Lopez & Rice 2018). Objects in and along the radius valley are excellent targets to study planetary formation and evaporation. For a detailed discussion of the radius–period diagram including the radius-gap, the radius desert at very short periods, and the cosmic shoreline separating bodies with and without atmospheres, see Zhang (2020).

- **What does the composition of rocky planets tell us about their formation?** The refractory component of low-mass planets should mirror that of the host star (Dorn et al. 2017; Hinkel & Unterborn 2018), thus setting the rock to iron core ratio to that of the star and reducing the degeneracy by one degree of freedom. However, the data on rocky planets do not necessarily support this idea (Plotnykov & Valencia 2020). In addition, planets more iron-rich than Mercury (e.g., super-Mercuries) are found (Plotnykov & Valencia 2020), as well as planets that if rocky, are depleted in iron with respect to Earth (e.g., super-Moons). We do not have a reliable formation theory that explains their existence (Scora et al. 2020). Figure 16.1 shows the current known exoplanets with measured mass and radius with an error below 25%. The rocky threshold line shows the maximum size a rocky planet has including all the relevant Mg–Si–O mineralogy. The curve for pure $Fe_{0.95}$-$Ni_{0.05}$ shows an extreme composition below which planets are expected to not exist. Any rocky planet is expected to lie between these lines. To compare to the stars' refractory composition, Plotnykov & Valencia (2020) translated the stars refractory ratios into $M-R$ relationships (red). Apparently, the detected planets span a wider distribution that needs to be explained by any successful formation theory.

## 16.2 Important Questions and Goals

**What are the structure parameters we are interested in?** For gas giants, the information we aim to obtain is foremost the bulk mass fraction of heavy elements $Z_p$ (elements that are heavier than hydrogen and helium), and the corresponding gas mass fraction (assuming H and He in roughly protosolar mass ratio of $M_{He}/(M_{He} + M_H) = 0.270$). A planet's $Z_p$ value yields a first classification as a terrestrial or super-Earth ($Z = 1$), sub-Neptune (high $Z_p$, but also with gaseous H/He and/or water), ice giant (high $Z_{H2O} > 0.5$ is a solution), or gas giant planet ($Z_p < 0.5$). In practice, theoretical mass–radius relations for different assumed compositions serve to invert an observed $M_p$–$R_p$ pair for possible composition. For an extensive set of $M_p$–$R_p$ relations for gaseous planets ranging from $M_p = 10\ M_E$ to $10\ M_J$ see Fortney et al. (2007), for rocky and water-rich planets in the 0.4–20 $M_E$ mass range and including thermal effects in the water EoS see Valencia et al. (2013), Thomas & Madhusudhan (2016), and with $S_E$ as a third parameter see Zeng et al. (2019).

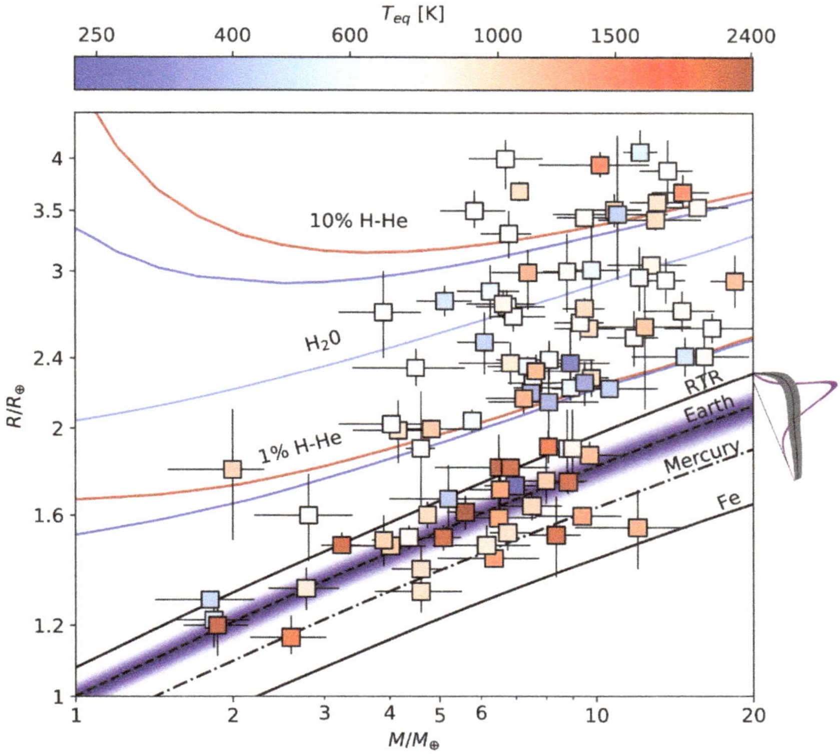


**Figure 16.1.** Masses and radii of supposedly rocky exoplanets with measurement uncertainties below 25%, color coded according to $T_{\rm eq}$ (in log-scale). $M-R$ relationships are shown for reference: pure Fe–Ni (core mass fraction cmf=1), Mercury-like (cmf=0.63), Earth (0.325), and the rocky threshold radius (RTR, cmf=0) above which planets have volatiles, earth-like rock with 1% and 10% by mass H–He with equilibrium temperatures of 300 K (blue) and 1000 K (red), a pure $H_2O$-planet at 1000 K equilibrium temperature. Shaded purple region show the $M-R$ of rocky planets with the same refractory ratio as stars (e.g., Fe/Mg and Fe/Si). Outlet figure: Comparison between probability density of stars (purple) and rocky exoplanets (planets within RTR and Fe, gray) in cmf space. Adapted from Plotnykov and Valencia (2020).

Second, we are interested in the distribution of heavy elements in the interior, such as the presence of a core. In the case of gas giants we usually refer to the core as the refractory material that can either be at the center, or diluted, meaning it is mixed into the envelope up to some radius but with high-Z at the center. In the case of terrestrial bodies, the core refers to the iron-rich central mass agglomeration below a silicate mantle. For gaseous planets, simple two-parameter models would quantify the distribution as $Z_{\rm env}$ and $M_{\rm core}$, with $Z_p = Z_{\rm env}(M_p - M_{\rm core}) + M_{\rm core}$. Mass and radius alone are not enough to provide $M_{\rm core}$ even if they would be exactly known, because $R_p(M_p)$ is only one constraining parameter. Uncertainties in the temperature profile, which underneath the radiative atmosphere is assumed to be adiabatic but the transition point is model-dependent (Thorngren et al. 2019), adds to the uncertainty in $Z_p$. The high observed heat flow out of Jupiter indeed suggests a convective interior with adiabatically stratified temperature profile, although Jupiter and Saturn may have super-adiabatic layers. The assumption of an adiabatic interior without extra heating (beyond insolation) allows us to estimate the minimum $Z_p$ value of a giant planet.

**What is the composition of the envelope of sub-Neptunes?** Sub-Neptunes are small planets that have a gaseous envelope. It may be light (few % $M_p$), but large enough to affect the planetary radius. This is in contrast to rocky planets, like Earth and Venus, that have an atmosphere that does not affect the radius in a considerable way. The most important science goal for sub-Neptunes is to determine what the composition of the envelope is, as it holds key information to their formation. Conventional formation theories predict that planets that form outside the snowline will have large water contents. These may or may not migrate to the close orbits where we find many sub-Neptunes. Instead, if formed *in situ*, close-in sub-Neptune planets should have negligible water. The problem is that from $M-R$ pair data it is not possible to discern between either scenario (water-rich versus water-poor). And even with good atmospheric characterization, we would need to connect the composition of the upper millibars of atmosphere to that of the bulk envelope.

**What is the origin of iron-poor or iron-rich rocky planet?** Planets that are sufficiently heated and compact enough to be well within the rocky region, should have no water and no H–He. This is because a steam atmosphere or any H–He enlarges the planet's radius considerably. Thus, compact and highly irradiated planets are presumed to be completely rocky, composed of silicate mantles and iron cores. Plotnykov & Valencia (2020) constrained the composition of $\sim$30 rocky planets and obtained the refractory ratios Fe/Mg and Fe/Si. Compared to planet-hosting stars, rocky planets span a wider range in composition, see Figure 16.1. They find a prevalence of super-Mercuries in the exoplanet data, as well as planets that are depleted in iron (super-Moons) should they truly be rocky. At the moment we lack a reliable theory that explains both iron enriched and iron-depleted massive rocky planets (Scora et al. 2020). Alternatively, the planets classified as iron-depleted instead could have very small amounts of volatiles. However, some of them are also highly irradiated ($\gtrsim$300 $S_E$) around FGK stars, and thus, expected to have lost their atmospheres. Either way, these planets are a mystery. Perhaps they have water-rich atmospheres that have undergone runaway greenhouse states (Turbet et al. 2019), which is the predicted fate of Earth in about 600 Myr due to the evolution of the Sun.

**We expect inferred metallicities from structure models to agree with those from planet formation models.** An assumption for transiting gaseous exoplanets is that they formed somewhere in the disk, migrated inward, and finalized mass accretion a few Myr after the young star reached the main sequence. Their inferred composition today should then be in agreement with predictions from planet formation models.

Core accretion formation (CA) assumes the formation of initial rocky core seeds from dust in the protostellar disk. The core then grows with some gas accretion. Captured planetesimals may dissolve in the gaseous envelope or sink to the core. Once the envelope mass equals about the core mass, which can occur for $M_{\rm core}$ = 1 to several 10 $M_E$, runaway gas accretion sets in. Mass of the core, bulk $Z_p$, and final mass are thereby dependent also on gas opacity and solids surface density. In any event, CA formation predicts an inhomogeneous structure with a slightly

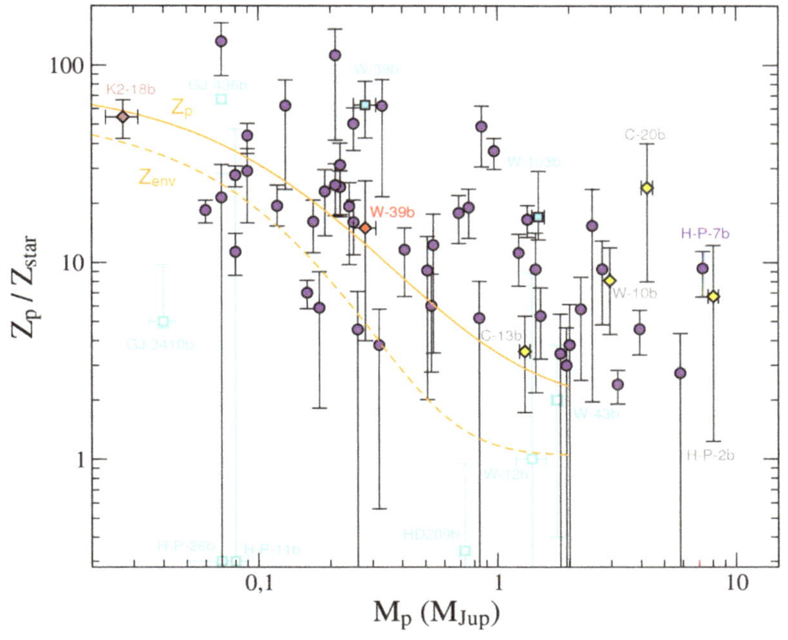


**Figure 16.2.** Compilation of heavy element enrichment by mass over planet mass; *violet, circles*: bulk $Z_p$ for the sample of 47 non-inflated planets in Thorngren et al. (2016); *yellow diamonds*: selection of inflated planets which are predicted to carry a large amount of heavy elements even though extra heating is not included in the interior model calculation; *red diamond*: WASP-39b based on models that account for extra heating and delay of cooling by clouds (Poser et al. 2019); *brown diamond*: super-Earth K2-18b (Madhusudhan et al. 2020); *orange curves* show fits to the prediction for $Z_{env}$ and $Z_p$ from CA formation (Venturini et al. 2016). *Cyan squares* are atmospheric metallicities, *open*: from Chachan et al. (2019) with observed values for HAT-P-11b and HAT-26b lower than shown here (placed at 0.2) and likely influenced by clouds; *filled*: WASP-39b (Wakeford et al. 2018) and WASP-103b (Kreidberg et al. 2018).

super-stellar heavy element abundance in the gaseous envelope. Jupiter is an example (Vazan et al. 2018). In the CA model, and also in the gravitational disk instability model (GI), planets the mass of Jupiter and higher are massive because they accreted so much gas from the disk. $Z_p$ should decrease the more massive the gaseous envelope is (Venturini et al. 2016).

The $Z_p$ value inferred from planetary interior modeling depends on internal temperatures. Suppose two planets of same mass and age, but one inflated and the other one not. The inflated one will be larger and warmer since an inflated planet experiences mechanisms that heat the interior (tidal, Ohmic) or delay its cooling (high opacity in the atmosphere). As a result, including extra heating/delay of cooling in the modeling and evolution runs leads to higher inferred $Z_p$ values.

In Figure 16.2 we plot the inferred heavy element enrichment of the sample of 47 non-inflated planets ($T_{eq}$ < 1000 K) with measured mass and radius known by 2016 according to the models in Thorngren et al. (2016). The predicted trend from CA formation of $Z_p$ decreasing with $M_p$ can clearly be seen. However, $Z_p(M_p)$ of planets more massive than Jupiter does not approach the stellar values $Z_{star}$ toward the brown dwarf regime ($M_{BD} = 13–80M_J$) as predicted by both CA and GI formation: the heavy element enrichment of massive planets tends to be higher. For the $\sim 7M_J$ planet HAT-P-20b, $Z_p \sim 9 \times Z_{star}$ implies the enormous amount of heavy elements of $550–750M_E$ (Thorngren et al. 2016). This is 7–9 times the mass of heavy elements confined to the planets of the solar system. There are also a couple of inflated and massive hot Jupiters for which models predict high enrichment even though extra heating has not even been taken into account (yellow diamonds). Huge amounts of hundreds of $M_E$ of heavy elements in a single planet are not predicted by any formation theory.

We also plot in Figure 16.2 atmospheric metallicities ($Z_{atm}$) obtained from transmission and emission spectra (Wakeford et al. 2018; Kreidberg et al. 2019; Chachan et al. 2019). Generally one expects $Z_{atm} \leqslant Z_p$ since $Z_{atm} > Z_p$ would trigger an instability that erases the inversion. The finding $Z_{atm} > Z_p$ for WASP-39b is a mystery.

**Tidal response can be used to infer internal density distribution.** In order to constrain both $M_{core}$ and $Z_{env}$, or for terrestrial planets the iron-core mass fraction, a further gravitational parameter is required. This may be the tidal Love number $k_2$ or the shape parameter $h_2$. In linear approximation, $h_2 = 1 + k_2$ (Love 1911).

A close-in planet suffers tidal deformation. In 1:1 spin–orbit resonance and neglecting librations around the equilibrium position, the tidal bulge of the planet is permanently elongated toward the star; the response is static. One can define a linear response coefficient for the tidal shape deformation with respect to the unperturbed shape, $h_2$, and an equivalent coefficient for the gravity field perturbation, $k_2 := -V_{22}/W_{22}$, where $V_{22}$ and $W_{22}$ are based on (tesseral) harmonics of the planetary and stellar gravity field evaluated at the planet's surface, respectively (Nettelmann 2019).

For fluid planets and relaxed solid planets, $k_2$ is dictated by the internal density distribution. By measuring $k_2$ (or $h_2$) an upper limit can be placed on the mass of the central core. For instance, if the sub-Neptune GJ1214b were a water world it would have $k_2 \sim 0.9$, whereas $k_2 < 0.1$ if it were composed of a rocky core with H/He atmosphere (Nettelmann et al. 2011). For predictions of $k_2$ values in the fluid regime of super-Earths see Kellermann et al. (2018), and for the warm Neptune 436b see Kramm et al. (2011) and Padovan et al. (2018).

Figure 16.3 shows $k_2$ values for planets and satellites. Among terrestrial bodies, the spread in $k_2$ is large as a result of different internal density distributions, rheologies (Moon versus Mercury), and possible internal oceans (Titan). For fluid exoplanets we highlight the potential of $k_2$ to reveal internal density distribution with the help of the two black curves. Both curves are for planets of same mass, same orbital distance of 0.1 AU around the Sun, same age of about 4–6 Gyr, and notably, of same composition $Z_p(M_p)$ following roughly the exoplanet distribution in Figure 16.2. The upper curve is for homogeneous irradiated planets with $M_{core} = 1$ and $Z_{env} = 0.78$ at $M_p = 20$ (Neptune-like), $Z_{env} = 0.3$ at $M_p = 100\ M_E$ (Saturn-like), $Z_{env} = 0.1$ at $M_p = 1–10\ M_J$ (gas giants). Models along the lower curve have that same amount of heavy elements but confined to the core ($Z_{env} = 0.01$),

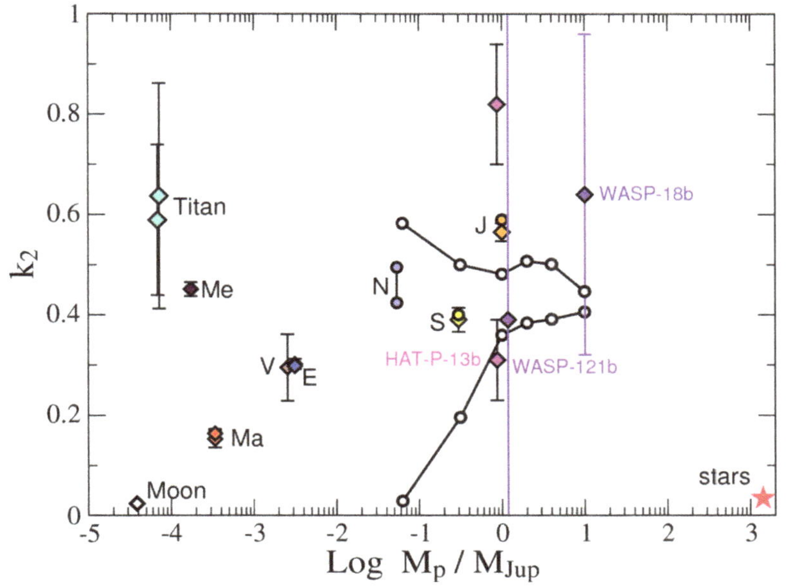


**Figure 16.3.** Observed Love number $k_2$ values *(diamonds)* in the solar system and of three exoplanets as labeled. The value for WASP-121b was derived from its $h_2$ estimate Hellard et al. 2020; for WASP-18b from RV variations (Csizmadia et al. 2019); for HAT-13b assuming locked apsidal motion between planets b and c (Buhler et al. 2016; Hardy et al. 2017), for Saturn from Cassini and ground-based long-term astrometric data (Lainey et al. 2017), and Jupiter's value is the current Juno observation (Durante et al. 2020). *Circles* show calculated values from interior models assuming static tidal response; the value for Saturn is for a rotation rate of 10 hr 32 minutes, for Neptune for respectively $16^h06^m40^s$ and $17^h27^m29^s$ (Helled et al. 2010); the two curves with *gray circles* are for irradiated exoplanets with $Z_p(M_p)$ relation after Thorngren et al. (2016); where the upper curve assumes homogeneous distribution (small $M_{core} = 1\ M_E$, high $Z_{env}$) and the lower curve heavy elements to be confined to the core (low $Z_{env} = 1x$ solar, high $M_{core}$). Since thermal structure matters, these exoplanet models have been evolved to 4–6 Gyr.

leading to a strong reduction in $k_2$ especially for the particularly degenerate class of (sub-)Neptune sized planets. Even the curve for a homogeneous planet underestimates the value of Jupiter, which has some central mass condensation: this is because Jupiter's atmosphere is cold and dense while that of an irradiated hot Jupiter is warm and more extended. Both bracketing curves fail to explain the high $k_2$ value of ~0.60 of WASP-18b (Csizmadia et al. 2019): this is a puzzle.

## 16.3 Challenges

**Formation and evolution of super-Mercuries and super-Moons**: A large compositional diversity in presumed rocky planets has been found (Plotnykov & Valencia 2020), larger than the Fe/Si of the planet-hosting stars. This diversity includes planets more iron-rich than Mercury (e.g., Super-Mercuries) that are difficult to form even by invoking giant planet collisions of differentiated massive embryos (Scora et al. 2020). Also, if truly rocky, there is subsample of planets 2-fold depleted in Fe/Mg with respect to stars. It is essential that we determine if these are truly bare rocks because at the moment we have no theory that predicts preferential loss of iron compared to Mg+Si rocks (Scora et al. 2020). One avenue is to look at the heat redistribution of their phase curves to rule out the presence of atmospheres (Kreidberg et al. 2019).

**One-to-one stellar-planet compositional comparison:** There is tremendous potential to learn about planet formation in systems where both the planets refractory composition and the host stars is known. At the moment, there is only a handful of planets for which this is the case (Plotnykov & Valencia 2020). Unfortunately, the error bars in planetary radius, but especially in planetary mass, as well as in stellar refractory ratios are too large to yield definite conclusions. It will be even more powerful to compare multiplanet systems, given the difference in planetary irradiation.

**Love number computation and measurements**: In the fluid regime, the Love numbers $k_2$ and $h_2$ can straightforwardly be computed from the internal density distribution. However, the relaxation time of solid planets or solid rocky cores into the fluid regime depends on the rheology (viscosity, rigidity) of the material and the frequency of tidal forcing. Mars and Earths are not in the fluid regime. Viscosity of solid materials and the unknown rotation rate are a major source of uncertainty for $k_2$ of solid planets. For Jupiter, Juno spacecraft measurement just revealed a deviation from the theoretical static response value of 0.5897 (Nettelmann 2019; Wahl et al. 2020) by −2% (Durante et al. 2020), teaching us that dynamic effects play a role, which are much more challenging to quantify.

Current measurement uncertainties for exoplanets are large and at present preclude a robust estimate of core mass upper limit. For TLC method, uncertainties in $h_2$ are predicted to be at best ± 20% with upcoming instrumentation like PLATO and JWST (Hellard et al. 2020). Nevertheless, long-term observations in this new field will help to reduce uncertainties.

**Internal temperature profile:** In 99% of Jupiter's mass, hydrogen is in a metallic, quantum-mechanically degenerate state. The opacity is high and thermal conductivity hampered by moderate temperatures (<20,000 K) and high densities (~1 g/cc). Similar conditions are predicted to occur in more massive planets and brown dwarfs (Becker et al. 2018), suggesting a largely adiabatic interior. However, observed oscillations in the C-ring of Saturn's indicate a large stably stratified core (Fuller 2014). If super-adiabatic, inferred metallicities could be significantly higher, and even explain some inflated planets (Chabrier & Baraffe 2007). Uncertainties in thermal structure and thus inferred metallicity could be large for sub-Neptune to Saturn-sized exoplanets, even in the absence of extra heating, rendering inferring their "nature" challenging.

Furthermore, the temperature structure of a planet has a profound effect on the state of the matter and related properties. The presence of liquid iron cores that may produce magnetic fields, surface magma oceans that can create specific emission signatures, or partially molten mantles that can affect the outgassing, and tidal dissipation of the planet, all depend on the mode of heat transport and temperature structure.

**Equations of State (EoS) and phase diagrams:** Interior structure models rely on adequate equations of state for the end-members (H–He, $H_2O$/ices, rocks, metallic iron) and their mixtures and phase diagrams. Uncertainties in melting lines, for instance, add to the uncertainty of whether super-Earths possess a magnetic field from a dynamo that operates in a partially molten silicate layer or in a liquid iron core. While the EoS and phase diagram of pure water under planetary interior conditions is well understood (uncertainties still persist regarding lattice structures in super-ionic and ice phases at high pressures outside the realm of planetary interiors), solubility of water with methane, ammonia, and refractory materials is not well known. Ab initio calculations (Wahl et al. 2013) predict solubility of water, iron, and also of Si and Mg at least at low concentrations of 1:256 in metallic hydrogen, which occurs at about $P > 0.5$ Mbar in H-rich planets. This suggests that planets more massive than Neptune, where pressures of 0.5 Mbar are reached at $\sim 0.7 R_{\mathrm{Nep}}$, have partially mixed interiors while low-mass planets may harbor segregated rock-iron cores. Of importance to habitability is the melting lines of rock material, (as it connects to outgassing ad tidal dissipation), and iron alloys (as it relates to magnetic field generation). More experimental and theoretical work is needed to understand the mixing behavior of multi-component planetary materials.

**Inflation mechanisms:** Mechanisms that transfer energy from the star to the planet are not well understood, in particular how much and where in the planet the heat is deposited as a function of $M_p$, and $T_{\mathrm{eq}}$. Tidal heating (from eccentricity or obliquity) scales with $1/Q_p$, offering the opportunity to estimate a planet's tidal quality factor $Q_p$. For gas giants, several studies come to the conclusion that the efficiency of Ohmic heating peaks at $T_{\mathrm{eq}} \sim 1500$–1600 K (Thorngren & Fortney 2018); Ohmic heating may also be relevant for sub-Neptunes (Pu & Valencia 2017) but has not systematically been studied for Neptune-to-Saturn mass planets.

## 16.4 Opportunities

**New observational facilities:** At the moment the TESS space telescope is surveying the closest M Dwarfs and measuring the radii of super-Earths and large exoplanets. The next leap into observation capabilities will come with the JWST scheduled to launch late 2021. This telescope is expected to revolutionize our understanding of the atmospheres of exoplanets by routinely obtaining transit spectroscopy for the exoJupiters, and sub-Neptunes. PLATO (launch date 2026) will survey more than 1 million stars and its focus is to find rocky planets in the habitable zone of Sun-like stars. Following, Ariel will launch in mid-2028 and will also obtain the composition and thermal structure of atmospheres, but contrary to JWST, it can dedicate more time for each planet observed. All these space missions are being supported by existing and in-construction ground facilities that measure planetary masses. With all these resources, the future of exoplanets research is bright.

**Exoplanets with liquid/icy water:** Due to observational biases, exoplanets with liquid or icy water on the surface have not been identified yet. However, soon we expect detections of compact long-period planets around M Dwarfs, some of which could harbor oceans. The high $k_2$ value of Saturn's moon Titan (Figure 16.3) is indicative of a deep water ocean, and ESA's JUICE mission (launch 2022) and NASA's Europa Clipper (launch 2024) will set out to measure the depths of sub-surface oceans on Ganymede and Europa by means of tidal distortion. Measuring the tidal response of distant low-mass planets is beyond the scope of near-future instrumentation and thus offers opportunities for future generations of exoplanetary scientists.

**Love number measurements:** Within just a decade, four methods (TLC, RVV, TTV, orbital fixed points) have been elaborated to observationally infer $k_2$ or $h_2$. While uncertainties are still large for each single method application of these independent methods to the same planet could greatly reduce the uncertainty, like the observable part of the universe is now known to be globally flat ($\Omega = 1$) due to independent observations of Type 1 supernovas, of the cosmic microwave background, and of baryon oscillations (Frieman et al. 2008). The RVV method most benefits from longer baseline observations (decades instead of years), and the TLC method from long exposure times for single planets (Hellard et al. 2020).

Since assumed static tidal response of a fluid body places a strict upper limit on the $k_2$ value, higher observed values offer the unique opportunity to study dynamic effects. Future extended RV observations of the ultra-hot Jupiter WASP-18b (Csizmadia et al. 2019), which has maximum static $k_2$ of only 0.43 may reveal if that era has already begun.

**Statistical analysis of increasingly large samples:** As the sample size of exoplanets has grown, so has the ability to extract information from the planets as a population. Statistical studies have aimed at uncovering any trends in the data: planetary mass and star metallicity (Miller & Fortney 2011; Mulders 2018; Neves et al. 2012), radius and irradiation (Thorngren & Fortney 2018), composition and irradiation (Plotnykov & Valencia 2020), etc. This trend will continue to grow, especially for planets around M dwarfs and evolved stars, and small planets in general, as the observational challenges are being overcome. There is power in numbers. Even if not as precise, sophisticated statistical analysis can yield quantitative constraints on planet formation and evolution if the sample size is large and informative enough. In parallel there is a shortcut to planetary understanding by focusing on the outliers that challenge conventional wisdom.

## References

Becker, A., Bethkenhagen, M., Kellermann, C., & Redmer, R. 2018, AJ, 156, 4
Buhler, P. B., Knutson, H. A., Batygin, K., et al. 2016, ApJ, 821, 26
Chabrier, G., & Baraffe, I. 2007, ApJ, 661, L81
Chachan, Y., Knutson, H. A., Gao, P., et al. 2019, AJ, 158, 244
Charbonneau, D., Brown, T. M., Latham, D. W., & Mayor, M. 2000, ApJ, 529, L45
Cloutier, R., & Menou, K. 2020, AJ, 159, 211
Csizmadia, S., Hellard, H., & Smith, A. M. S. 2019, A&A, 623, A45
Dorn, C., Hinkel, N. R., & Venturini, J. 2017, A&A, 597, A38
Durante, D., Parisi, M., Serra, D., et al. 2020, GeoRL, 47, e86572
Figueira, P., Pont, F., Mordasini, C., et al. 2009, A&A, 493, 671
Fortney, J. J., Marley, M. S., & Barnes, J. W. 2007, ApJ, 659, 1661
Frieman, J. A., Turner, M. S., & Huterer, D. 2008, ARA&A, 46, 385
Fuller, J. 2014, Icar, 242, 283
Fulton, B. J., Petigura, E. A., Howard, A. W., et al. 2017, AJ, 154, 109
Guillot, T., Burrows, A., Hubbard, W. B., Lunine, J. I., & Saumon, D. 1996, ApJ, 459, L35

Guillot, T., & Showman, A. P. 2002, A&A, 385, 156
Hardy, R. A., Harrington, J., Hardin, M., et al. 2017, ApJ, 836, 143
Hellard, H., Csizmadia, S, Padovan, S., Sohl, F., & Rauer, H. 2020, ApJ, 889, 66
Helled, R., Anderson, J. D., & Schubert, G. 2010, Icar, 210, 446
Hinkel, N. R., & Unterborn, C. T. 2018, ApJ, 853, 83
Hsu, D. C., Ford, E. B., Ragozzine, D., & Ashby, K. 2019, AJ, 158, 109
Kellermann, C., Becker, A., & Redmer, R. 2018, A&A, 615, A39
Kramm, U., Nettelmann, N., Redmer, R., & Stevenson, D. S. 2011, A&A, 528, A18
Kreidberg, L., Bean, J. L., & Désert, J.-M. 2014, ApJL, 793, L27
Kreidberg, L., Koll, D. D. B., Morley, C., et al. 2019, Natur, 573, 87
Kreidberg, L., Line, M. R., Parmentier, V., et al. 2018, AJ, 156, 17
Lainey, V., Jacobson, R. A., Tajeddine, R., et al. 2017, Icar, 281, 286
Lopez, E. D., & Fortney, J. J. 2013, ApJ, 776, 2
Lopez, E. D., & Rice, K. 2018, MNRAS, 479, 5303
Love, A. E. H. 1911, Some Problems of Geodynamics (Cambridge: Cambridge University Press), Chapter IV
Madhusudhan, N., Nixon, M. C., Welbanks, L., Piette, A. A. A., & Booth, R. A. 2020, ApJL, 891, L7
Mayor, M., & Queloz, D. 1995, Natur, 378, 355
Miller, N., & Fortney, J. J. 2011, ApJ, 736, L29
Morley, C. V., Knutson, H., Line, M., et al. 2017, AJ, 153, 86
Mulders, G. D. 2018, in Handbook of Exoplanets, ed. H. J. Deeg, & J. A. Belmonte (Cham: Springer)
Nettelmann, N. 2019, ApJ, 874, 156
Nettelmann, N., Fortney, J. J., Kramm, U., & Redmer, R. 2011, ApJ, 733, 2
Nettelmann, N., Kramm, U., Redmer, R., & Neuhäuser, R. 2010, A&A, 523, A26
Neves, V., Bonfils, X., Santos, N. C., et al. 2012, A&A, 538, A25
Owen, J. E., & Wu, Y. 2013, ApJ, 775, 105
Padovan, S., Spohn, T., Baumeister, P., et al. 2018, A&A, 620, A178
Plotnykov, A., & Valencia, D. 2020, MNRAS, 499, 932
Poser, A. J., Nettelmann, N., & Redmer, R. 2019, Atmos, 10, 664
Pu, B., & Valencia, D. 2017, ApJ, 846, 47
Scora, J., Valencia, D., Morbidelli, A., & Jacobson, S. 2020, MNRAS, 493, 4910
Stevenson, D. J. 2020, AREPS, 48, 465
Thomas, S. W., & Madhusudhan, N. 2016, MNRAS, 458, 1330
Thorngren, D., Gao, P., & Fortney, J. J. 2019, ApJL, 884, L6
Thorngren, D. P., & Fortney, J. J. 2018, AJ, 155, 214
Thorngren, D. P., Fortney, J. J., Murray-Clay, R. A., & Lopez, E. D. 2016, ApJ, 831, 64
Turbet, M., Ehrenreich, D., Lovis, C., Bolmont, E., & Fauchez, T. 2019, A&A, 628, A12
Valencia, D., Guillot, T., Parmentier, V., & Freedman, R. S. 2013, ApJ, 775, 10
Vazan, A., Helled, R., & Guillot, T. 2018, A&A, 610, L14
Venturini, J., Alibert, Y., & Benz, W. 2016, A&A, 596, 90
Wahl, S. M., Parisi, M., Folkner, W. M., Hubbard, W. B., & Militzer, B. 2020, ApJ, 891, 42
Wahl, S. M., Wilson, H. F., & Militzer, B. 2013, ApJ, 773, 95
Wakeford, H. R., Sing, D. K., Deming, D., et al. 2018, AJ, 155, 29
Zeng, L., Jacobsen, S. B., Sasselov, D. D., et al. 2019, PNAS, 116, 9723
Zhang, X. 2020, RAA, 20, 99

# Chapter 17

## Atmospheric Modeling and Retrieval

**Jonathan J. Fortney, Joanna K. Barstow and Nikku Madhusudhan**

### 17.1 Introduction

We seek to learn about the atmospheres of exoplanets both to understand their physical and chemical processes at work, as well as to potentially tie atmospheric abundances to theories of planet formation and evolution. Such work relies on spectroscopic investigation. The interpretation of spectra of exoplanetary atmospheres is heavily dependent on atmospheric models. The accuracy and reliability of these models is especially critical in the absence of *in situ* measurements that in the solar system provide us with the "ground truth" to anchor remote sensing observations.

Atmospheric models for exoplanets vary in type and complexity (Figure 17.1). Models attempt to understand the state of an atmosphere, including the temperature structure, atmospheric abundances, and cloud composition and distribution. "Self-consistent" models, where a wide range of atmospheric physics and chemistry are specified, are typically either 1D and iterate to radiative-convective equilibrium, (e.g., Fortney et al. 2008), or 3D, general circulation models (GCMs), which use the equations of fluid dynamics (or some simplifications thereof) to describe the atmospheric dynamics, (e.g., Showman et al. 2009; Heng & Showman 2015). Either in 1D or 3D, such models need to include sub-models that treat chemical abundances as a function of pressure and temperature (Lodders & Fegley 2002), the chemical effects of incoming stellar UV flux and vertical mixing (Moses et al. 2011), as well as cloud microphysics and opacities (Ackerman & Marley 2001; Helling & Casewell 2014). With all physical effects either explicitly treated within some framework (or explicitly ignored) one calculates a solution for a given atmosphere's state, and may then generate an emission, reflection, or transmission spectrum, which is then compared to observations.

By contrast, parametric models are normally incorporated into a data-driven "retrieval" approach (Madhusudhan 2018). A variety of atmospheric parameters that control, for instance, the temperature structure, atmospheric abundances, or a small number of cloud parameters, can be tuned in order to provide the best match

doi:10.1088/2514-3433/abfa8fch17  

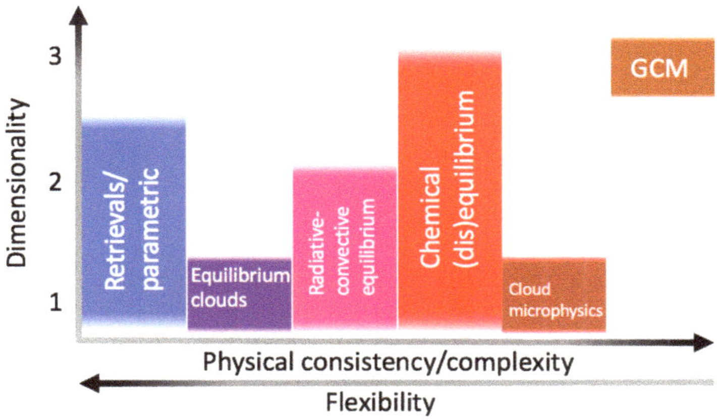


**Figure 17.1.** This schematic illustrates the typical complexity/flexibility and dimensionality of various types of exoplanet atmospheric modeling tool. Retrieval models can contain varying degrees of physical constraint. Equilibrium clouds includes models such as Ackerman & Marley (2001), microphysical clouds refers to models used by, e.g., Gao et al. (2020).

to a spectrum, typically with minimal physical assumptions. Within this framework one performs a wide exploration of a range of possible atmosphere models that can yield a best-fit to an observed spectrum. Such models can find solutions outside the confines of self-consistent models. They are most important in providing a sound statistical assessment of retrieved parameters, including mixing ratios of atmospheric gases and the shape of the atmospheric temperature profile. These models have mostly been developed in the 1D context, but recent work has shown the importance of 2D and 3D frameworks for inhomogeneous atmospheres (Caldas et al. 2019; MacDonald et al. 2020; Taylor et al. 2020).

These families of different model types represent a compromise between computation speed/flexibility and physical accuracy; as such they have different uses. For example, highly physics-based GCMs are typically used to simulate detailed conditions on a specific planet, with a pre-determined set of inputs. 1D self-consistent models can be calculated across large "grids" of parameter space (as is often done in classic stellar atmospheres modeling) to investigate the role of surface gravity, metallicity, and intrinsic fluxes (Burrows et al. 1997; Fortney et al. 2008; Mollière et al. 2015; Gandhi & Madhusudhan 2017). Retrievals can be run on individual planets, or larger samples of spectra for many planets, to search for trends (Barstow et al. 2017; Line et al. 2017; Welbanks et al. 2019), which can be compared to expectations from self-consistent models.

Atmospheric modeling and retrieval methods ultimately seek to answer central questions regarding the diversity of physical and chemical processes in exoplanetary atmospheres and their possible connections to planetary formation conditions. Some of the major questions include:

- How do atmospheric properties and processes in exoplanets depend on the macroscopic planetary conditions, such as the irradiation, elemental composition, bulk properties (masses and radii), and stellar environment?

- How do different atmospheric processes (e.g., chemistry, energy transport, dynamics and escape) interact with each other across the diversity of exoplanets?
- How representative are atmospheric properties of exoplanets of their interior conditions? For giant planets, are chemical abundances in the observable atmosphere representative of the entire H/He envelope? For low-mass planets what can observable compositions tell us about atmosphere-surface interactions?
- How well can we extract information about 3D atmospheric properties of exoplanets from their 1D disc- and terminator averaged observations?
- What are the limits on accuracy and precision for exoplanet spectroscopy, in emission, transmission, and reflection?

## 17.2 State of the Art

Exoplanet atmosphere modeling emerged as an offshoot both of solar system planetary atmosphere modeling (Burrows et al. 1997; Goukenleuque et al. 2000) as well as stellar atmosphere modeling (Seager & Sasselov 1998; Barman et al. 2001). As such, a very wide variety of methods, with heritages, assumptions, strengths, and pitfalls, from a diverse array of modeling frameworks have long come into play, and continue to do so. This kind of merging of cultures and disciplines can be seen in many aspects of modeling. An example is the technique, "correlated-k," of tabulating strongly non-gray molecular opacity with hundreds of millions of individual lines (Goody et al. 1989), commonplace for decades in the solar system, which has gradually been adopted by many exoplanet atmosphere codes (Amundsen et al. 2017; but not without healthy pushback and discussion).

Pioneering methods of retrieval, to quantify the quality of model fits and robustly assess uncertainties on atmospheric abundances and temperature structures, were independently developed in an exoplanet astrophysics context (Madhusudhan & Seager 2009; Madhusudhan et al. 2011) alongside modification and updating of methodologies created in Earth and solar system remote sensing (Rodgers 2000; Irwin et al. 2008). The inherent similarities in these methods was investigated (Line et al. 2013), and there is now considerable work to understand and quantify the role of different modeling choices in affecting accuracy and precision on constraints (Barstow et al. 2020).

While retrieval codes were originally created as 1D "average" model atmospheres, to compare to both transmission and emission spectra, there has been significant thought in the past five years about limitations of this 1D framework, in particular for hot Jupiters that have strongly inhomogeneous atmospheres from day to night. Complications could arise from partly cloudy atmospheres (Line & Parmentier 2016), temperature inhomogeneities (Feng et al. 2016; Blecic et al. 2017; MacDonald et al. 2020; Taylor et al. 2020), day-to-night temperature differences (Caldas et al. 2019; Lacy & Burrows 2020). 2D and 2D+ retrieval scenarios are now being investigated for interpreting thermal emission phase curves (Irwin et al. 2020; Feng et al. 2020).

Three-dimensional GCM simulations of exoplanet atmospheres mostly have a heritage in Earth and solar system atmosphere modeling, both for giant planets (Showman & Guillot 2002; Showman et al. 2009; Rauscher & Menou 2010; Mayne et al. 2014) and for terrestrial worlds (Joshi et al. 1997; Leconte et al. 2013; Carone et al. 2014). However, there is also work underway to create new general GCM tools specifically for exoplanets (Deitrick et al. 2020). Bringing realistic radiative transfer to these simulations is mostly now a solved problem (Showman et al. 2009; Amundsen et al. 2014). However, significant work is currently ongoing on treatments of cloud particle formation and transport via a wide variety of schemes (Lee et al. 2016; Lines et al. 2018, 2019; Roman & Rauscher 2019), as well as explorations of simplified chemical networks that can be used within GCMs (Drummond et al. 2018; Zhang & Showman 2018; Venot et al. 2020).

## 17.3 Important Questions/Goals

- **What is the impact of photochemistry on exoplanet temperature structures and spectra?** The solar system's giant planets are incredibly cold, leading to a dearth of gases that can readily be photolyzed to drive further chemistry (e.g., $CH_4$). For warmer planets, a much richer variety of gases will be found, which may drive extremely complex pathways of photochemical reactions, and create novel pathways for haze formation, that are not seen in the solar system (Zahnle et al. 2009; Hörst et al. 2018; He et al. 2020). For terrestrial planets, which unlike giant planets, do not have a vast reservoir of a deep atmosphere in thermochemical equilibrium, photochemistry takes on an ever greater importance, likely with a very wide phase space of possible atmospheric compositions. It is possible, perhaps even likely, that a wide variety of necessary rate constants and dissociation cross-sections have not yet been determined in the laboratory (Fortney et al. 2019).
- **What will we be able to learn about secondary atmospheres from JWST?** Secondary atmospheres—generated by outgassing from a rocky planet—are more challenging to observe in transmission than primary, $H_2$–He dominated atmospheres due to their higher mean molecular weight resulting in a reduced scale height. Molecular absorption signatures in these atmospheres will be relatively low amplitude. However, with sufficient coadded observations it has been demonstrated (Barstow & Irwin 2016; Morley et al. 2017) that JWST would be capable of detecting various molecules in the atmospheres of favorable targets, such as the TRAPPIST-1 planets. Due to the small signals, likely challenges will be distinguishing atmospheric signatures from those of stellar activity (Rackham et al. 2018), or identifying muted features in the presence of clouds.
- **How are atmospheric abundances impacted by surface/atmosphere interactions for thin primary atmospheres and for secondary atmospheres?** The role of surface/atmosphere exchange is very uncertain for the atmospheres that will be characterized in the coming decade. The high temperature and surface pressures are mostly outside the realm of previous work in the solar system.

For instance, the role of a long-lived molten surface altering the abundances in thin H/He dominated atmospheres is just now beginning to be investigated (Kite et al. 2020).

- **What is the appropriate treatment for cloud and haze in retrieval models, and what can we learn from microphysics simulations?** Clouds and hazes are complex atmospheric phenomena that are challenging to represent, especially in parametric retrieval models that aim to be as computationally efficient as possible. Recent work (Mai & Line 2019; Barstow 2020; Min et al. 2020) has been done to explore the effects of different parameterizations on retrievals of both simulated and real spectra. As yet, only very basic properties of clouds can be directly inferred from observations, and as a consequence assumptions about characteristics such as composition can only come from microphysical simulations (e.g., Helling et al. 2016, 2017; Gao et al. 2020). Symbiosis between such simulations and retrievals will be necessary as we enter the JWST era, such that detailed models are used to benchmark parameterizations, and information gained from retrievals can provide observational constraints that allow microphysical models to be refined.
- **For what data sets is the use of a 1D retrieval framework sufficient for determining atmospheric properties?** Recent work has demonstrated that using 1D retrievals, with the inherent assumption of atmospheric homogeneity, may introduce bias when they are applied to planets with significant variations in temperature and cloud coverage around or across the terminator (Line & Parmentier 2016; MacDonald et al. 2020; Lacy & Burrows 2020) or the disk-averaged emitting hemisphere (Feng et al. 2016). The sensitivity to inhomogeneity and degree of bias, and therefore the necessity of including more complex parameterizations, is dictated both by the wavelength coverage and signal to noise of the observation. Tests on simulated observations with different signal to noise properties, such as those conducted by Lacy & Burrows (2020), will be necessary to determine this.
- **How do we build 2D/3D frameworks capable of extracting information from transit/disc averaged spectra?** Knowledge gained from phase curve observations of hot Jupiters (e.g., Stevenson et al. 2014; Wong et al. 2020, Kreidberg et al. 2018) tells us how we might expect conditions to vary across the dayside and terminator regions of tidally locked, highly irradiated exoplanets. While spectroscopic observations are available for only a handful of these objects, 3D GCMs provide an alternative source of information about expected atmospheric inhomogeneity. Using existing, detailed observations and simulated outputs from GCMs will allow us to benchmark ways of parameterizing variations in temperature, cloud coverage and chemistry.
- **What are *unknown* unknowns, beyond known unknowns?** So far, we know so little about exoplanetary atmospheres that it is natural to wonder, up until now, if we are even asking the right questions. It is quite plausible that the first high-precision broad-wavelength-coverage spectra obtained with JWST may highlight important issues of atmospheric physics and chemistry that have not yet been highlighted in previous work.

## 17.4 Opportunities

- Scheduled for launch in late 2021, JWST will enable breakthrough transiting planet spectroscopy from 0.6 to 5 $\mu$m via several instruments, and longward of 5 $\mu$m with the MIRI instrument. The high signal-to-noise observations across a very broad-wavelength range will be transformational for the field, across the range of planets from gas giants orbiting Sun-like stars (Beichman et al. 2014; Greene et al. 2016; Lacy & Burrows 2020), to Earth-size planets in orbit around M dwarfs (Barstow & Irwin 2016; Morley et al. 2017; Koll et al. 2019).
- Scheduled for launch in 2029, Atmospheric Remote-sensing Infrared Exoplanet Large-survey (Ariel), is the fourth medium-class mission of the European Space Agency. Using photometry from 0.4 to 2 $\mu$m and spectroscopy from 2 to 8 $\mu$m, Ariel will focus on transmission spectra, emission spectra, and phase curves of $\sim$1000 transiting planets across a wide range of planet masses (Tinetti et al. 2018). Ariel will provide a statistical assessment of exoplanet atmospheres, to complement the detailed view of a smaller number of targets investigated by JWST (Zellem et al. 2019; Changeat & Al-Refaie 2020).
- As multiple-node machines become increasingly cost effective, retrievals in 2D or even 3D will become more computationally viable. Optimizing codes to run on GPU nodes can also improve efficiency for retrievals (e.g., Kitzmann et al. 2020.)
- A decade after the first detection of an exoplanet atmosphere via high spectral resolution spectroscopy (Snellen et al. 2010), the field is now coming into its own (Birkby 2018). A bounty of high spectral resolution ($R > 25,000$ platforms are just now coming online) that will enable a broad array of atmospheric characterization in the optical and near-infrared. In addition to detecting atoms and molecules, wind speeds (Snellen et al. 2010; Louden & Wheatley 2015) and planetary rotation (Brogi et al. 2016) can be characterized. The development of spectrographs with a broad simultaneous wavelength coverage will be especially powerful.
- Comparative exoplanetary spectroscopy between transiting planets and directly-imaged planets. While the number of directly-imaged planets with measured emission spectra is relatively low, the quality of these observations has been dramatically increasing (e.g., Gravity Collaboration et al. 2020). In the Extremely Large Telescope (ELT) era, spectra of $\sim$100 directly-imaged gas giants, all with known dynamical masses, will be possible (Brandt et al. 2019). Contrasting the atmospheric abundances for these planets on multi-AU orbits, compared to close-in transiting planets, may prove critical in unraveling how atmospheric abundances are tied to planetary formation location (Öberg et al. 2011; Madhusudhan et al. 2014; Mordasini et al. 2016).
- As exoplanet observations reveal more detail about their atmospheric structure, techniques and findings from solar system research will become increasingly applicable. Over the last several decades, observations of solar system planets have improved and expanded from observations taken from the ground to detailed imagery from orbiting satellites, and even in some

cases *in situ* measurements using descent probes. As observational constraints have improved, solar system planet models have been forced to develop in complexity and detail. With the exoplanet community facing a similar dramatic change in observational data, we can benefit from the experience of solar system researchers during this process.

## 17.5 Challenges

- Inference from exoplanet observations is hampered by a lack of ground truth. In the solar system, interpretation of remote sensing observations is often anchored by a direct, *in situ* measurement of conditions at some time or place on the planet (for example, directly measured temperature and cloud profiles were obtained by the Pioneer Venus mission, which has allowed informative priors to be applied for subsequent analyses of the Venusian atmosphere). Similar constraints could be obtained from GCMs or chemical models for exoplanets, but this is potentially risky as our understanding of atmospheric processes under extreme conditions may well be incomplete.
- A large sample size of atmospheres may be needed to disentangle the role of a variety of complex physical and chemical effects. The phase space of exoplanet atmospheres is large. While stellar atmospheres can largely be understood in terms of the three factors of $T_{\rm eff}$, surface gravity, and metallicity, a much larger number of factors influences planetary atmospheres (Fortney et al. 2020). While dramatic progress will certainly be made in the 2020s, it will likely open up many new questions that can only be solved from spectra of numerous, possibly hundreds, of exoplanets.
- It is becoming increasingly apparent that heterogeneities on the stellar surface can substantially affect transmission spectra, and if not accounted for can result in biased interpretation (e.g., Rackham et al. 2018, 2019). Further study and modeling of starspots and faculae, requiring input from the fields of heliophysics and stellar astrophysics, will be required to support transit spectroscopy with JWST and beyond.
- The challenges for atmospheric retrievals lie in expanding the current paradigm to higher dimensional models and higher quality observations. Robust retrievals involve computing $\sim 10^4$–$10^9$ model evaluations on a $\sim$10–20 dimensional parameter space, depending on the data at hand, even for 1D models. Therefore, multi-D retrievals are associated with a number of practical challenges. These include increased computation time, requirement of realistic and efficient parameterizations for temperature profiles, compositions and clouds/hazes, and efficient statistical inference and parameter estimation algorithms.
- New advances are also being made in atmospheric retrievals using a combination of high-resolution and low-resolution spectra. However, there are significant challenges in adapting both modeling and retrieval algorithms for this purpose, including computational costs as well as needs for robust likelihood metrics for the retrievals with multi-instrument, multi-resolution data.

- Retrievals of directly-imaged and isolated planets/sub-stellar objects are typically limited by a lack of measured masses and radii, and hence gravities, which need to be retrieved from the spectra in addition to the atmospheric properties. The spectra of such objects are also sensitive to temperature structure in the deep atmosphere as well as clouds, which make their retrievals more challenging than those of thermal emission from irradiated objects.
- Since exoplanetary atmospheres probe regimes not seen on Earth or in the solar system, the needed physical and chemical inputs into models can be missing or incomplete. Needed improvements in areas such as the opacities of molecules, reaction rate constants, and optical properties of aerosols can come from increased investments in laboratory work and first-principle simulations (Fortney et al. 2019).
- Ultimately, interpretation of atmospheric observations will require a holistic approach of using both robust atmospheric retrieval algorithms as well as detailed self-consistent models in tandem to both constrain the atmospheric properties and the underlying physical processes. Thus the ultimate challenge in this area is to consider the wide range of atmospheric processes possible (energy transport, clouds/hazes, disequilibrium chemistry, dynamics, escape) and the broad range of observations (UV to infrared) that can probe the different processes in a unified framework.

## References

Ackerman, A. S., & Marley, M. S. 2001, ApJ, 556, 872
Amundsen, D. S., Baraffe, I., Tremblin, P., et al. 2014, A&A, 564, A59
Amundsen, D. S., Tremblin, P., Manners, J., Baraffe, I., & Mayne, N. J. 2017, A&A, 598, A97
Barman, T. S., Hauschildt, P. H., & Allard, F. 2001, ApJ, 556, 885
Barstow, J. K., Aigrain, S., Irwin, P. G. J., & Sing, D. K. 2017, ApJ, 834, 50
Barstow, J. K., & Irwin, P. G. J. 2016, MNRAS, 461, L92
Barstow, J. K. 2020, MNRAS, 497, 4183
Barstow, J. K., Changeat, Q., Garland, R., et al. 2020, MNRAS, 493, 4884
Beichman, C., Benneke, B., Knutson, H., et al. 2014, PASP, 126, 1134
Birkby, J. L. 2018, arXiv:1806.04617
Blecic, J., Dobbs-Dixon, I., & Greene, T. 2017, ApJ, 848, 127
Brandt, T., Briesemeister, Z., Savransky, D., et al. 2019, BAAS, 51, 269
Brogi, M., de Kok, R. J., Albrecht, S., et al. 2016, ApJ, 817, 106
Burrows, A., Marley, M., Hubbard, W. B., et al. 1997, ApJ, 491, 856
Caldas, A., Leconte, J., Selsis, F., et al. 2019, A&A, 623, A161
Carone, L., Keppens, R., & Decin, L. 2014, MNRAS, 445, 930
Changeat, Q., Al-Refaie, A., & Mugnai, L. V. 2020, AJ, 160, 80
Deitrick, R., Mendonça, J. M., Schroffenegger, U., et al. 2020, ApJS, 248, 30
Drummond, B., Mayne, N. J., Manners, J., et al. 2018, ApJ, 869, 28
Feng, Y. K., Line, M. R., & Fortney, J. J. 2020, arXiv:2006.11442
Feng, Y. K., Line, M. R., Fortney, J. J., et al. 2016, ApJ, 829, 52
Fortney, J. J., Lodders, K., Marley, M. S., & Freedman, R. S. 2008, ApJ, 678, 1419
Fortney, J. J., Visscher, C., & Marley, M. S. 2020, AJ, 160, 288

Fortney, J., Robinson, T. D., Domagal-Goldman, S., et al. 2019, in Astro2020: Decadal Survey on Astronomy and Astrophysics, science white papers, 146
Gandhi, S., & Madhusudhan, N. 2017, MNRAS, 472, 2334
Gao, P., Thorngren, D. P., Lee, G. K. H., et al. 2020, NatAs, 4, 951
Goody, R., West, R., Chen, L., & Crisp, D. 1989, JQSRT, 42, 539
Goukenleuque, C., Bézard, B., Joguet, B., Lellouch, E., & Freedman, R. 2000, Icar, 143, 308
Gravity CollaborationNowak, M., Lacour, S., et al. 2020, A&A, 633, A110
Greene, T. P., Line, M. R., Montero, C., et al. 2016, ApJ, 817, 17
He, C., Hörst, S. M., Lewis, N. K., et al. 2020, NatAs, 4, 986
Helling, C., Tootill, D., Woitke, P., & Lee, G. 2017, A&A, 603, A123
Helling, C., Lee, I., Dobbs-Dixon, G., et al. 2016, MNRAS, 460, 855
Helling, C., & Casewell, S. 2014, A&ARv, 22, 80
Heng, K., & Showman, A. P. 2015, AREPS, 43, 509
Hörst, S. M., He, C., Lewis, N. K., et al. 2018, NatAs, 2, 303
Irwin, P. G. J., Teanby, N. A., de Kok, R., et al. 2008, JQSRT, 109, 1136
Irwin, P. G. J., Parmentier, V., Taylor, J., et al. 2020, MNRAS, 493, 106
Joshi, M. M., Haberle, R. M., & Reynolds, R. T. 1997, Icar, 129, 450
Kite, E. S., Fegley, B. Jr+, Schaefer, L., & Ford, E. B. 2020, ApJ, 891, 111
Kitzmann, D., Heng, K., Oreshenko, M., et al. 2020, ApJ, 890, 174
Koll, D. B., Malik, M., Mansfield, M., et al. 2019, ApJ, 886, 140
Kreidberg, L., Line, M. R., Thorngren, D., Morley, C. V., & Stevenson, K. B. 2018, ApJ, 858, L6
Lacy, B. I., & Burrows, A. S. 2020, ApJ, 905, 131
Leconte, J., Forget, F., Charnay, B., Wordsworth, R., & Pottier, A. 2013, Natur, 504, 268
Lee, G., Dobbs-Dixon, I., Helling, C., Bognar, K., & Woitke, P. 2016, A&A, 594, A48
Line, M. R., Marley, M. S., Liu, M. C., et al. 2017, ApJ, 848, 83
Line, M. R., Wolf, A. S., Zhang, X., et al. 2013, ApJ, 775, 137
Line, M. R., & Parmentier, V. 2016, ApJ, 820, 78
Lines, S., Mayne, N. J., Boutle, I. A., et al. 2018, A&A, 615, A97
Lines, S., Mayne, N. J., Manners, J., et al. 2019, MNRAS, 488, 1332
Lodders, K., & Fegley, B. 2002, Icar, 155, 393
Louden, T., & Wheatley, P. J. 2015, ApJ, 814, L24
MacDonald, R. J., Goyal, J. M., & Nikole, N. K. 2020, ApJ, 893, L43
Madhusudhan, N., Amin, M. A., & Kennedy, G. M. 2014, ApJ, 794, L12
Madhusudhan, N., Harrington, J., Stevenson, K. B., et al. 2011, Natur, 469, 64
Madhusudhan, N., & Seager, S. 2009, ApJ, 707, 24
Madhusudhan, N. 2018, in Handbook of Exoplanets (Berlin: Springer), 104
Mai, C., & Line, M. R. 2019, ApJ, 883, 144
Min, M., Ormel, C. W., Chubb, K., Helling, C., & Kawashima, Y. 2020, A&A, 642, A28
Mayne, N. J., Baraffe, I., Acreman, D. M., et al. 2014, A&A, 561, A1
Mollière, P., van Boekel, R., Dullemond, C., Henning, T., & Mordasini, C. 2015, ApJ, 813, 47
Mordasini, C., van Boekel, R., Mollière, P., Henning, T., & Benneke, B. 2016, ApJ, 832, 41
Morley, C. V., Kreidberg, L., Rustamkulov, Z., Robinson, T., & Fortney, J. J. 2017, ApJ, 850, 121
Moses, J. I., Visscher, C., Fortney, J. J., et al. 2011, ApJ, 737, 15
Öberg, K. I., Murray-Clay, R., & Bergin, E. A. 2011, ApJ, 743, L16
Rackham, B. V., Apai, D., & Giampapa, M. S. 2018, ApJ, 853, 122
Rackham, B. V., Apai, D., & Giampapa, M. S. 2019, AJ, 157, 96

Rauscher, E., & Menou, K. 2010, ApJ, 714, 1334
Rodgers, C. D. (ed.) 2000, in Inverse Methods for Atmospheric Sounding: Theory and Practice, Series on Atmospheric Oceanic and Planetary Physics, Vol. 2 (Singapore: World Scientific Publishing), 2
Roman, M., & Rauscher, E. 2019, ApJ, 872, 1
Seager, S., & Sasselov, D. D. 1998, ApJ, 502, L157
Showman, A. P., Fortney, J. J., Lian, Y., et al. 2009, ApJ, 699, 564
Showman, A. P., & Guillot, T. 2002, A&A, 385, 166
Snellen, I. A. G., de Kok, R. J., de Mooij, E. J. W., & Albrecht, S. 2010, Natur, 465, 1049
Stevenson, K. B., Desert, J. M., Line, M. R., et al. 2014, Sci, 346, 838
Taylor, J., Parmentier, V., Irwin, P. G. J., et al. 2020, MNRAS, 493, 4342
Tinetti, G., Drossart, P., Eccleston, P., et al. 2018, ExA, 46, 135
Venot, O., Cavalié, T., Bounaceur, R., et al. 2020, A&A, 634, A78
Welbanks, L., Madhusudhan, N., & Allard, N. F. 2019, ApJ, 887, L20
Wong, I., Shporer, A., Morris, B. M., et al. 2020, AJ, 160, 88
Zahnle, K., Marley, M. S., Freedman, R. S., Lodders, K., & Fortney, J. J. 2009, ApJ, 701, L20
Zellem, R. T., Swain, M. R., Cowan, N. B., et al. 2019, PASP, 131, 094401
Zhang, X., & Showman, A. P. 2018, ApJ, 866, 2

# Chapter 18

# Atmospheric Chemistry

Julianne I. Moses

## 18.1 Introduction

Two decades of exoplanet observations have taught us that planets in our galaxy are ubiquitous and diverse. By studying a statistically large population of exoplanets, we can constrain theories of planetary formation and evolution and place our own solar system in context. The atmospheres of exoplanets are critical in this regard. They provide potentially observable relics of the planetary formation era and highlight the evolutionary processes that have shaped the planet over time. Because giant planets are more likely to have held onto their primordial atmospheres, the bulk elemental composition of giant-exoplanet atmospheres reflects conditions within the formation region of the protoplanetary disk, reveals the extent to which the planets have migrated, and informs about the relative contribution of gas versus solids during accretion, thus providing critical constraints on planetary formation theories. The atmospheres of smaller terrestrial exoplanets and super-Earths are more likely to have been affected by atmospheric loss and secondary outgassing from the planetary interior, thus providing a window into the evolutionary history and the bulk composition of the solid planet. If life were to develop on such smaller planets, the atmospheric composition would provide vital hints to its presence. Atmospheres of widely diverse exoplanets therefore provide clues needed for unraveling such fundamental questions as: **How do planets form? How do atmospheres arise on planets? How do atmospheres evolve with time? What processes currently affect atmospheric properties? Does life exist elsewhere in the universe?**

Because atmospheric chemistry controls the observable composition of exoplanets, understanding how chemistry operates in exoplanet atmospheres is necessary for furthering our knowledge of the big-picture questions described above. However, atmospheric chemistry cannot be considered in isolation. Chemistry is highly coupled to physical processes within the planet, and both the physics and chemistry are affected by external radiation from the host star (and other star–planet interactions), by the age and evolutionary history of the planet, and by potential

doi:10.1088/2514-3433/abfa8fch18  

gravitational interactions with other planets in the system. In order to interpret observations of exoplanetary atmospheres, we need to look at the entire stellar and planetary system and consider the various feedbacks between stellar irradiation, atmospheric chemistry, dynamics, and climate that can affect the three-dimensional atmospheric composition. We also need to understand how various elements can be tied up in condensed phases (clouds/hazes) or in spectrally inactive gas-phase species to fully understand how to relate the observed composition to the bulk elemental composition.

Below, we briefly review the state of the art of observations and theoretical modeling of exoplanet atmospheric chemistry, and then we discuss the major questions, challenges, and opportunities at the forefront of the field.

## 18.2 Background and State of the Art in the Field

### 18.2.1 Observations of Atmospheric Composition

Exoplanet transit and eclipse observations have led to detections of numerous atmospheric constituents. For example, absorption lines from neutral and ionized atoms such as H, He, Li, $C^+$, O, Na, Mg, $Mg^+$, $Si^{2+}$, Ca, K, $Sc^+$, $Ti^+$, $Cr^+$, Fe, $Fe^+$, and $Y^+$ have been detected on various exoplanets through electronic transitions at ultraviolet and visible wavelengths (e.g., Charbonneau et al. 2002; Vidal-Madjar et al. 2003, 2004; Redfield et al. 2008; Linsky et al. 2010; Fossati et al. 2010; France et al. 2011; Sing et al. 2011, 2012; Colón et al. 2012; Astudillo-Defru & Rojo 2013; Jensen et al. 2012; Spake et al. 2018; Chen et al. 2018; Hoeijmakers et al. 2018, 2019; Cauley et al. 2019). The cores of these atomic lines probe the planet's extended high-altitude thermosphere, where temperatures can be many thousands of kelvins and the gas is often actively escaping. For some of the neutral atomic species that can be present below the thermosphere, such as Na and K, the line wings can probe deeper into the stratosphere of the planet. At near-infrared and visible wavelengths that typically probe the planet's stratosphere and/or upper troposphere between ~0.1–1000 mbar, molecular bands from $H_2O$, and (less definitively) CO, TiO, VO, and AlO have been identified in transit and/or eclipse (e.g., Deming et al. 2013; Huitson et al. 2013; Wakeford et al. 2013; Mandell et al. 2013; Kreidberg et al. 2014; Crouzet et al. 2014; Haynes et al. 2015; Evans et al. 2017; Sheppard et al. 2017; Sedaghati et al. 2017; von Essen et al. 2019). The atomic species can provide information about thermospheric processes, atmospheric escape, and atmospheric metallicities on exoplanets, as well as indirectly provide information about cloud condensation (or lack thereof) deeper in the atmosphere, photochemical destruction of molecules in the stratosphere, and potential "pollution" from impacting solid bodies. Observations of molecular species provide important information on bulk elemental abundances in the atmosphere (and so indirectly constrain formation scenarios), as well as information about cloud condensation, atmospheric energetics, and chemical processes in the atmosphere. See Chapter 7 for more details about the transit technique and key observational results obtained to date.

Young, hot, giant planets that orbit far from their stars have also been imaged directly (e.g., Marois et al. 2008, 2010). Near-infrared spectra from such directly imaged exoplanets have revealed the presence of $H_2O$, CO, and $CH_4$ in the planet's atmosphere (Konopacky et al. 2013; Lee et al. 2013; Oppenheimer et al. 2013; Janson et al. 2013; Barman et al. 2015; Macintosh et al. 2015; Todorov et al. 2016; Chilcote et al. 2017; Samland et al. 2017). The relative abundance of these species points to the effectiveness of disequilibrium chemical processes such as transport-induced quenching (e.g., Bowler et al. 2010; Barman et al. 2011, 2015; Konopacky et al. 2013). See Chapter 5 for a discussion of the technique and key observational results to date.

Even if a planet itself is unresolved in observations, using high-resolution spectroscopy to track the Doppler shift from multiple lines within a molecular band as their position varies with time during the planet's orbit, combined with cross-correlation techniques, can be used to detect atmospheric constituents in the planet's atmosphere (e.g., Snellen et al. 2010; de Kok et al. 2013; Birkby et al. 2013). This technique has worked well for transiting, close-in, giant planets, but can also work for emission from non-transiting planets (e.g., Brogi et al. 2012) and eventually smaller transiting planets. One important advantage of the high-resolution observations is that multiple individual lines within a molecular band are observed, eliminating ambiguities, and allowing for robust detections of molecular species. Some disadvantages include difficulties in accurately removing telluric contamination, difficulties in constraining atmospheric abundances, and limitations on the wavelength range and species that can be explored using this technique from ground-based facilities. Combining this technique with lower-resolution transit observations to obtain abundances can also prove useful (Brogi et al. 2017; Hoeijmakers et al. 2018; Hawker et al. 2018). Atmospheric species detected to date from high-resolution Doppler spectroscopy include CO, $H_2O$, TiO, HCN, Fe, $Fe^{+}$, Ti, $Ti^{+}$, Na, $Cr^{+}$, $Sc^{+}$, and $Y^{+}$ (Snellen et al. 2010; Brogi et al. 2012; Birkby et al. 2013; Nugroho et al. 2017; Hawker et al. 2018; Hoeijmakers et al. 2018, 2019). See Chapter 8 in this volume for more details about this observing technique and key observational results to date.

Bailey (2014), Burrows (2014), Madhusudhan et al. (2014, 2016), Crossfield (2015), Deming & Seager (2017), and Madhusudhan (2019) review what we currently know about exoplanet atmospheric composition from observations, as well as discuss the degeneracies, complications, and pitfalls involved with the attempt to characterize exoplanet atmospheres. These reviews also contain discussions of what the future holds in terms of relevant space missions and ground-based observing campaigns. A more comprehensive list of the individual species detected on various exoplanets to date can be found in Madhusudhan (2019).

### 18.2.2 Theoretical Predictions for Atmospheric Chemistry

To first order, the composition of an exoplanet atmosphere can be predicted assuming *thermochemical equilibrium*, where the abundance of the chemical constituents achieves a steady-state balance representing a minimum in the total Gibbs free energy of the system (e.g., Burrows & Sharp 1999; Lodders & Fegley 2002). Thermochemical-equilibrium calculations, which depend solely on temperature, pressure, and bulk elemental abundances, are particularly useful for identifying the likely dominant atmospheric form of different elements, such as $CH_4$ versus CO, $CO_2$, C, or $C^+$ for carbon (see Figure 18.1). As long as equilibrium can be kinetically maintained, as in the hot, deep atmospheres of isolated giant planets and brown dwarfs (where chemical timescales are very short and no other processes compete to disturb this balance), thermochemical equilibrium should prevail. However, in real planetary atmospheres, certain processes can act to drive the composition away from equilibrium.

*Transport-induced quenching* is one such disequilibrium process that occurs when dynamical transport of atmospheric gases operates more rapidly than chemical reactions can maintain equilibrium. In this situation, the mixing ratios of the different constituents can be "frozen in" at the point at which the dynamical timescale begins to dominate over the chemical timescale. A common example is the quenching of CO and $CH_4$ in the strongly convecting tropospheres of giant planets and brown dwarfs, allowing the less thermodynamically favored constituent to be carried to higher altitudes, where it can be observed (e.g., Prinn & Barshay 1977; Lodders & Fegley 2002; Visscher & Moses 2011). *Photochemistry* is another disequilibrium process that occurs when atmospheric constituents interact with stellar ultraviolet photons or stellar/magnetospheric high-energy particles, causing atmospheric species to be ionized or photolyzed (broken apart), initiating further chemical reactions. Photochemistry can lead to the production and maintenance of species at abundances well in excess of chemical equilibrium, as with ozone in the Earth's atmosphere or complex hydrocarbons on Titan and our solar-system giant planets (e.g., Yung & DeMore 1999). Photochemistry is likely important in the middle and upper atmospheres of most terrestrial and giant exoplanets (e.g., Liang et al. 2003; Segura et al. 2005; Muñoz 2007; Moses et al. 2011). Transport-induced quenching can be important on planets with

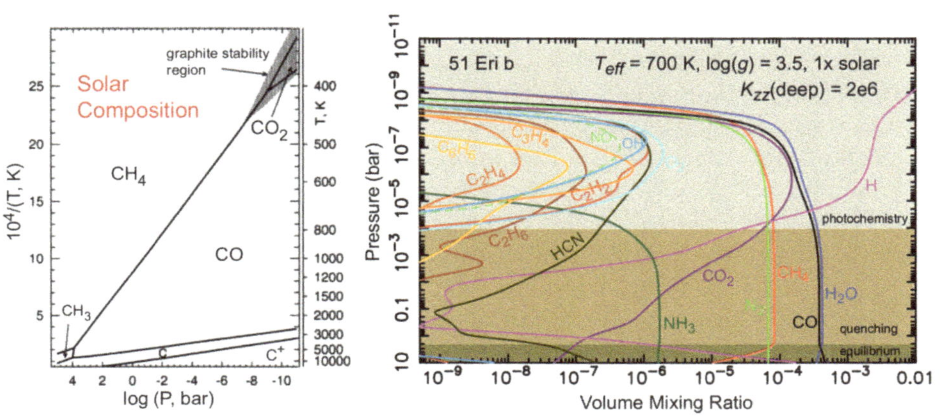


**Figure 18.1.** (Left) Thermodynamic-equilibrium stability regions for the major carbon gases in a solar-composition atmosphere, as a function of pressure and temperature (figure adapted from Lodders & Fegley 2002; Copyright (2002), with permission from Elsevier). (Right) Mixing ratios predicted for 51 Eri b from a thermo/photochemical kinetics and transport model; shaded colors highlight the regions of the atmosphere where thermochemical equilibrium, transport-induced quenching, and photochemistry dominate the chemistry (figure adapted from Moses et al. 2016, © 2016. The American Astronomical Society. All rights reserved).

warm-to-hot atmospheres that exhibit large temperature gradients (either vertically or horizontally) and efficient atmospheric dynamics (e.g., Visscher & Moses 2011; Zahnle & Marley 2014; Agúndez et al. 2014a; Tsai et al. 2018). Thermochemical equilibrium can be important for planets with very hot, thick atmospheres (e.g., Moses et al. 2011, 2013; Venot et al. 2012; Kopparapu et al. 2012). All three processes can dominate in giant-exoplanet atmospheres at different depths (see Figure 18.1).

## 18.3 Important Questions/Goals

The following are major science questions related to exoplanet atmospheric chemistry.

### 18.3.1 What Can the Current Exoplanet Atmospheric Composition Tell Us about Formation Conditions and Evolutionary Pathways?

The elemental abundances within the atmospheres of giant planets provide a window into the conditions encountered during the planetary formation era (e.g., Helled & Lunine 2014). The current composition of an exoplanet atmosphere depends on the composition of the protoplanetary disk within which the planet formed (assumed to be similar to that of the star), the structure and evolutionary history of the disk, the planet's formation location and migration history within the disk, the relative influence of solids and gases being accreted over time, the composition of those solids and gases, the physical properties of the solids as they interact with both the disk and planetary atmosphere (e.g., pebbles versus larger bodies; strong versus friable), the efficiency of mixing and exchange between the interior and atmosphere, and atmospheric evolutionary processes such as escape, weathering, or sequestration of volatiles into solids/liquids (e.g., Mordasini et al. 2016). Although these steps are complex individually and collectively, comparisons of a giant-planet's atmospheric composition with that of its star can help put limits on some of the processes. With compositional information from a statistically significant population of giant exoplanets, a more robust understanding of planet formation can begin to emerge. Here we summarize the ways in which exoplanet composition can reflect back on formation conditions; see Chapters 12, 13, and 14 for more in-depth discussions of planetary formation topics.

As is illustrated on the left side of Figure 18.2 (Mousis et al. 2018), the overall atmospheric metallicity and the ratios of key elements in the atmosphere can help distinguish between competing giant-planet formation theories, including core accretion (e.g., Mizuno et al. 1978; Pollack et al. 1996) and disk instability (e.g., Cameron 1978; Boss 1997), and the source of any heavy-element enrichment (e.g., disk photoevaporation versus trapping of volatiles in clathrate hydrates versus low-temperature amorphous ice). Moreover, as a result of condensation in the evolving protoplanetary disk, fractionation of the elements between gas and solid phases occurs in the disk (e.g., Stevenson & Lunine 1988; Cyr et al. 1998; Dodson-Robinson et al. 2009; Öberg et al. 2011), and the relative abundance of the elements in an exoplanet atmosphere could potentially reveal where in the disk the planet formed. For example, $H_2O$, $CO_2$, and CO have condensation fronts or "snow lines"

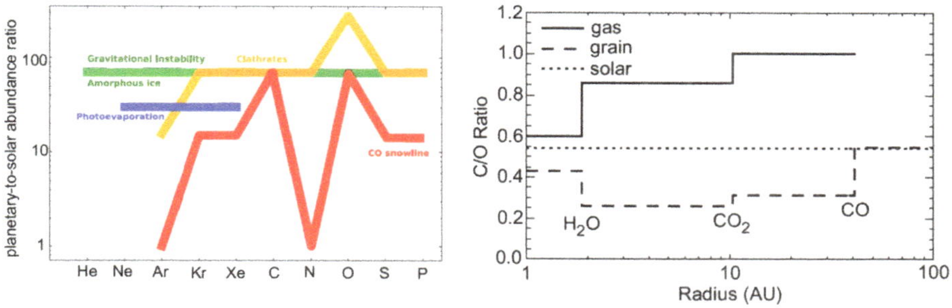


**Figure 18.2.** (Left) Volatile abundances (normalized to carbon) in Uranus and Neptune predicted from different formation scenarios. Note that the different scenarios leave different imprints on the elemental abundances; the same will be true for exoplanets (figure adapted from Mousis et al. 2018; Copyright (2018), with permission from Elsevier). (Right) The C/O ratio in the gas (solid curve) versus grains (dashed curve) in a "typical" protoplanetary disk around a solar-type star, including snow lines demarking $H_2O$, $CO_2$, and CO condensation (figure adapted from Öberg et al. 2011 © 2011. The American Astronomical Society. All rights reserved).

progressively farther out in the disk (see Figure 18.2), and as oxygen and carbon become tied up in condensates, the C/O ratio in the gas and solids changes as a function of distance from the star. The C/O ratio in the planetary atmosphere (in comparison to that of the star) and its overall metallicity can help constrain the formation location, potential migration history, and the relative influence of solids during accretion (e.g., Lodders 2004; Mousis et al. 2009; Fortney et al. 2013; Madhusudhan et al. 2014; Ali-Dib et al. 2014; Venturini et al. 2016; Espinoza et al. 2017). Some studies even suggest that exoplanet atmospheric composition can help reveal what type of migration (e.g., Type II, planet–planet scattering) occurred and whether the volatiles were delivered through pebble or planetesimal accretion (Madhusudhan et al. 2014, 2017; Venturini et al. 2016).

### 18.3.2 Can We Determine Bulk Elemental Compositions from Atmospheric Observations?

We clearly have many reasons for wanting to know the bulk elemental composition of exoplanets, but is such a goal even achievable? The atmospheric region we can probe remotely may not be representative of the whole atmosphere or the whole planet. Molecular absorbers, clouds, Rayleigh scattering, and refraction limit the extent to which we can probe the deep atmospheres of exoplanets (e.g., de Kok & Stam 2012; Benneke & Seager 2013; Misra et al. 2014), and condensation at depth can remove elements from our view (e.g., Morley et al. 2012; Marley et al. 2013). During various stages of planetary evolution, elements may become permanently sequestered in the planetary interior (Stevenson & Salpeter 1977; Wilson & Militzer 2010), convection may erode the core and add heavy elements to the atmosphere (e.g., Guillot et al. 2004), or there may be concentration gradients with altitude or radiative layers limiting convection such that the atmospheric envelope is not well mixed (e.g., Leconte & Chabrier 2012; Vazan et al. 2018). Some elements may be tied up in species that are not spectrally active, or disequilibrium processes may convert spectrally observable molecules to unobservable ones (e.g., Moses et al. 2011, 2016). To answer the question of how to relate the observed composition to the bulk composition, we need to know the fate of the volatile elements that were supplied to the planet during different stages of formation and evolution. Remote observations of exoplanets themselves provide only few clues on this topic, other than mass, radius, and the presence of clouds and gases in the observable regions of the atmosphere, so we must rely on theoretical models, on spacecraft data from our own giant planets, and on observations of protoplanetary disks at various stages of evolution to inform the models.

Theoretical models play a critical role in constraining the bulk elemental composition of an exoplanet based on observations. Interior and evolution models can help connect what is observed in the outer layers of the planet to properties of the planet as a whole (e.g., Baraffe et al. 2008; Vazan et al. 2013; Espinoza et al. 2017; Helled & Guillot 2017; Lozovsky et al. 2017; Thorngren & Fortney 2019; Valletta & Helled 2019). Models that track the formation of condensate clouds help determine what fraction of any element is sequestered in condensed phases versus gas phases at various depths in the atmosphere (see Figure 18.3)—information that is needed to back out the relative abundance of elements from observations (e.g., Visscher et al. 2010a; Marley et al. 2013; Helling 2019). As an example, ~20% of the atmospheric oxygen inventory will be tied up in silicate and metal-oxide clouds at depth in the atmospheres of cooler solar-composition transiting exoplanets (Visscher et al. 2010b). The fraction of any element sequestered in condensates will depend strongly on the overall metallicity and initial relative elemental abundances, as well as the atmospheric thermal structure. Photochemical models can provide sanity checks on abundances derived from spectral retrievals, identifying any unlikely

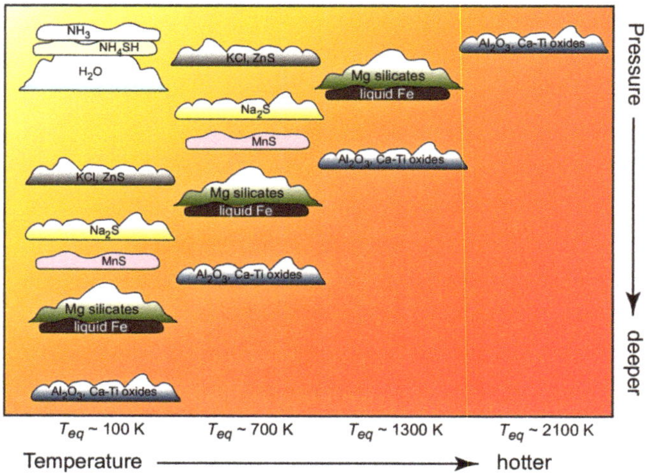


**Figure 18.3.** Cartoon illustrating the equilibrium condensation cloud layers expected in hydrogen-dominated giant-exoplanet atmospheres, for planets with equilibrium temperatures ranging from ~100 K to 2100 K. Refractory elements are tied up in condensates at greater depths in colder atmospheres (figure concept from Lodders & Fegley 2006).

relative abundances that result from low signal-to-noise data or instrument systematics. The photochemical models can also help identify situations where elements might be partitioned into gas-phase species that are not observable, thereby indicating cases in which observed elemental budgets represent upper limits only (e.g., Zahnle et al. 2009; Line et al. 2011; Moses et al. 2011, 2013, 2016; Venot et al. 2012, 2014; Agúndez et al. 2014a, 2014b; Miguel & Kaltenegger 2014; Lavvas et al. 2014; Lupu et al. 2016; Drummond et al. 2016, 2018; Tsai et al. 2018; Mendonça et al. 2018).

### 18.3.3 How Diverse Are Atmospheric Compositions of Exoplanets?

Observations to date show a wide diversity of planetary sizes, densities, and atmospheric properties. Both observers and theoreticians seek to identify trends in atmospheric properties based on physical parameters of the planet and its host star, with the goal of better understanding the physical and chemical processes at work. Given observable properties of the planet—such as its mass, radius, and orbital parameters—and the host star—such as stellar type, effective temperature, spectral energy distribution, metallicity, age, activity level—is it possible to accurately predict the composition of an exoplanet atmosphere? For example, young, newly formed giant planets that reside far from their stars will have hotter deep atmospheres and colder upper atmospheres than much older giant planets that reside close to their stars, even if the planets have similar observed "effective" temperatures and masses. The volatile elements carbon, nitrogen, oxygen, and sulfur will be partitioned into molecular forms differently on these planets based on their thermal profiles. Moreover, planets that are closer to their host stars receive a higher flux of ultraviolet radiation and thus may have a greater abundance of photochemical products than planets at larger orbital distances (see Figure 18.4).

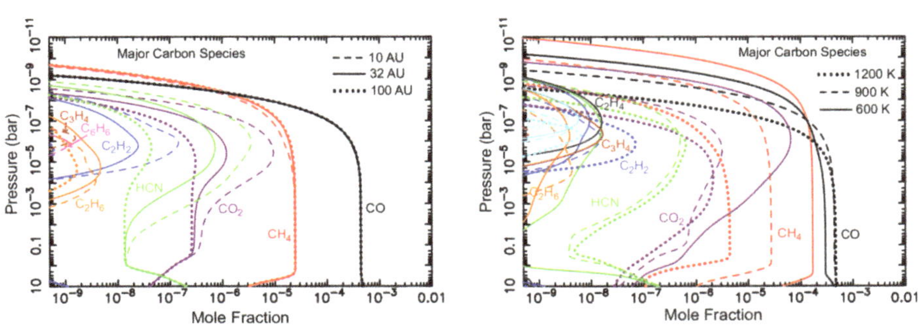


**Figure 18.4.** Predicted mixing-ratio profiles of several carbon-containing constituents on a young Jovian exoplanet being irradiated by an HR 8799-like star: (left) as a function of orbital distance, for a planet with an effective temperature of $T_{\rm eff} = 1000$ K and an assumed gravity $g = 10^4$ cm s$^{-2}$, and (right) as a function of $T_{\rm eff}$, for a planet with an assumed gravity $g = 10^{3.5}$ cm s$^{-2}$ at an orbital distance of 68 au. Species that are produced from photochemistry, such as HCN, $CO_2$ and complex hydrocarbons, are more abundant for planets at smaller orbital distances, due to the higher ultraviolet flux received. Methane, which is very sensitive to temperature, is much more abundant on cooler planets (figure adapted from Moses et al. 2016 © 2016. The American Astronomical Society. All rights reserved).

The abundance of heavy elements within the atmosphere of a hydrogen-dominated planet may depend the planet's mass, with smaller planets exhibiting greater atmospheric metallicities (e.g., Fortney et al. 2013). The atmospheres of some planets in multiple-planet systems could potentially be tidally heated due to orbital resonances with other planets and/or forced eccentricities, affecting the chemical makeup of the atmosphere. Small planets that are highly irradiated may have lost their atmospheres or have lost lighter elements, preferentially enriching heavy elements (Owen & Wu 2013; Hu et al. 2015; Malsky & Rogers 2020). Any identified trends in the atmospheric composition as a function of planetary and stellar properties can provide clues about current processes operating in the atmosphere, and past formation and evolution.

### 18.3.4 How Does the Composition of an Exoplanet Atmosphere Evolve with Time?

The composition we observe now in an exoplanet atmosphere may not reflect the initial volatile budget of the planet. Volatiles can be lost through atmospheric escape as a result of hydrodynamic flow, traditional thermal escape (Jean's escape), impact erosion, as well as a variety of non-thermal processes that lead to excited neutral species (e.g., charge exchange, dissociative recombination, ion-neutral reactions, sputtering) or ion escape (polar wind, solar-wind pickup; see the reviews of Chamberlain & Hunten 1987; Catling & Kasting 2017). Large impacts can either "erode" the atmosphere, or deliver volatiles, depending on impact and atmospheric conditions (e.g., Griffith & Zahnle 1995). Volatiles can be supplied through interior degassing or volcanic outgassing, and geological and biological processes can serve as either sources or sinks. Variations in stellar properties with time and/or planetary orbital properties with time can lead to time-variable atmospheric composition. Observations constraining the atmospheric mean molecular mass, metallicity, relative elemental abundances, and isotopic ratios therefore shed light not only on current processes, but also on past evolutionary processes. Trends in these properties with planetary size, bulk density, orbital properties, and stellar properties serve as useful constraints on models of atmospheric evolution. Ultraviolet and visible transit observations that probe a planet's extended thermosphere and even escaping atmosphere can also provide clues to atmospheric evolution over time (e.g., Sanz-Forcada et al. 2011).

### 18.3.5 How and Why Does Exoplanet Atmospheric Composition Vary with Altitude, Latitude, and Longitude?

Atmospheres are not simple, one-dimensional constructs. Stellar insolation, temperatures, and cloud cover will vary across the planet, potentially affecting the gas-phase composition. Atmospheric dynamics can also affect local and global atmospheric properties. Dynamics, chemistry, and radiative processes are highly coupled, and we must fully understand the various mechanisms involved—with all their feedbacks—to accurately predict or interpret the three-dimensional atmospheric structure and composition. Due to their unique orbital geometry as seen from Earth, transiting planets are able to provide some insights into the three-dimensional (3D) properties of their atmospheres. The transits themselves probe the atmosphere at the dawn/dusk terminators, while eclipse observations probe the dayside atmosphere. Observations across the full orbital phase of the planet sample emission and reflection changes as a function of longitude (and potentially latitude), which can provide information about the 3D temperature structure, wind fields, and cloud distribution (e.g., Knutson et al. 2007; Demory et al. 2013).

One current outstanding problem in exoplanet atmospheric science is the degree to which the atmospheric composition changes with location on the planet. Close-in transiting exoplanets, in particular, will have an atmospheric temperature structure that is highly variable with location, as a result of the presumed tidal locking that keeps one hemisphere of the planet facing the star, with subsequent strong, non-uniform, radiative forcing. If the atmospheric constituents were to remain in thermochemical equilibrium, the composition would be expected to vary with longitude and latitude as a result of these temperatures differences. The non-uniform radiative forcing, however, can also produce strong horizontal winds that rapidly transport gases across the planet. If the horizontal transport occurs more rapidly than the rate at which the constituents can react to maintain chemical equilibrium, then the composition can be "quenched" horizontally (Cooper & Showman 2006; Agúndez et al. 2014a; Tsai et al. 2018), similar to the vertical transport-induced quenching discussed in Section 18.2.2. In that situation, the atmospheric composition would be expected to be more uniform with location (see Figure 18.5), although differences may still exist in the photochemical regime at high altitudes. The strong horizontal temperature gradients can also affect cloud formation, leading to differences in cloud cover and/or cloud heights between the dayside, terminators, and nightside of the planet. All of these differences affect the observed spectra.

To better understand the 3D composition/structure and current processes at work in exoplanet atmospheres, we need spectral observations covering a full planetary orbit at a wide range of wavelengths from ultraviolet to mid-infrared. Shorter wavelengths are needed to better constrain the scattering properties of clouds and hazes, and to probe the

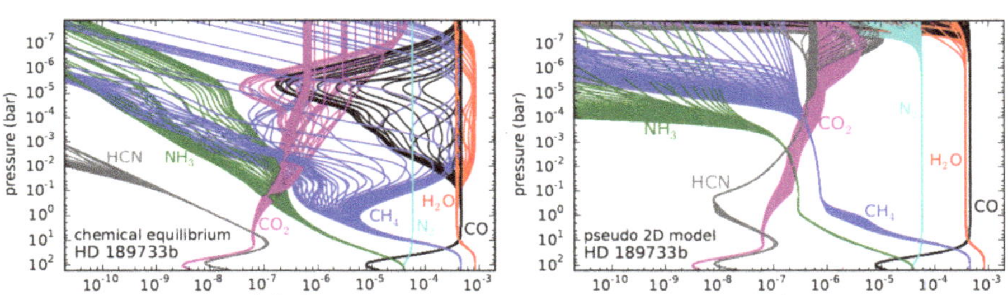


**Figure 18.5.** Predicted vertical mixing-ratio profiles for some major atmospheric constituents on the hot Jupiter HD 189733b at multiple longitudes (in the 0–360° range), from (left) thermochemical-equilibrium models and (right) pseudo-2D disequilibrium chemistry-transport models. Note that the atmospheric composition changes much more strongly with longitude (and thus orbital phase) if the atmosphere remains in thermochemical equilibrium (figure from Agúndez et al. 2014a; reproduced with permission from Astronomy & Astrophysics, © ESO).

thermosphere/exosphere; the visible and near-infrared region tends to have high signal-to-noise and can help diagnose key gas-phase absorbers, as well as inform about the presence of aerosols; longer-wavelength infrared is useful because several key molecules have diagnostic absorption bands in this region, and aerosols have less of an influence, reducing degeneracies in interpretation of the spectra. Because exoplanets are diverse, spectra from a statistically large sample of planets is desirable, and multi-dimensional models such as general climate models that include chemistry will be needed to truly help further our understanding of the processes that shape the 3D composition (e.g., Drummond et al. 2018; Mendonça et al. 2018).

### 18.3.6 How Does Life Affect the Composition of Exoplanet Atmospheres?

The potential detection of life on other worlds has always been an important motivation for exoplanet studies. Concerns about how we might be able to definitively detect life, if it exists on other planets, have a similarly long history (e.g., Lovelock 1965). On Earth, life is currently altering the atmospheric composition in observable ways, and presumably that could be true for other biologically active planets. The main problem is that we have only one known example—that of terrestrial life—on one known life-hosting planet, the Earth. There is no reason to believe that all life would resemble that on Earth, and even if it did, the host exoplanet and its environment may differ, causing remotely sensed "biosignatures" to differ. See Chapter 23 on Astrobiology in this volume for a more thorough discussion.

Spectroscopic observations will be our main tool for identifying any biosignatures on exoplanets, whether they consist of biogenic gases that affect atmospheric composition (such as $O_2$ on the Earth) or spectral changes on the planet's surface due to the presence of life (such as the "vegetation red edge" in terrestrial spectra). In their discussion of biosignature candidates for exoplanets, Des Marais et al. (2002) identify atmospheric gases such $O_2$, $O_3$, $CH_4$, $N_2O$, and $CH_3Cl$ (and/or combinations thereof) as potentially signaling the presence of life. Schwieterman et al. (2018) expand on this list in their comprehensive recent review of the topic, and they emphasize the importance of evaluating any potential biosignatures in the context of the planetary environment and climate. For example, as discussed by Meadows et al. (2018), $O_2$ can be produced in abundance abiotically under a variety of atmospheric conditions, and for a large portion of Earth's history, life existed without the presence of significant amounts of $O_2$. Therefore, understanding the processes affecting $O_2$ (and other potential biosignatures) in the context of the planetary orbit, stellar properties, atmospheric properties, and surface/interior properties is key to evaluating whether it is an appropriate biosignature or not. Photochemistry and climate modeling can help provide some context, but there will remain many unknown biogeochemical factors about the planetary environment that provide uncertainty. We must have some way of quantitatively evaluating the probability that observed atmospheric signatures actually signal life. Catling et al. (2018) and Walker et al. (2018) discuss this topic in detail.

The search for biosignatures from exoplanets has commenced. Atmospheric characterization plays an important role in these efforts. Fujii et al. (2018) review current and future technological capabilities and concepts that could enable the characterization of potentially habitable exoplanets, along with specific observing strategies for biosignatures.

## 18.4 Opportunities

In the next decade and beyond, opportunities will abound for characterizing exoplanet composition and chemistry. Synergies between observations, theoretical modeling, and laboratory experiments will propel us forward in the upcoming years, helping us better understand what exoplanet atmospheric composition can tell us about planetary formation, atmospheric evolution, and the possibility of life elsewhere in the universe.

***Observations:***

- Current transit surveys such as K2, TESS, and CHEOPS (from space) and SuperWASP, SPECULOOS, KELT, NGTS, and MEARTH (ground-based telescope campaigns) are identifying large numbers of exoplanet targets amenable to atmospheric characterization; the upcoming PLATO mission will do the same.
- Ground-based surveys such as GPI, SPHERE, and SCExAO are providing direct imaging and spectral characterization of young planets orbiting far from their host stars.
- HST and ground-based observations are continuing to provide useful information about exoplanet atmospheric composition and structure.
- Our understanding of atmospheric chemistry promises to be transformed with the 2021 launch of the JWST mission, a space-based telescope that will enable atmospheric spectra to be obtained for dozens of exoplanets at wavelengths ranging from 0.6 to 28.5 $\mu$ m.
- Looking beyond the next ten years, the ARIEL mission will be dedicated to acquiring visible-through-near-infrared spectra of exoplanets, providing the large statistical sample (500–1000 planets) needed to address some of the major outstanding questions described in the previous sections.
- Future extremely large (30 m class) ground-based telescopes (ELTs) will perform high-sensitivity transit spectroscopy and/or high-contrast direct imaging of exoplanets
- Future spaced-based mission concepts that are currently under study (e.g., HabEX and LUVOIR) will enable potential detection of biospheres on terrestrial exoplanets.

Reviews of the upcoming observational opportunities for exoplanet atmospheric characterization can be found in Cowan et al. (2015), Beichman & Greene (2018), Tinetti et al. (2018), Fujii et al. (2018), and Madhusudhan (2019).

***Theory:***

Opportunities also exist on the theoretical side of exoplanet atmospheric chemistry. A hierarchy of models can be used to address the major science questions facing the field.

- One-dimensional (1D) atmospheric chemistry and radiative-transfer models can highlight the important physical and chemical processes affecting atmospheric composition and spectra. Complex 1D thermo/photochemical models can accommodate large lists of species and reactions, helping to identify the important chemical mechanisms taking place under the diversity of conditions encountered in exoplanet atmospheres. Such 1D models can also more quickly and efficiently explore a large range of parameter space of atmospheric environments. While a new generation of sophisticated 3D general climate models will be needed to fully understand the complex coupling between atmospheric dynamics, chemistry, cloud formation, and radiative processes occurring in real exoplanet atmospheres, such models are computationally expensive and require various simplifications, including reduced chemical networks or chemical relaxation schemes (e.g., Drummond et al. 2018; Tsai et al. 2018; Mendonça et al. 2018). One-dimensional models will be needed to define these appropriate simplifications.
- Progress in theoretical modeling in the next decade will include 3D global climate models that fully couple the dynamics with chemistry (taking into account key disequilibrium chemical processes such as transport-induced quenching and photochemistry), and realistic radiative transfer, cloud formation, and climate feedbacks.
- Chemical models should expand to include elements besides H, C, N, O, and S, and should consider ion chemistry and realistic photochemical haze production, as well as the formation of equilibrium condensate clouds.
- More sophisticated models for exoplanet interior structure, atmosphere–interior exchange, atmospheric evolution, star–planet interactions, atmospheric escape, and (on potentially habitable planets) biogeochemical couplings could help further our understanding of the processes that can affect the observable atmospheric composition of exoplanets.

***Experiments:***

New laboratory measurements and ab initio calculations will also advance the field of exoplanet chemistry in the next decade. The exoplanet models described above need accurate laboratory and theoretical inputs relevant to a wide range of temperature, pressure, and redox conditions. Opportunities for future advancements include improvements in our understanding of kinetic reaction rates, thermodynamic parameters, and photoabsorption cross sections at conditions different from those encountered on Earth; aerosol formation and the microphysical/optical properties of "exotic" hazes relevant to exoplanets; spectroscopic line-shape parameters for mixtures of appropriate gases at relevant temperatures and pressures; and definitive identification of the dominant quench reactions for different key species.

## 18.5 Challenges

Many observational challenges impede our ability to reliably determine the composition and bulk elemental abundances of exoplanet atmospheres, and to address the questions described above. Exoplanet observations are difficult, often plagued by instrument systematics, and typically of limited wavelength range, signal-to-noise, and/or spectral resolution. Stellar activity can compromise the results. Degeneracies exist between thermal structure, species abundances, and cloud properties when attempting to derive abundances from spectral retrievals from limited data sets, especially given the horizontal and vertical gradients within real atmospheres, and the simplifications needed for retrievals. Elements can be tied up in condensates that are removed from view or in spectrally inactive molecules, complicating estimates of bulk abundances. Some elements may have escaped over time. Clouds and hazes can obscure molecular absorption features, and it is not clear we can accurately predict where, when, and how aerosols will affect atmospheric observations. The atmosphere may be variable due to changing cloud coverage or atmospheric dynamics. All of these issues complicate the attempts to constrain atmospheric composition.

Developing appropriate theoretical models is also challenging. Models rely on accurate laboratory and theoretical inputs relevant to the physical and chemical conditions of the exoplanets being observed, but many of our current inputs are based on conditions appropriate to the Earth and other solar-system planets, having little resemblance to exoplanet atmospheric conditions. There are many cases in which our knowledge of spectroscopic line-shape parameters, species thermodynamic parameters, reaction rate-coefficients, temperature-dependent UV cross sections, microphysical and optical properties of potential condensates, and haze-formation mechanisms is limited, impeding our ability to accurately retrieve or theoretically predict atmospheric composition. The complicated 3D behavior of short-period exoplanets requires 3D models, but such models are computationally expensive, and adding fully coupled chemistry to 3D general climate models is typically not feasible, necessitating approximations. Clouds are notoriously difficult to include accurately in 3D dynamical models, even in terrestrial cases where abundant high-quality data exist. Certain properties of the host star are not always well known, complicating the models of star–planet interactions.

Telescope time is also limited, so we need to go after targets with the best characterization possibilities, and make trade-offs when planning observations. For example, we desire to obtain as much 3D information as possible, which can be provided with full phase curve observations, but such observations can take a significant fraction of available observing time. Many of the big-picture science questions described in the previous section will require a large statistical sample, which can be problematic without dedicated exoplanet missions or telescopes.

Overcoming these challenges will be a major focus of exoplanet science in the next decade and beyond.

## References

Agúndez, M., Parmentier, V., Venot, O., Hersant, F., & Selsis, F. 2014a, A&A, 564, A73

Agúndez, M., Venot, O., Selsis, F., & Iro, N. 2014b, ApJ, 781, 68

Ali-Dib, M., Mousis, O., Petit, J.-M., & Lunine, J. I. 2014, ApJ, 785, 125

Astudillo-Defru, N., & Rojo, P. 2013, A&A, 557, A56

Bailey, J. 2014, PASA, 31, e043
Baraffe, I., Chabrier, G., & Barman, T. 2008, A&A, 482, 315
Barman, T. S., Konopacky, Q. M., Macintosh, B., & Marois, C. 2015, ApJ, 804, 61
Barman, T. S., Macintosh, B., Konopacky, Q. M., & Marois, C. 2011, ApJ, 733, 65
Beichman, C. A., & Greene, T. P. 2018, in Handbook of Exoplanets, ed. H. Deeg, & J. Belmonte (Berlin: Springer), 1283
Benneke, B., & Seager, S. 2013, ApJ, 778, 153
Birkby, J. L., de Kok, R. J., Brogi, M., et al. 2013, MNRAS, 436, L35
Boss, A. P. 1997, Sci, 276, 1836
Bowler, B. P., Liu, M. C., Dupuy, T. J., & Cushing, M. C. 2010, ApJ, 723, 850
Brogi, M., Line, M., Bean, J., Désert, J. M., & Schwarz, H. 2017, ApJ, 839, L2
Brogi, M., Snellen, I. A. G., de Kok, R. J., et al. 2012, Natur, 486, 502
Burrows, A., & Sharp, C. M. 1999, ApJ, 512, 843
Burrows, A. S. 2014, Natur, 513, 345
Cameron, A. G. W. 1978, M&P, 18, 5
Catling, D. C., & Kasting, J. F. 2017, Atmospheric Evolution on Inhabited and Lifeless Worlds (Cambridge: Cambridge Univ. Press)
Catling, D. C., Krissansen-Totton, J., Kiang, N. Y., et al. 2018, AsBio, 18, 709
Cauley, P. W., Shkolnik, E. L., Ilyin, I., et al. 2019, AJ, 157, 69
Chamberlain, J. W., & Hunten, D. M. 1987, Theory of Planetary Atmospheres (2nd edn; Orlando, FL: Academic)
Charbonneau, D., Brown, T. M., Noyes, R. W., & Gilliland, R. L. 2002, ApJ, 568, 377
Chen, G., Pallé, E., Welbanks, L., et al. 2018, A&A, 616, A145
Chilcote, J., Pueyo, L., & Rosa, R. D. 2017, AJ, 153, 182
Colón, K. D., Ford, E. B., Redfield, S., et al. 2012, MNRAS, 419, 2233
Cooper, C. S., & Showman, A. P. 2006, ApJ, 649, 1048
Cowan, N. B., Greene, T., Angerhausen, D., et al. 2015, PASP, 127, 311
Crossfield, I. J. M. 2015, PASP, 127, 941
Crouzet, N., McCullough, P. R., Deming, D., & Madhusudhan, N. 2014, ApJ, 795, 166
Cyr, K. E., Sears, W. D., & Lunine, J. I. 1998, Icar, 135, 537
de Kok, R. J., Brogi, M., Snellen, I. A. G., et al. 2013, A&A, 554, A82
de Kok, R. J., & Stam, D. M. 2012, Icar, 221, 517
Deming, D., & Seager, S. 2017, JGRE, 122, 53
Deming, D., Wilkins, A., McCullough, P., et al. 2013, ApJ, 774, 95
Demory, B.-O., de Wit, J., Lewis, N., et al. 2013, ApJ, 776, L25
Des Marais, D. J., Harwit, M. O., Jucks, K. W., et al. 2002, AsBio, 2, 153
Dodson-Robinson, S. E., Veras, D., Ford, E. B., & Beichman, C. A. 2009, ApJ, 707, 79
Drummond, B., Mayne, N. J., Manners, J., et al. 2018, ApJ, 855, L31
Drummond, B., Tremblin, P., Baraffe, I., et al. 2016, A&A, 594, A69
Espinoza, N., Fortney, J. J., Miguel, Y., Thorngren, D., & Murray-Clay, R. 2017, ApJ, 838, L9
Evans, T. M., Sing, D. K., Kataria, T., et al. 2017, Natur, 548, 58
Fortney, J. J., Mordasini, C., Nettelmann, N., et al. 2013, ApJ, 775, 80
Fossati, L., Haswell, C. A., Froning, C. S., et al. 2010, ApJ, 714, L222
France, K., Linsky, J. L., Yang, H., Stocke, J. T., & Froning, C. S. 2011, Ap&SS, 335, 25
Fujii, Y., Angerhausen, D., Deitrick, R., et al. 2018, AsBio, 18, 739
Muñoz, A. G. 2007, P&SS, 55, 1426

Griffith, C. A., & Zahnle, K. 1995, JGR, 100, 16907
Guillot, T., Stevenson, D. J., Hubbard, W. B., & Saumon, D. 2004, in Jupiter. The Planet, Satellites and Magnetosphere, ed. F. Bagenal, T. E. Dowling, & W. B. McKinnon (Cambridge: Cambridge Univ. Press), 35
Hawker, G. A., Madhusudhan, N., Cabot, S. H. C., & Gandhi, S. 2018, ApJ, 863, L11
Haynes, K., Mandell, A. M., Madhusudhan, N., Deming, D., & Knutson, H. 2015, ApJ, 806, 146
Helled, R., & Guillot, T. 2017, in Handbook of Exoplanets, ed. H. J. Deeg, & J. A. Belmonte (Berlin: Springer), 44
Helled, R., & Lunine, J. 2014, MNRAS, 441, 2273
Helling, C. 2019, AREPS, 47, 583
Hoeijmakers, H. J., Ehrenreich, D., Kitzmann, D., et al. 2019, A&A, 627, A165
Hoeijmakers, H. J., Ehrenreich, D., Heng, K., et al. 2018, Natur, 560, 453
Hu, R., Seager, S., & Yung, Y. L. 2015, ApJ, 807, 8
Huitson, C. M., Sing, D. K., Pont, F., et al. 2013, MNRAS, 434, 3252
Janson, M., Brandt, T. D., Kuzuhara, M., et al. 2013, ApJ, 778, L4
Jensen, A. G., Redfield, S., Endl, M., et al. 2012, ApJ, 751, 86
Knutson, H. A., Charbonneau, D., Allen, L. E., et al. 2007, Natur, 447, 183
Konopacky, Q. M., Barman, T. S., Macintosh, B. A., & Marois, C. 2013, Sci, 339, 1398
Kopparapu, R. k., Kasting, J. F., & Zahnle, K. J. 2012, ApJ, 745, 77
Kreidberg, L., Bean, J. L., & Désert, J.-M. 2014, ApJ, 793, L27
Lavvas, P., Koskinen, T., & Yelle, R. V. 2014, ApJ, 796, 15
Leconte, J., & Chabrier, G. 2012, A&A, 540, A20
Lee, J.-M., Heng, K., & Irwin, P. G. J. 2013, ApJ, 778, 97
Liang, M.-C., Parkinson, C. D., Lee, A. Y.-T., Yung, Y. L., & Seager, S. 2003, ApJ, 596, L247
Line, M. R., Vasisht, G., Chen, P., Angerhausen, D., & Yung, Y. L. 2011, ApJ, 738, 32
Linsky, J. L., Yang, H., France, K., et al. 2010, ApJ, 717, 1291
Lodders, K. 2004, ApJ, 611, 587
Lodders, K., & Fegley, B. 2002, Icar, 155, 393
Lodders, K., & Fegley, B. Jr. 2006, in Astrophysics Update 2, ed. J. W. Mason (Chichester: Praxis), 1
Lovelock, J. E. 1965, Natur, 207, 568
Lozovsky, M., Helled, R., Rosenberg, E. D., & Bodenheimer, P. 2017, ApJ, 836, 227
Lupu, R. E., Marley, M. S., Lewis, N., et al. 2016, AJ, 152, 217
Macintosh, B., Graham, J. R., Barman, T., et al. 2015, Sci, 350, 64
Madhusudhan, N. 2019, ARA&A, 57, 617
Madhusudhan, N., Agúndez, M., Moses, J. I., & Hu, Y. 2016, SSRv, 205, 285
Madhusudhan, N., Amin, M. A., & Kennedy, G. M. 2014, ApJ, 794, L12
Madhusudhan, N., Bitsch, B., Johansen, A., & Eriksson, L. 2017, MNRAS, 469, 4102
Madhusudhan, N., Knutson, H., Fortney, J. J., & Barman, T. 2014, in Protostars and Planets VI, ed. H. Beuther, R. S. Klessen, C. P. Dullemond, & T. Henning (Tucson, AZ: Univ. Arizona Press), 739
Malsky, I., & Rogers, L. A. 2020, ApJ, 896, 48
Mandell, A. M., Haynes, K., Sinukoff, E., et al. 2013, ApJ, 779, 128
Marley, M. S., Ackerman, A. S., Cuzzi, J. N., & Kitzmann, D. 2013, in Comparative Climatology of Terrestrial Planets, ed. S. J. Mackwell, A. A. Simon-Miller, J. W. Harder, & M. A. Bullock (Tucson, AZ: Univ. Arizona Press), 367
Marois, C., Macintosh, B., Barman, T., et al. 2008, Sci, 322, 1348

Marois, C., Zuckerman, B., Konopacky, Q. M., Macintosh, B., & Barman, T. 2010, Natur, 468, 1080
Meadows, V. S., Reinhard, C. T., Arney, G. N., et al. 2018, AsBio, 18, 739
Mendonça, J. M., Tsai, S.-M., Malik, M., Grimm, S. L., & Heng, K. 2018, ApJ, 869, 107
Miguel, Y., & Kaltenegger, L. 2014, ApJ, 780, 166
Misra, A., Meadows, V., & Crisp, D. 2014, ApJ, 792, 61
Mizuno, H., Nakazawa, K., & Hayashi, C. 1978, PThPh, 60, 699
Mordasini, C., van Boekel, R., Mollière, P., Henning, T., & Benneke, B. 2016, ApJ, 832, 41
Morley, C. V., Fortney, J. J., Marley, M. S., et al. 2012, ApJ, 756, 172
Moses, J. I., Line, M. R., Visscher, C., et al. 2013, ApJ, 777, 34
Moses, J. I., Madhusudhan, N., Visscher, C., & Freedman, R. S. 2013, ApJ, 763, 25
Moses, J. I., Marley, M. S., Zahnle, K., et al. 2016, ApJ, 829, 66
Moses, J. I., Visscher, C., Fortney, J. J., et al. 2011, ApJ, 737, 15
Mousis, O., Atkinson, D. H., Cavalié, T., et al. 2018, P&SS, 155, 12
Mousis, O., Marboeuf, U., Lunine, J. I., et al. 2009, ApJ, 696, 1348
Nugroho, S. K., Kawahara, H., Masuda, K., et al. 2017, AJ, 154, 221
Öberg, K. I., Murray-Clay, R., & Bergin, E. A. 2011, ApJ, 743, L16
Oppenheimer, B. R., Baranec, C., Beichman, C., et al. 2013, ApJ, 768, 24
Owen, J. E., & Wu, Y. 2013, ApJ, 775, 105
Pollack, J. B., Hubickyj, O., Bodenheimer, P., et al. 1996, Icar, 124, 62
Prinn, R. G., & Barshay, S. S. 1977, Sci, 198, 1031
Redfield, S., Endl, M., Cochran, W. D., & Koesterke, L. 2008, ApJ, 673, L87
Samland, M., Mollière, P., Bonnefoy, M., et al. 2017, A&A, 603, A57
Sanz-Forcada, J., Micela, G., Ribas, I., et al. 2011, A&A, 532, A6
Schwieterman, E. W., Kiang, N. Y., Parenteau, M. N., et al. 2018, AsBio, 18, 663
Sedaghati, E., Boffin, H. M. J., MacDonald, R. J., et al. 2017, Natur, 549, 238
Segura, A., Kasting, J. F., Meadows, V., et al. 2005, AsBio, 5, 706
Sheppard, K. B., Mandell, A. M., Tamburo, P., et al. 2017, ApJ, 850, L32
Sing, D. K., Désert, J.-M., Fortney, J. J., et al. 2011, A&A, 527, A73
Sing, D. K., Huitson, C. M., Lopez-Morales, M., et al. 2012, MNRAS, 426, 1663
Snellen, I. A. G., de Kok, R. J., de Mooij, E. J. W., & Albrecht, S. 2010, Natur, 465, 1049
Spake, J. J., Sing, D. K., Evans, T. M., et al. 2018, Natur, 557, 68
Stevenson, D. J., & Lunine, J. I. 1988, Icar, 75, 146
Stevenson, D. J., & Salpeter, E. E. 1977, ApJS, 35, 239
Thorngren, D. P., & Fortney, J. J. 2019, ApJL, 874, L31
Tinetti, G., Drossart, P., Eccleston, P., et al. 2018, ExA, 46, 135
Todorov, K. O., Line, M. R., Pineda, J. E., et al. 2016, ApJ, 823, 14
Tsai, S.-M., Kitzmann, D., Lyons, J. R., et al. 2018, ApJ, 862, 31
Valletta, C., & Helled, R. 2019, ApJ, 871, 127
Vazan, A., Helled, R., & Guillot, T. 2018, A&A, 610, L14
Vazan, A., Kovetz, A., Podolak, M., & Helled, R. 2013, MNRAS, 434, 3283
Venot, O., Agúndez, M., Selsis, F., Tessenyi, M., & Iro, N. 2014, A&A, 562, A51
Venot, O., Hébrard, E., Agúndez, M., et al. 2012, A&A, 546, A43
Venturini, J., Alibert, Y., & Benz, W. 2016, A&A, 596, A90
Vidal-Madjar, A., Désert, J.-M., & Lecavelier des Etangs, A. 2004, ApJ, 604, L69
Vidal-Madjar, A., Lecavelier des Etangs, A., & Désert, J.-M. 2003, Natur, 422, 143

Visscher, C., Lodders, K., & Fegley, B. Jr. 2010a, ApJ, 716, 1060
Visscher, C., & Moses, J. I. 2011, ApJ, 738, 72
Visscher, C., Moses, J. I., & Saslow, S. A. 2010b, Icar, 209, 602
von Essen, C., Mallonn, M., Welbanks, L., et al. 2019, A&A, 622, A71
Wakeford, H. R., Sing, D. K., Deming, D., et al. 2013, MNRAS, 435, 3481
Walker, S. I., Bains, W., Cronin, L., et al. 2018, AsBio, 18, 779
Wilson, H. F., & Militzer, B. 2010, PhRvL, 104, 121101
Yung, Y. L., & DeMore, W. B. 1999, Photochemistry of Planetary Atmospheres (Oxford: Oxford University Press)
Zahnle, K., Marley, M. S., Freedman, R. S., Lodders, K., & Fortney, J. J. 2009, ApJ, 701, L20
Zahnle, K. J., & Marley, M. S. 2014, ApJ, 797, 41

# Chapter 19

## Atmospheric Dynamics

**Adam P. Showman**

### 19.1 Introduction

Exoplanet atmospheres are expected to exhibit great diversity. Exoplanets span a wide range of planetary masses, radii, interior composition, rotation rates, incident stellar fluxes, and orbital eccentricities. And within this diverse range, atmospheres range from massive to tenuous, exhibit a wide range of chemical compositions, redox states, evolutionary histories, and suitability (or lack thereof) for life. Observations, while in their infancy and so far emphasizing the hottest and most massive planets, already place important constraints on the properties of exoplanetary atmospheres. Despite their limited nature, these data sets already speak to wide diversity among such atmospheres (Figure 19.1).

Significant progress is being made in understanding the dynamical processes and three dimensional (3D) structure of exoplanetary atmospheres through such observations and theoretical studies. Extensive reviews on this subject can be found in Showman et al. 2010, 2013b, 2020, Heng & Showman 2015, Read et al. 2018, Pierrehumbert & Hammond 2019. In this chapter we focus on the major questions, challenges and opportunities on the path to understanding atmospheric dynamics in exoplanets.

### 19.2 Important Questions and Goals

In this section we survey major unsolved problems in atmospheric dynamics in exoplanets. These questions fall into four main classes:

- What is the basic atmospheric structure, that is, the three-dimensional (3D) distribution of temperature, wind, cloudiness, and trace gaseous species?
- What are the dynamical mechanisms that govern the atmospheric circulation/climate?
- What are the climates of terrestrial planets?
- What are the interior dynamics of fluid exoplanets and how do the atmosphere and interior interact?

doi:10.1088/2514-3433/abfa8fch19  

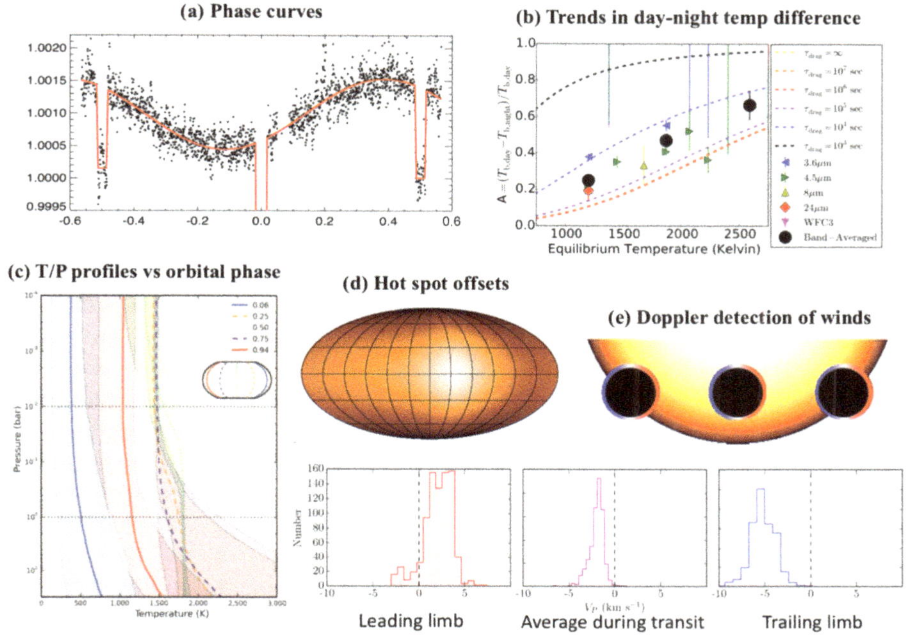


**Figure 19.1.** Examples of observations constraining the atmospheric circulation of hot Jupiters (adapted from Showman et al. 2020). (a) Thermal phase curves, providing constraints on day–night temperature differences and longitudinal offsets of hot and cold spots. This example is HD 189733b at 4.5 μm (reproduced from Zellem et al. 2014. © 2014. The American Astronomical Society. All rights reserved). (b) Trends of fractional day–night brightness temperature differences vs. planetary equilibrium temperature, suggesting that hotter planets have larger day–night temperature differences, along with theoretical predictions to explain this trend (reproduced from Komacek et al. 2017. © 2017. The American Astronomical Society. All rights reserved). (c) Temperature–pressure profiles at various orbital phases for WASP-43b; such inferences are possible when light curves at many wavelengths are obtained (from Stevenson et al. 2014. Reprinted with permission from AAAS.). (d) Thermal map inferred from phase curves, indicating eastward offset of dayside hot spot (reproduced from Knutson et al. 2007. © 2007, Nature Publishing Group). (e) Doppler shift of atmospheric spectral lines observed during transit place constraints on winds at the terminator, and indicate an eastward, superrotating jet stream on HD 189733b (reproduced from Louden & Wheatley 2015. © 2015. The American Astronomical Society. All rights reserved.).

For each of the above classes of questions, we wish to understand how the answers vary across the known population of exoplanets, and how the answers relate to the known behavior of atmospheres in the solar system.

### 19.2.1 Basic Atmospheric Structure

A major goal for exoplanet atmosphere science over the next 10–20 years is to characterize the 3D distribution of temperature, wind, and cloudiness on a wide range of exoplanets. This is interesting in its own right, but, perhaps more importantly, provides the observational foundation for understanding the physical and chemical *processes* that govern planetary atmospheres generally.

What is the vertical temperature profile, on the dayside, nightside, and terminator regions? Most models—and our limited observations—lead to an expectation that temperatures decrease with increasing altitude throughout most of the observable atmosphere, with the gradients exhibiting greatest amplitude on the nightside. We wish to determine observationally what these temperature profiles are. Thermal inversions—regions where temperature rises with altitude over some range—have been repeatedly suggested in the literature for several hot Jupiters (e.g. Haynes et al. 2015, Mikal-Evans et al. 2019), but confirmation has proved difficult in some cases (e.g. Line et al. 2016). On what types of exoplanets do thermal inversions exist in their atmospheres, over what pressure range do they occur, and what is their longitudinal and latitudinal extent?

It is critical to determine the horizontal temperature structure as well, including the day–night temperature differences and how those temperature differences depend on pressure. These observations have much to say about the dynamical mechanisms that regulate the thermal structure in planetary atmospheres. Observations from infrared (IR) light curves of hot Jupiters to date indicate a wide range of day–night temperature differences, tentatively encompassing a trend wherein the most strongly irradiated of the hot Jupiters exhibit the greatest fractional day–night temperature differences, and more poorly irradiated hot Jupiters exhibit smaller fractional day–night temperature differences (e.g. Cowan & Agol 2011, Komacek et al. 2017). Determining how such day–night temperature differences vary with pressure is important for constraining the dynamical mechanisms and understanding how heat transport in the atmosphere links to that in the interior; this requires obtaining high signal-to-noise light curves in a variety of wavebands that probe both absorption bands (which sample high in the atmosphere) and spectral windows (which probe more deeply).

The specific horizontal thermal structure across the globe is also important, in particular the longitudes and latitudes of the hottest and coldest points (the "hot spots" and "cold spots"). A fundamental theoretical prediction for hot, tidally locked exoplanets has been that the hot and cold spots are shifted to the east of the substellar and antistellar points, respectively, due to eastward advection by a superrotating (eastward) equatorial jet stream centered near the equator (e.g. Showman & Guillot 2002, Showman et al. 2009; Heng et al. 2011, Rauscher & Menou 2012, Dobbs-Dixon & Agol 2013). For the dayside hot spots of hot Jupiters, this prediction has been borne out from IR light curve measurements for a variety of hot Jupiters (e.g. Knutson et al. 2007, Zellem et al. 2014), but the situation with cold-spot offsets is less clear, partly because the nightside flux is lower and thus the signal-to-noise ratio is smaller. Such data have not yet been obtained for many mini Neptunes or super-Earths, but doing so is important in the JWST era and beyond. Light curves only provide longitudinal information, but eclipse mapping (during the ingress and egress of secondary eclipse) allows 2D mapping of the thermal structure across the entire dayside, including the latitudinal offset (if any) of the hot spot from the equator (de Wit et al. 2012; Majeau et al. 2012; Rauscher et al. 2018). So far, such ingress/egress mapping—performed for one planet, HD 189733b—suggests the hot spot is near the equator, but this needs to be repeated for dozens of hot Jupiters, super-Earths, and other planets to better understand the thermal structure and provide constraints on the underlying mechanisms.

Similarly, we wish to determine the 3D circulation patterns in the atmosphere—e.g., characterize the existence and properties of zonal jet streams, equatorial superrotation/subrotation, day-to-night flow, large coherent vortices, storms, and other features. Determining the wind patterns is difficult, but knowledge of the thermal structure provides valuable constraints on wind patterns, given that winds advect the thermal structure. For transiting planets, direct detection of winds is easiest on the terminator using Doppler measurements (Miller-Ricci Kempton & Rauscher 2012; Showman et al. 2013a); to date, this has only been performed for a few hot Jupiters—indicating fast, multi-km $s^{-1}$ wind speeds and equatorial superrotation (Snellen et al. 2010; Louden & Wheatley 2015)—but should be repeated for a wide range of hot Jupiters, mini Neptunes, and super-Earths to understand trends with incident stellar flux, atmospheric composition, and other factors.

Current observations suggest clouds and/or hazes on a wide range of hot Jupiters and several smaller exoplanets (mini Neptunes and super-Earths), but the data remain sparse (Sing et al. 2016; Crossfield & Kreidberg 2017). We need to better understand the 3D distribution of clouds and hazes, their composition, and trends in these quantities across the wide population of exoplanets. Clouds couple to the circulation in critical ways (Parmentier et al. 2013; Helling et al. 2019)—they exist only because of transport by the atmospheric circulation, and can also significantly affect the atmospheric circulation through radiative feedbacks and latent heating. Because of the circulation, clouds are likely to exhibit a patchy pattern at global scales—for example, clouds on the nightside and morning terminator, but not near the substellar point. Data sets required to constrain cloud properties—detailed IR spectra at many orbital phases—are essentially similar to those necessary for extracting the thermal structure, an exercise that needs to be conducted jointly. Polarization could provide important insights into the properties of cloud/haze particles.

A question of great interest is whether exoplanet atmospheres are time variable, and if so, over what timescales and amplitudes. Given the difficulty of resolving the spatial structure on exoplanets, observations of variability contain rich information difficult to obtain any other way; an analogy is the sound produced when a bell is rung, which can provide information on the size and even the shape of the bell, even if the sound is heard from far enough away that the bell itself cannot easily be imaged. For exoplanets, variability searches are most easily carried out by determining whether the secondary eclipse depth varies from one orbit to the next. Possible variability in the transit signature are also ripe for searches—for example, wavelength-dependent variation in the transit depth from one transit to the next could occur in response to changes to the thermal and cloud structure at the terminator; moreover, any changes in zonal wind speeds at the terminator would cause orbit-to-orbit changes in the Doppler shifts of spectral lines observed during transit. Variability is expected because atmospheres are unstable against various fluid dynamical instabilities. Thus, observational characterization of the variability amplitudes at a range of frequencies could provide information on the basic dynamical structure and its stability.

Another goal is to characterize the atmospheric structure in "exotic" regimes, such as planets with high orbital eccentricity. Planets on eccentric orbits experience time variable stellar heating, which will drive the atmospheric circulation in unclear ways that need to be understood (e.g. Kataria et al. 2013). Such exotic regimes are of interest in their own right but can also provide insights into how atmospheres in general respond to strong perturbations, thus yielding insights into physical and chemical mechanisms at play. Likewise, it will be of interest to characterize the orbital obliquity of exoplanets, a challenging task that may be possible with Doppler measurements, ingress/egress mapping, or by characterizing seasonal cycles via IR light curves.

Most of our discussion so far has emphasized close-in, highly irradiated transiting exoplanets, but a long-term goal is to characterize a much wider range of planetary atmospheres, including brown dwarfs, directly imaged giant planets, and irradiated exoplanets that are more distant from their stars than typical hot Jupiters and Neptunes (e.g. Showman et al. 2019). Within the overall broad population of giant planets, there are several, distinct observationally booming subfields that will yield constraints at the extreme ranges of key parameters (see Figure 19.2). Solar system giant planets exhibit weak irradiation, weak interior heat flux, and rotate rapidly (a factor that strongly influences the dynamical regime). Hot Jupiters receive intense radiation, but exhibit weak interior flux (perhaps within an order of magnitude of Jupiter), and because of tidal locking have modest rotation rates. Brown dwarfs and directly imaged giant planets receive negligible irradiation from their stars, but have

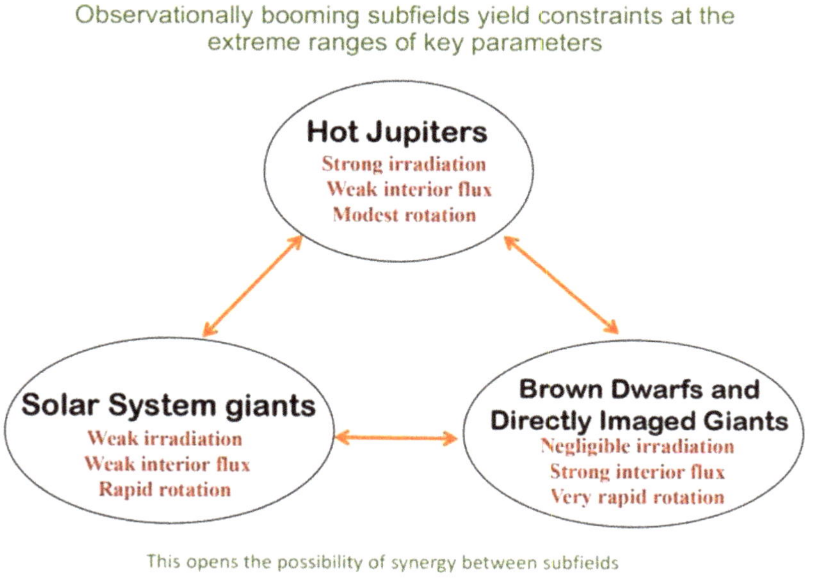


**Figure 19.2.** Three distinct fields providing important constraints on atmospheric structure and circulation of giant planets—hot Jupiters, solar system giant planets, and brown dwarfs—together spanning a wide range of physical properties.

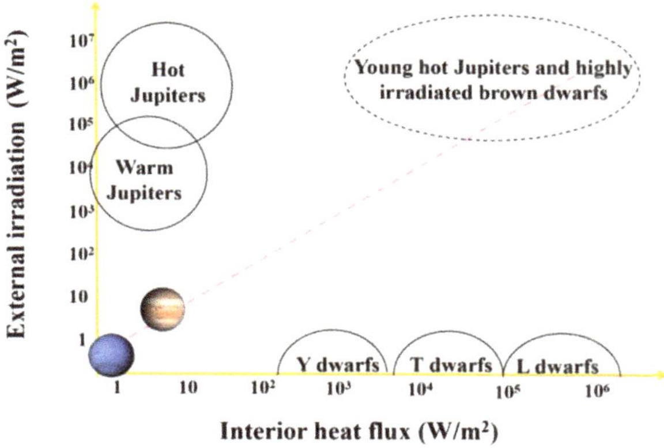


**Figure 19.3.** Regime diagram of various giant planet populations as a function of interior heat flux (abscissa) and global-mean heat flux received from the star (ordinate). Solar system giant planets lie in the lower left, corner, hot Jupiters in the upper left, corner, and brown dwarfs in the lower right, corner. A handful of highly irradiated brown dwarfs are also providing information and reside in the upper right, corner. Observational constraints from these diverse populations place tight constraints on how fundamental theories for the atmospheric circulation scale with these heat fluxes over many orders of magnitude. (From Showman et al. 2020)

enormous interior heat fluxes (at least for the hot, massive planets we are able to characterize), and are rapidly rotating—in many cases faster than Jupiter.

All three of these subfields (Figure 19.2) are yielding a wealth of new observations, and provide constraints on the dynamical behavior at the extreme ranges of these key parameters. Taking advantage of recent advances in all three of these fields opens the possibility of synergy between subfields.

To illustrate this range more concretely, Figure 19.3 shows these diverse populations as a. function of the incident stellar heat flux and the internal (convected) heat flux lost from the interior. Jupiter, Saturn, and Neptune lie near the lower left corner, with external and interior heat fluxes that are comparable, and very low. Hot Jupiters reside in the upper left corner, with large external and smaller interior heat fluxes. Brown dwarfs lie in the lower right corner, with large interior fluxes and typically negligible external fluxes. Finally there exists a small population of highly irradiated brown dwarfs for which the external and internal fluxes are both large, and comparable to order of magnitude. This latter population includes both brown dwarfs orbiting main-sequence stars (such as KELT-lb; Siverd et al. 2012) at distances typical of hot Jupiters, but also massive brown dwarfs orbiting on ~l–3-hour orbits around white dwarfs (e.g. Steele et al. 2013). Continued characterization of all four of these populations is critical, and will provide tight constraints on how the atmospheric circulation varies with these heat fluxes over many orders of magnitude.

### 19.2.2 Dynamical Mechanisms Governing Atmospheric Circulation and Climate

Perhaps the key goal in studying atmospheres is not only to characterize observationally their structure, or even to have models that match the observed behavior, but to *understand* the interplay of physical and chemical mechanisms that act to cause the behavior we see. Obtaining such an understanding requires a detailed program of modeling and theory to explore a hierarchy of models spanning wide variation in physical, dynamical, and chemical complexity, and that explore the response when various effects are turned on or off in any given class of model. In atmospheric dynamics, such models can range from sophisticated 3D full nonlinear general circulation models (GCMs) of the atmospheric circulation that include state-of-the-art prescriptions for dynamics, radiative transfer, cloud physics (if relevant), and other processes—to more idealized 3D GCMs that still capture the full dynamics but represent one or more processes (e.g., radiative transfer) using simplified means—to models with idealized dynamics, including models that filter various classes of wave modes (such as sound waves or gravity waves), as well as models that simplify the vertical structure (such as the one-layer shallow-water system); see e.g. reviews listed at the beginning of this chapter. Linearized versions of all these 3D and shallow-water systems provide great insight that complements our understanding of the full nonlinear response, and in idealized settings such linear calculations can be performed analytically. Note that within this hierarchy it is misguided to view any of these classes of models as “superior” to another; generally, one can obtain a deep understanding of a given phenomenon only when it has been studied robustly using a wide range of models across this hierarchy. This “modeling hierarchy” approach is a standard technique that has been used for decades to investigate Earth’s atmospheric circulation and climate, but the approach is still greatly underutilized in the context of exoplanet atmospheres.

Thus as observations of exoplanets are obtained, and their 3D distributions of temperature, wind, cloudiness and other properties are better understood, numerous puzzles will surely arise that call for deep theoretical work to understand the underlying mechanisms. Although our field is observationally driven, this theoretical program will represent a major challenge for exoplanet atmosphere research over the next 10–20 years and should not be neglected, for only after such robust modeling/theoretical hierarchies are explored can we claim any true understanding of the phenomena we observe. Below we list a few major areas for theoretical exploration at the present time, a picture that will need to be revisited in 5–10 years as better data are obtained in the JWST era.

- Equatorial superrotation and atmospheric circulation: As discussed in Section 19.2.1, theoretical work has identified basic dynamical mechanisms (e.g. Showman & Polvani 2011; Tsai et al. 2014; Cho et al. 2015; Mayne et al. 2017), but numerous questions remain about what controls the speed to which the jet equilibrates, the depth to which it extends, its role in advecting tracer species, and possible interaction with other dynamical processes occurring in the atmosphere. Superrotation may occur under only a portion of the transiting exoplanet parameter space, and the conditions under which it does—or does not—occur need to be better investigated. More broadly, numerous questions remain about the overall circulation regime on tidally locked planets, and how it depends on stellar irradiation, planetary rotation rate, gravity, atmospheric composition, and other parameters.

- Spatial temperature structure, including day–night temperature differences: Finite nightside temperatures on synchronously rotating planets indicate the transport of heat from the dayside, and the dynamics controlling the heat transport and the resulting day–night temperature difference need to be better investigated. Theories exist (Perez-Becker & Showman 2013; Komacek & Showman 2016), but additional work is needed. In particular we need to determine the extent to which observed trends of day–night temperature differences (versus incident stellar flux and other parameters) can be quantitatively explained, how the day–night temperature difference varies with depth, and the role of rotation rate, clouds, differing atmospheric composition, and other processes in influencing the longitudinal temperature variations.
- What are the dynamical processes that control vertical and horizontal (e.g., day to night) mixing of cloud/haze particles and gaseous tracers on exoplanets? We need to understand the processes that cause vertical mixing, and determine how the mixing efficiency scales with incident stellar flux, planetary rotation rate, and other parameters (e.g. Parmentier et al. 2013; Drummond et al. 2018; Helling et al. 2019). The mixing behavior may differ greatly depending on where the particles are produced (e.g., deep in the atmosphere versus at high altitudes) and their spatial pattern of production (e.g., production everywhere over the globe versus on the dayside only).
- More poorly irradiated planets (e.g., warm and cool Jupiters and super-Earths): Planets in this regime will not be tidally locked and will have a circulation driven primarily by the equator-to-pole gradient in zonal-mean heating, rather than by the day–night heating gradient as occurs on hot Jupiters. This population of exoplanets greatly outnumbers the closer-in planets, and although they are harder to observe, this observational frontier will open up over the next decade. Understanding the expected dynamical behavior in this regime is therefore critical.
- Regime transitions: The vast parameter space occupied by exoplanets implies that exoplanet atmospheric circulation and climate will encompass a variety of fundamentally different regimes, each governed by differing dynamics. Even among the hottest population of gas-giant planets where observations are best, several such transitions have already been suggested (see e.g. Ohno & Zhang 2019; Showman et al. 2020), such as a transition from strong day–night flow at extreme irradiation levels to a more banded flow at more "typical" hot-Jupiter irradiation levels, and from a flow dominated by day–night forcing—leading to equatorial superrotation—to a flow dominated by equator-to-pole heat transport and multiple mid-to-high latitude jets for more weakly irradiated planets (e.g., beyond ~0.1–0.2 au for a typical Sun-like star). The regime of Titan and Venus differs greatly from that of Earth and Mars, and this transition is also likely present among exoplanets. Significantly more work is needed before the full scope of possible circulation regimes is understood. Such work is important not only for the fundamental insight it supplies into the underlying dynamics, but will also yield observational

predictions, as there may be changes in observable properties that occur when regime transitions are crossed.

### 19.2.3 Climates of Terrestrial Planets

The atmospheres of terrestrial planets are far harder to observe than those of gas giants, but because of the interest in habitability and life, understanding their climate is nevertheless critical. In principle a better understanding of their behavior may also help to place our understanding of Earth, Venus, Mars, and Titan into context as specific examples among a wide range of possibilities. This applies both to the mechanisms maintaining global-mean climate (largely controlled by radiative processes) as well as dynamical regimes controlling the circulation (see e.g. Showman et al. 2013b and Read et al. 2018 for recent reviews on this topic).

A key goal is to better understand the global-scale feedbacks that control climate on terrestrial planets. This includes transitions from temperate, Earth-like climate regimes to fully glaciated snowball states (wherein the uppermost km or more of the ocean globally freezes) and transitions to runaway greenhouses of the type that may have occurred for Venus. We also need to better understand long-term surface/atmosphere feedbacks like the carbonate-silicate feedback cycle, which is a negative feedback thought to be critical in having helped Earth remain habitable over billions of years despite a gradually brightening Sun. In particular how this feedback behaves on planets differing from Earth—for example planets with global oceans lacking any continents standing above-water—is unclear and needs further work. Long-term exchange of volatiles such as nitrogen, carbon, and water between the mantle and atmosphere/ocean system is critical in determining the equilibrium mass and composition of a planet's atmosphere and/or ocean, and significant more research is needed in these areas.

We need to better understand how these feedbacks interact with the circulation in 3D to control the mean climate. Much of the prior work on global-scale climate feedbacks has been in the context of 1D radiative–convective–equilibrium climate models, which will continue to be useful, but this provides only a partial look at how these feedbacks behave. For example, the ice-albedo feedback that is responsible for causing transitions to the snowball regime depends on the temperature structure near the poles, which in turns depends sensitively on the latitudinal heat transport by the atmospheric circulation. Otherwise identical models that have differing rates of atmospheric heat transport (due to different planetary rotation rates or other factors) exhibit transitions to snowballs under different conditions. Likewise, transitions to runaway greenhouse atmospheres depend on the horizontal distribution of temperature, humidity, and clouds in a way that remains incompletely understood. Water–rock interactions at the base of the ocean depend on ocean temperature and chemistry, which in turn depend on the atmospheric circulation and the way it communicates with the overlying atmosphere. Careful numerical work with full 3D models that resolve the dynamics can lead to significant new insights on these interactions.

We need to understand the extent to which climates are stable or unstable, whether multiple equilibria exist, and the conditions needed to switch between these multiple equilibria. This is related to the topics described above—for example, under certain conditions an Earth-like planet receiving the same stellar flux can potentially be in temperate, relatively ice-free states or ice-covered snowballs—but other feedbacks besides the ice-albedo feedbacks may allow the existence of additional multiple equilibria that have not yet been identified or explored.

As our understanding of the various climate processes improves, we need to continually reassess the implications for the width of the habitable zone around various classes of stars; this will likely have important implications for observational searches for terrestrial planets, as well as which planets are prioritized for observational follow-up.

At the end of the day, the long-term goal is to understand the actual climates on various discovered exoplanets, and whether these exoplanets might be habitable.

### 19.2.4 Interior Dynamics and Atmosphere-Interior Interactions

We now turn to the interior dynamics and atmosphere—interior interactions of fluid exoplanets. Because giant planets are fluid throughout, and exhibit a continuous gradation of the atmosphere into the interior, the key issues differ fundamentally from the corresponding issues of atmosphere/interior interactions on terrestrial planets.

Like almost all giant planets, hot Jupiters are expected to have convective interiors, but the strong stellar irradiation generates a stably stratified layer that can extend as deep as 1000 bars. The day—night heating contrast and strong atmospheric circulation lie within this stratified layer. Because the deep radiative zone (from ∼10–1000 bars) has a long radiative time constant, and is not directly subjected to strong insolation gradients, the overturn timescales inside this layer are expected to be extremely long. As a result, many studies have assumed that the layer is relatively quiescent and provides a barrier of sorts that discourages direct dynamical interaction of the deeper convection zone with the atmospheric circulation in the overlying photospheric layers (e.g. Showman et al. 2009, Heng et al. 2011, Rauscher & Menou 2012, Mayne et al. 2017). Nevertheless, this assumption has not been well tested; current GCMs have difficulty accurately simulating this deep radiative layer, because the timescales required for this layer to reach equilibrium in a GCM simulation are far longer than computational resources typically allow.

Significant work is needed to understand the possible types of flows that may exist in the convective interior, and how they may (or may not) couple to the atmospheric flows driven by the day-night heating contrast near photosphere levels. A fundamental question is whether the interiors of hot Jupiters differentially rotate, that is, whether they exhibit zonal flows, and if so, to understand the typical wind speed, meridional scale (e.g., latitudinal jet width), and vertical structure of these flows. Numerical simulations of convection in spherical shells, and analogy with Jupiter and Saturn, would suggest that hot-Jupiter interiors may well have alternating zonal flows that extend into the interior, but no models of this type that emphasise the hot-Jupiter regime have yet been performed. These dynamics are linked to those causing the convective dynamo that drive planetary magnetic fields, so future observations providing constraints on hot-Jupiter magnetic fields may yield insights into the interior flow dynamics. More generally, we seek to understand how the dynamics influences the structure of the radiative zone. This class of questions has to date received little attention in the literature; attacking it will require use of a new generation of coupled interior/atmosphere models run on modern supercomputers.

It is also worthwhile to investigate the interior/atmosphere interactions of mini Neptunes and super Earths whose interiors are largely comprised of higher-molecular-weight materials such as water. Uranus and Neptune provide solar system analogs, and exhibit the interesting property that their interiors are a mix of "icy" volatiles (fluid, partially ionised water, ammonia, and other species) while their atmospheres are predominantly hydrogen. This compositional stratification suggests the possibility that convective mixing is inhibited at the interface. This problem is not well understood for Uranus and Neptune, calling for new modelling and theoretical work. This type of investigation will naturally have major implications for the conditions under which such compositional stratification can occur more generally. Exoplanet observations can constrain this problem, since mean planet density provides constraints on interior composition, while transit and secondary eclipse spectroscopy provide constraints on atmosphere composition.

## 19.3 Opportunities and Challenges

Most of the progress will come in the area of observations; this has been generally true in planetary sciences over the past 40 years and will continue to be true in exoplanets, indeed perhaps even more so. A fundamental question concerns the extent to which observational resources should be focused on a small number of benchmark planets (such as HD 189733b, HD 209458b, and WASP-43b) versus being spread more thinly over a wider range of planets. Recent experience has highlighted that for understanding the atmospheric circulation, the broad array of observations obtainable for benchmark planets is absolutely critical to building a robust understanding—including full-orbit light curves at a variety of IR band-passes, secondary eclipse and transit spectra from the ultraviolet through mid-IR, Doppler constraints on wind speeds at the planet's terminator, dayside thermal maps from secondary-eclipse ingress/egress mapping, and constraints (upper limits) on secondary eclipse variability over time. Each of these observations place constraints on complementary aspects of the atmospheric dynamics. If any major aspect of them is missing for the benchmark planets, it would significantly hinder our ability to pin down how the circulation operates on those worlds.

At the same time, the hot-Jupiter population embodies an enormous variation in parameters like incident stellar flux, rotation (orbital) period, and atmospheric composition, and the atmospheric circulation of exoplanets likewise will vary significantly across this regime. Thus to understand trends—such as how the day–night temperature difference, longitudinal offset of the dayside hot spot, existence of clouds/hazes, and vertical temperature structure depend on external stellar irradiation, planetary rotation (orbital) period, and other parameters—we need good observations across a wide diversity of planets. This has already started to bear fruit, particularly in understanding the dynamics governing the day–night temperature differences (e.g. Perez-Becker & Showman 2013; Komacek & Showman 2016). Obtaining good observations across a wide range of exoplanets—for purposes of cleanly identifying trends—will require significant observational resources, but is going to be absolutely critical over the next 10–20 years.

There is a strong need for dedicated spacecraft missions beyond JWST to study exoplanet atmospheres. JWST will be an outstanding platform for exoplanet atmosphere observation, but progress on exoplanet atmosphere characterization will likely be bottlenecked by the fierce competition for time, given that JWST is one only telescope serving the entire astronomy/astrophysics community. For observing the hotter exoplanets (e.g., hot Jupiters and super-Earths), large telescope collecting area is less critical than extreme pointing stability of the spacecraft, and modest-cost missions specifically designed for extreme stability with mirror diameters in the range ~1–2 m can do far better than the Hubble or Spitzer Space Telescopes. ARIEL, recently selected by ESA, will play a critical role.

On the modeling and theory side, there is need in GCMs for continued development of schemes for radiative transfer, cloud and haze formation, coupling to chemistry, and dynamical cores that can handle sounds waves, shocks, and allow for subgrid scale parameterizations. Coupling to magnetic fields will be critical for the hottest of the hot Jupiters, and is another area for model development. In each of these areas, the goal is not simply to develop the most sophisticated scheme possible—such models will be computationally expensive and their results difficult to understand in detail—but rather a range of schemes varying from simple (hence fast and easy to understand, but sacrificing realism) to complex (slower and harder to understand, but more realistic). More realistic does not imply better. Stated another way, the concept of a "modeling hierarchy" commonly used in GFD should be extended to these various modeling components as well.

As models build up, inter-model comparisons will become increasingly important. These have become a standard exercise in the Earth climate community, but in exoplanet atmospheres, little effort has to date been devoted to this issue. Part of the goal is to test differing numerical approaches of distinct codes solving essentially similar equations, but part of the goal is to better understand the extent to which so-called "reduced" equation sets, such as the primitive equations, capture (or not) various aspects of the dynamics. The intercomparison approach will become particularly critical once many different codes include schemes for clouds/hazes, radiative transfer, and chemistry, since these parameterization schemes involve many modeling choices, and simulation results may be affected by parameter choices in these schemes that are not obvious at first glance.

There will be a continued need to simulate a broad range of exoplanets using a broad range of different codes, especially as observations become better, trends become apparent, and more sophisticated models come online. This should include efforts to explore the deep atmosphere and interactions with the interior.

Computational resources will continue to be a major challenge to simulating the deep atmosphere especially, but this will (hopefully) be surmounted over the next decade by a combination of faster computers, numerical solvers better suited to modern computer architecture, and the introduction of solution techniques that attempt to iteratively determine the statistical equilibrium structure (e.g., of the deep atmosphere) without having to directly simulate it using a time-stepping algorithm. Idealized models to identify dynamical mechanisms is an area that has received far too little attention. Most progress in true understanding will come from this area.

Editor's note: This chapter is published in honor and fond memory of the author Prof. Adam Showman who passed away while this book was in preparation. Prof. Showman pioneered the study of atmospheric dynamics of exoplanets, with numerous seminal contributions, and was a great visionary, colleague, mentor and friend to many in the field of exoplanetary science. We thank his family for facilitating his manuscript to be edited and published in the present book.

## References

Cho, J. Y.-K., Polichtchouk, I., & Thrastarson, H. Th. 2015, MNRAS, 454, 3423
Cowan, N. B., & Agol, E. 2011, ApJ, 729, 54
Crossfield, I. J. M., & Kreidberg, L. 2017, AJ, 154, 261
de Wit, J., Gillon, M., Demory, B.-O., & Seager, S. 2012, A&A, 548, A128
Dobbs-Dixon, I., & Agol, E. 2013, MNRAS, 435, 3159
Drummond, B., Mayne, N. J., Manners, J., et al. 2018, ApJ, 869, 28
Haynes, K., Mandell, A. M., Madhusudhan, N., Deming, D., & Knutson, H. 2015, ApJ, 806, 146
Helling, C., Iro, N., Corrales, L., et al. 2019, A&A, 631, A79
Heng, K., Menou, K., & Phillipps, P. J. 2011, MNRAS, 413, 2380
Heng, K., & Showman, A. P. 2015, AREPS, 43, 509
Kataria, T., Showman, A. P., Lewis, N. K., et al. 2013, ApJ, 767, 76
Knutson, H. A., Charbonneau, D., Allen, L. E., et al. 2007, Natur, 447, 183
Komacek, T. D., & Showman, A. P. 2016, ApJ, 821, 16
Komacek, T. D., Showman, A. P., & Tan, X. 2017, ApJ, 835, 198
Line, M., Stevenson, K. B., Bean, J., et al. 2016, AJ, 152, 203
Louden, T., & Wheatley, P. J. 2015, ApJ, 814, L24
Majeau, C., Agol, E., & Cowan, N. B. 2012, ApJ, 747, L20
Mayne, N., Debras, F., Baraffe, I., et al. 2017, A&A, 604, 79
Mikal-Evans, T., Sing, D. K., Goyal, J. M., et al. 2019, MNRAS, 488, 2222
Miller-Ricci Kempton, E., & Rauscher, E. 2012, ApJ, 751, 117
Ohno, K., & Zhang, X. 2019, ApJ, 874, 1
Parmentier, V., Showman, A. P., & Lian, Y. 2013, A&A, 558A, 91P
Perez-Becker, D., & Showman, A. P. 2013, ApJ, 776, 134
Pierrehumbert, R. T., & Hammond, M. 2019, AnRFM, 51, 275
Rauscher, E., & Menou, K. 2012, ApJ, 750, 96
Rauscher, E., Suri, V., & Cowan, N. B. 2018, AJ, 156, 235
Read, P. L., Lewis, S. R., & Vallis, G. K. 2018, in Handbook of Exoplanets (Berlin: Springer), 50
Showman, A. P., & Guillot, T. 2002, A&A, 385, 166S

Showman, A., Fortney, J. J., Lian, Y., et al. 2009, ApJ, 699, 564
Showman, A., Cho, J. Y.-K., & Menou, K. 2010, in Exoplanets, ed. S. Seager (Tucson, AZ: Univ. Arizona Press), 471
Showman, A. P., & Polvani, L. M. 2011, ApJ, 738, 71
Showman, A., Fortney, J. J., Lewis, N. K., & Shabram, M. 2013a, ApJ, 762, 24
Showman, A. P., Wordsworth, R. D., Merlis, T. M., & Kaspi, Y. 2013b, in Comparative Climatology of Terrestrial Planets, ed. S. J. Mackwell, A. A. Simon-Miller, J. W. Harder, & M. A. Bullock (Tucson, AZ: Univ. Arizona Press), 277
Showman, A. P., Tan, X., & Zhang, X. 2019, ApJ, 883, 4 arXiv:1807.08433
Showman, A., Tan, X., & Parmentier, V. 2020, SSRv, 216, 139
Sing, D., Fortney, J. J., Nikolov, N., et al. 2016, Natur, 529, 59
Siverd, R. J., Beatty, T. G., Pepper, J., et al. 2012, ApJ, 761, 123
Snellen, I. A. G., de Kok, R. J., de Mooij, E. J. W., & Albrecht, S. 2010, Natur, 465, 7301
Steele, P. R., Saglia, R. P., Burleigh, M. R., et al. 2013, MNRAS, 429, 3492
Stevenson, K. B., Désert, J.-M., Line, M. R., et al. 2014, Sci, 346, 838
Tsai, S.-M., Dobbs-Dixon, I., & Gu, P.-G. 2014, ApJ, 793, 141
Zellem, R. T., Lewis, N. K., Knutson, H. A., et al. 2014, ApJ, 790, 53

AAS | IOP Astronomy

ExoFrontiers
Big questions in exoplanetary science
Nikku Madhusudhan

# Chapter 20

# Clouds in Exoplanetary Atmospheres

**Christiane Helling**

## 20.1 Introduction

Today, we know over 4300 exoplanets orbiting their host stars in over 3200 planetary systems.[1] The diversity of these exoplanets is huge, and none of the known exoplanets is a twin to any of the solar system planets, nor is any of the known extrasolar planetary systems a twin of the solar system. Such diversity on many scales and structural levels requires fundamental theoretical approaches. Large efforts are underway to develop fundamental theoretical approaches for individual aspects of exoplanet sciences, like exoplanet atmospheres, cloud formation, disk chemistry, planet system dynamics, mantle convection, and mass loss of planetary atmospheres.

The diversity of exoplanets calls for a fundamental, for an interdisciplinary and for an intelligent approach in order to fully understand their atmospheric composition. Many harbor a dynamic atmosphere where cloud particles can form in abundance. Some planets will be fully covered in clouds, some have clouds on the nightside but are largely cloud-free on the dayside. The basic processes of cloud formation are understood, and the main challenge lays in the fundamental physics and chemistry of material properties. The simultaneous development of different cloud models (e.g., Helling & Woitke 2006; Kawashima & Ikoma 2018) has enabled great insight into individual processes that now need to be combined. Such a combination requires an intelligent approach for tackling the micro-physical complexity involved. This includes the modeling of cluster formation, particle-particle growth/destruction, ionization and photo-processes, diffusion and turbulent transport consistently coupled. Cloud models are a key part of the virtual laboratories that enable us to study the large parameter range of exoplanets, and to decipher the combination of individual observations.

Several of the known exoplanets orbit their host star very closely such that a rocky surface turns into magma ocean (e.g., CoRot-7b or 55 Cnc e) or Jupiter-like exoplanets

[1] http://exoplanet.eu

doi:10.1088/2514-3433/abfa8fch20  

expand their atmosphere considerably (e.g., HD 189733b, HD 209456b) and asymmetrically (e.g., HAT-P-7b, WASP-18b). Inside the atmospheres of exoplanets, clouds form. The cloud particles (*aerosols*) are made of a mix of minerals, for example, silicates and iron oxide in giant gas planets (Helling 2019); and sulfur hazes (Gao et al. 2017). The global cloud composition changes with changing element abundance which can be affected by outgasing (e.g., for 55 Cnc e; Mahapatra et al. 2017) and pre-determined by the location of planet formation within the disk (Helling et al. 2014; Cridland et al. 2019; Gravity Collaboration et al. 2020). The composition of the upper crust of a rocky planet does determine the atmospheric element abundances, and eventually, the stability of liquid water (Herbort et al. 2020). Hydrocarbon hazes have been shown to form in the upper atmospheres of super-Earths and mini-Neptunes (e.g., GJ 1214b and GJ436b; Kawashima et al. 2019) under the effect of the UV radiation field of the M-dwarf type host stars. Similar effects have been proposed for the dayside of the giant gas planet WASP-43b where the UV photons of a K star enable the formation of hydrocarbon hazes in tandem with mineral cloud particles (Helling et al. 2005). Spectroscopic studies of exoplanets are repeatedly frustrated by clouds and/or hazes blocking the view into the underlying atmosphere to reveal the existence of potential biosignatures, signature of their evolutionary state, or traces of planet formation. Clouds on Earth are linked to the support of the terrestrial cycle of life as they can preserve the right temperature of the biosphere and enable the transport of water through the atmosphere. Can we expect exoplanet clouds to be of similar significance?

The following challenges need to be addressed in order to answer this question, and, simultaneously provide the opportunity to progress our understanding of exoplanets and their atmospheres by exploring our models as virtual laboratories to fill gaps in observational data from different instruments and missions, and taken at different instances of time:

Challenge (a) Building *complex models based on theoretical rigor* that aim to understand the interactions of atmospheric processes, to treat cloud formation and its feedback onto the gas-phase chemistry and the energy budget of the planetary atmosphere moving away from solar-system inspired parameterizations.

Challenge (b) Enabling *cloud modeling based on fundamental physio-chemical insights* in order to be applicable to the large and unexplored chemical, radiative and thermodynamical parameter range of exoplanets in the Universe. Challenge (b) will be explored in what follows.

## 20.2 State of the Art—Cloud Formation Modeling

The results of atmosphere models (Fortney et al. 2013; Juncher et al. 2017; Witte et al. 2009; Showman et al. 2009; Dobbs-Dixon & Agol 2013; Mayne et al. 2014; Carone et al. 2015; Lee et al. 2016; Mendonça et al. 2016), or retrieval approaches (Madhusudhan & Seager 2011; Barstow et al. 2017; Blecic et al. 2017; Min et al. 2020) will depend on their individual components, their ability to model physics and chemistry and to solve it numerically. Cloud formation modeling in Earth and solar system atmosphere modeling, e.g., Rossow (1978); Hellmuth (2006), has a long tradition. It is essential that different modeling approaches are followed in order to enable a benchmarking process. Cloud formation modeling is also a key component in atmosphere modeling and retrieval for exoplanets where the atmospheric gas is chemically and dynamically very different to what is known from our solar system. Cloud particles are tracers of the gas from which they form, through which they move, which they deplete or enrich with respect to elements abundances, and which they heat or cool due their opacity. In order to understand the chemical composition of exoplanet atmosphere, to link this composition to their evolutionary state by interpreting observations from CHEOPS, HST, JWST, Ariel, PLATO, etc., cloud formation needs to be understood in enough detail to be applicable to the diversity of known exoplanets. Such a cloud model needs to be applicable to the wide ranges of thermodynamics, radiation, and chemical regimes that exoplanets have presented us with. Here we lay out the challenge of building such a cloud model, starting from our kinetic modeling philosophy. This is not a speedy exercise and the retrieval modeler may therefore still wish to hold on to representing a cloud by a set of parameters like particle size, or by reducing complexity by keeping physical consistency (Ormel & Min 2019).

### 20.2.1 Basic Processes to Form a Cloud Particle

**Seed formation:** Unless an exoplanet has dust particles or other condensation seeds swapped into its atmosphere (e.g., condensing meteoritic dust, volcano outbreaks, sand storms), the formation of a cloud particles begins in the gas phase by chemical reactions of those atoms/molecules/ions that are available in the atmosphere of a certain element composition. The element composition and the local thermodynamic conditions determine the first condensates. The formation of condensation seeds can be understood as a chain of chemical gas–gas reactions that progresses to molecules and clusters with increasing size and complexity. The challenge is to determine the required thermodynamic cluster data, the free energy change associated with the formation of a cluster of size $N$, $\Delta G(N)$. Assuming that no activation energy is required, one can formulate a rate network for all cluster sizes (Patzer et al. 1998) or approximate $\Delta G(N)$ based on the concepts of classical nucleation theory (Lee et al. 2018). If cluster formation requires an activation energy, all individual chemical reaction will need consideration. Such fundamental concepts enable the modeling of the formation of mineral and organic seeds for a wide range of chemical situations if the required material data (e.g., $\Delta G(N)$) are available. Further chemical paths open up in highly irradiated planets where photochemistry leads to the formation of hydrocarbon molecules of increasing complexity after carbon is released by photochemically destroying CO or $CH_4$ (Kawashima & Ikoma 2018; Rimmer et al. 2014). Two chemically independent ensembles of seed particles maybe expected in highly irradiated exoplanets, as shown for WASP-43b (Helling et al. 2005). It needs to be understood, however, that a seed formation rate will depend on how much elements are consumed by further bulk growth.

**Bulk growths:** Once seed particles are present, the gas phase can easily condense on these surfaces because non-photochemically-driven seed formation requires a supercooling of the gas phase which is then highly supersaturated. The kinetic bulk growth of cloud particles of mixed composition by gas-surface reactions can efficiently be modeled through a moment method that provides consistent information about cloud particle composition, particles sizes and cloud extension (Helling & Woitke 2006). A challenge are the details of the hundreds of surface reactions that remain largely unexplored by laboratory studied. A further process of increasing particles sizes is coagulation, i.e., the collisional growth of existing cloud particles. This can occur for hydrocarbon monomers or for mineral cloud particles if the relative velocities are appropriate. Collisions between cloud particles may also lead to shattering causing a decrease of the mean particle size but an increase in particles number (Güttler et al. 2010). The arising challenge is that binning methods work well for solving Smoluchowski's coagulation equation in a mass conserving scheme but are challenged by particles of mixed materials. Modeling the kinetic surface growth and the collisional growth of existing dust particles furthermore requires to determine the cloud-particle velocity in potentially different frictional regimes.

**Gravitational settling, frictional coupling, turbulent mixing and diffusion:** Cloud particles move. Frictional coupling of cloud particles to the gas phase determine how far cloud particles of certain sizes and masses fall into the deeper atmospheric layers (and hence, keep growing or begin to evaporate). They stand still (relative to the surrounding gas) in the gas of complete friction (i.e., position) coupling with the gas. Frictional coupling does also determine in how far turbulent gas motion affects their further growth or even their ionization. In case of complete coupling, turbulence will cause the formation of intermittent cloud structures due to the emergence of vorticity (Helling et al. 2004). Diffusion is often used as overarching term for transport against the atmospheric pressure gradient and turbulent mixing as a driver of such diffusion. Per definition, diffusion is driven by concentration gradients of gas-species but also of cloud particles. Each cloud-particle size would diffuse differently. Incorporating cloud-particle diffusion in cloud models has proven to be challenging, not the least because cloud-particle diffusion coefficients are unknown (Woitke et al. 2020).

**Neutral gas-phase chemistry:** Modeling cloud formation requires knowledge about the local thermodynamic conditions, including the chemical composition for a set of element abundances. This, in turn, will depend on the presence of a condensed phase (Herbort et al. 2020). The most complete set of information about the gas-phase composition for almost all elements that we know from the Sun's atmosphere (including O, Mg, Fe, Si, Ca, Na, Al, Ti, $\cdots$) is provided by chemical equilibrium calculations (e.g., Woitke et al. 2018). Large gas-kinetic or photochemical calculations remain limited to molecules made of C/H/N/O and S and can, because of their complexity, be complete only unto a certain number of elements per molecule (Rimmer & Helling 2016). C/H/N/O are used to study the formation of hydrocarbon hazes and the triggering of possibly biology related molecules like HCN. Such constrained chemical networks are very efficient for the study of specific processes, like for example, lightning chemistry (Rimmer & Helling 2016) and biomolecules on the surface of charged cloud particles (Stark et al. 2014). The challenge is far more fundamental, namely, to build kinetic rate networks that link to the need of the nucleation modeling outlines above and to model spectacular events like lightning in the exoplanet context.

### 20.2.2 Ionization Process

Atmospheres of exoplanets can be highly irradiated such that the atmospheric gas is photoionized and/or ionized thermally. The ionization efficiency may vary from day to nightside leading to a highly ionized and cloud-free dayside and a cloudy nightside with just an ionosphere on top on hot rocky planets like 55 Cnc e (Mahapatra et al. 2017) as on super-hot Jupiters (Helling et al. 2019). The photoionization can reach the upper cloud layers, but a more realistic process is that currents from the ionized atmosphere form and transport charges onto the surface of the cloud particles, similar to Earth. Figure 20.1 demonstrate that for the ultra-hot Jupiter HAT-P7b the day/night difference of the thermal degree of ionization (color coded). An ionosphere is extending deep into the dayside atmosphere but remains rather shallow on the nightside. The nightside, in contrast is covered in clouds (solid black lines show the

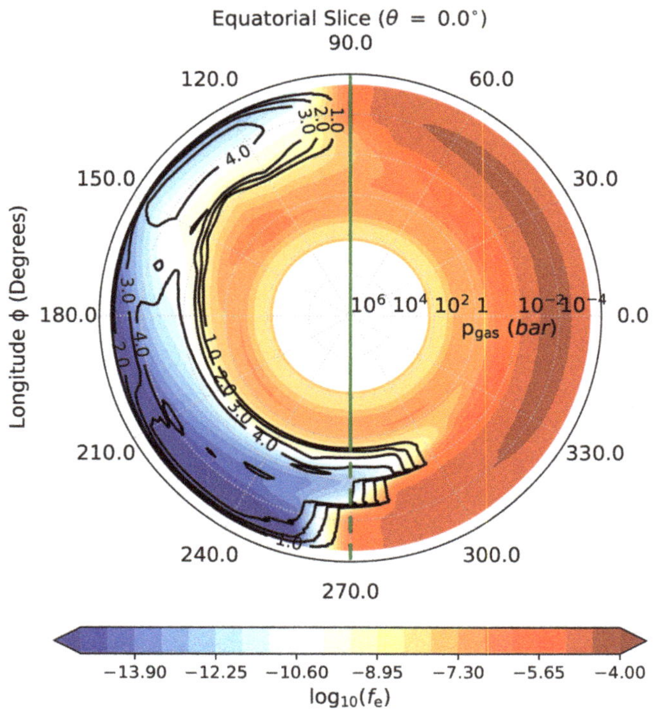


**Figure 20.1.** The exoplanet atmosphere challenge: ultra-hot Jupiters combine an ultra-hot dayside forming an ionosphere with a high degree of ionization (thermal degree of ionization, $f_e$—blue–yellow–red color coded) with a cloudy nightside ($\rho_d/\rho_{gas}$—black surface contour lines), shown here for HAT-P-7b as slice plot through the equatorial plane. The atmospheric pressure levels are shown as concentric rings. Credit: courtesy to D. Samra.

cloud-particle mass loss in term of the dust-to-gas mass density ratio) that also asymmetrically cover the two terminator regions.

## 20.3 Opportunities

Understanding if the atmosphere of an exoplanet enables biological activity requires us to understand the interaction of every-day processes that occur on Earth in a far larger parameter range than available on Earth. This is why laboratory experiments on Earth are unfortunately only of limited value for exoplanet research in its whole diversity of exoplanet conditions. The exoplanet parameter range is increased by being linked to the evolutionary state of the planet, and its formation history. We therefore require complex models to study in detail the physico-chemistry of exoplanet atmospheres; what is more, we require several of those models to be developed in parallel to be able to follow the fundamental research ethics of testing and comparing results by different means and approaches. The use of only one model defeats the purpose of research and the necessary falsification of results.

Building virtual laboratories by virtue of complex models is the only path to understand the vast exoplanet diversity in depth, and to decode observations that can only be taken for selected times (or time intervals) and for selected wavelengths (or wavelength bands) for selected objects. Monitoring exoplanets of different kinds, that orbit different stellar types, in wavelengths and in time is unfeasible in the foreseeable future. The purpose of such monitoring could be as far-fetched as finding a proper landing side or as fundamental as understanding the different planets as global objects. The EAS Voyage 2050 white papers on exoplanet research remain focused by necessity, including spectropolarimetry (Fossati et al. 2019), spatial resolved mid-IR of thermal emission (Quanz et al. 2019), and high-contrast imaging (Snellen et al. 2019) missions as next steps beyond JWST, Ariel, and the ELT-class telescopes. Modeling the underpinning exoplanet atmosphere physics and chemistry will be essential. Virtual laboratories already enable us to explore exoplanet atmosphere with respect to specific processes (like the effect of cloud formation), or as the sum of all known processes (like the emerging of weather pattern), and in particular beyond the singled-out terrestrial conditions (e.g., for a diversity of element abundances).

## 20.4 Challenges

Outstanding challenges to enable virtual laboratories to meet the observational challenge include obtaining:

- the element abundances that determine the planet's atmosphere chemistry
- chemical and cluster data for all possible nucleation species
- thermochemical data for reaction rates for elements other than C/N/O/H/(S)
- thermochemical data for gas-surface reactions, including sticking probabilities
- opacity data for condensates across large wavelengths ranges and crystalline species
- opacity modeling for non-spherical and charged cloud particles.

Meeting these challenges will enable us to decipher the evolutionary state and the planet's formation mechanisms spectroscopically, and to determine with confidence the state of habitability of an extrasolar planet.

## References


Barstow, J. K., Aigrain, S., Irwin, P. G. J., & Sing, D. K. 2017, ApJ, 834, 50
Blecic, J., Dobbs-Dixon, I., & Greene, T. 2017, ApJ, 848, 127
Carone, L., Keppens, R., & Decin, L. 2015, MNRAS, 453, 2412
Cridland, A. J., van Dishoeck, E. F., Alessi, M., & Pudritz, R. E. 2019, A&A, 632, A63
Dobbs-Dixon, I., & Agol, E. 2013, MNRAS, 435, 3159
Fortney, J. J., Mordasini, C., Nettelmann, N., et al. 2013, ApJ, 775, 80
Fossati, L., Rossi, L., Stam, D., et al. 2019, BAAS, 51, 33
Gao, P., Marley, M. S., Zahnle, K., Robinson, T. D., & Lewis, N. K. 2017, AJ, 153, 139
Güttler, C., Blum, J., Zsom, A., Ormel, C. W., & Dullemond, C. P. 2010, A&A, 513, A56
Helling, C., Gourbin, P., Woitke, P., & Parmentier, V. 2019, A&A, 626, A133
Helling, C., Kawashima, Y., Graham, V., et al. 2020, A&A, 641, A178
Helling, C., Klein, R., Woitke, P., Nowak, U., & Sedlmayr, E. 2004, A&A, 423, 657
Helling, C., & Woitke, P. 2006, A&A, 455, 325
Helling, C. 2019, AREPS, 47, 583
Helling, C., Woitke, P., Rimmer, P. B., et al. 2014, Life, 4, 142
Hellmuth, O. 2006, ACP, 6, 4175
Herbort, O., Woitke, P., Helling, C., & Zerkle, A. 2020, A&A, 636, A71
Juncher, D., Jørgensen, U. G., & Helling, C. 2017, A&A, 608, A70
Kawashima, Y., Hu, R., & Ikoma, M. 2019, ApJ, 876, L5
Kawashima, Y., & Ikoma, M. 2018, ApJ, 853, 7
Lee, G., Dobbs-Dixon, I., Helling, C., Bognar, K., & Woitke, P. 2016, A&A, 594, A48
Lee, G. K. H., Blecic, J., & Helling, C. 2018, A&A, 614, A126
Madhusudhan, N., & Seager, S. 2011, ApJ, 729, 41
Mahapatra, G., Helling, C., & Miguel, Y. 2017, MNRAS, 472, 447
Mayne, N. J., Baraffe, I., Acreman, D. M., et al. 2014, A&A, 561, A1
Mendonça, J. M., Grimm, S. L., Grosheintz, L., & Heng, K. 2016, ApJ, 829, 115
Min, M., Ormel, C. W., Chubb, K., Helling, C., & Kawashima, Y. 2020, A&A, 642, A28
Nowak, M., Lacour, S., et al. 2020, Gravity Collaboration, A&A, 633, A110
Ormel, C. W., & Min, M. 2019, A&A, 622, A121
Patzer, A. B. C., Gauger, A., & Sedlmayr, E. 1998, A&A, 337, 847
Quanz, S. P., Absil, O., Angerhausen, D., et al. 2019, arXiv:1908.01316
Rimmer, P. B., & Helling, C. 2016, ApJS, 224, 9
Rimmer, P. B., Helling, C., & Bilger, C. 2014, IJAsB, 13, 173
Rossow, W. B. 1978, Icar, 36, 1
Showman, A. P., Fortney, J. J., Lian, Y., et al. 2009, ApJ, 699, 564
Snellen, I., Albrecht, S., Anglada-Escude, G., et al. 2019, arXiv:1908.01803
Stark, C. R., Helling, C., Diver, D. A., & Rimmer, P. B. 2014, IJAsB, 13, 165
Witte, S., Helling, C., & Hauschildt, P. H. 2009, A&A, 506, 1367
Woitke, P., Helling, C., Hunter, G. H., et al. 2018, A&A, 614, A1
Woitke, P., Helling, C., & Gunn, O. 2020, A&A, 634, A23

# Chapter 21

# Atomic and Molecular Line Data

**Sergei N. Yurchenko and Jonathan Tennyson**

## 21.1 Introduction

The main technique for characterizing exoplanet atmospheres is via spectroscopy. Interpreting these spectra either directly or via models requires detailed knowledge of the spectroscopic properties of the atoms and molecules (likely to be) present in the atmosphere under consideration. This chapter gives a brief overview of the available sources of data.

## 21.2 State of the Art

Broadly speaking databases for spectroscopic studies of exoplanets divide into two categories. Databases such as HITRAN (Gordon 2017) and GEISA (Jacquinet-Husson et al. 2016) have been constructed for detailed spectroscopic studies of the Earth's and solar system planetary atmosphere. As such they have been extensively validated against detailed observational data; these databases provide an excellent starting point for studies of Earth-like exoplanets as well for other temperate exoplanets. However, molecular spectra become hugely more complicated with increases in temperature (Tennyson & Yurchenko 2018) and these databases are only designed to be used for atmospheres at about 296 K.

For higher temperatures specialist databases have been developed specifically to address the requirements of studies of (hot) exoplanets. Specialist activities, such as the ExoMol project (Tennyson & Yurchenko 2012) have been launched for this reason. These activities are under continuous development as the demands of exoplanetary science grows. Table 21.1 summarizes the major spectroscopic databases providing atomic and molecular line lists for atmospheric studies of exoplanets and cool stars. Below we briefly outline the attributes of the key databases.

The main sources of the atomic data for exoplanetary applications include NIST (Ralchenko & Kramida 2020) and VALD (Ryabchikova et al. 2015) databases. For most of the atoms and atomic ions of interest for studies of exoplanet, both the quality of the line positions and the quality of intensities in these databases are

doi:10.1088/2514-3433/abfa8fch21

**Table 21.1.** Databases providing spectroscopic data which can be used for exoplanet atmospheres.

| Database | url | Purpose | Contents | Type |
|---|---|---|---|---|
| CDMS (Endres et al. 2016) | cdms.astro.uni-koeln.de | interstellar medium | microwave spectra | empirical |
| ExoMol (Tennyson et al. 2020) | exomol.com | exoplanets, cool stars | molecules | theoretical |
| JPL (Pickett et al. 1998) | spec.jpl.nasa.gov | interstellar medium | microwave spectra | empirical |
| GEISA (Jacquinet-Husson et al. 2016) | geisa.aeris-data.fr | Earth, planets | molecules | empirical |
| HITRAN (Gordon 2017) | HITRAN.org | Earth, planets | molecules | empirical |
| HITEMP (Rothman et al. 2010) | hitran.org/hitemp | hot atmospheres | molecules | mixed |
| Kurucz (Kurucz 2011) | kurucz.harvard.edu | stellar atmospheres | atomic/molecular | mixed |
| MolLIST (Bernath 2020) | bernath.uwaterloo.ca/molecularlists.php | exoplanets, cool stars | small molecules | empirical |
| NASA Ames | huang.seti.org | planets, exoplanets | polyatomics | theoretical |
| NIST (Ralchenko & Kramida 2020) | nist.gov/pml/atomic-spectra-database | general | atomic | experimental |
| SPECTRA (Mikhailenko & Babikov 2005) | spectra.iao.ru | general | polyatomics | theoretical |
| TheoReTS (Rey et al. 2016) | theorets.tsu.ru | planets, exoplanets | polyatomics | theoretical |
| VALD3 (Ryabchikova et al. 2015) | vald.astro.uu.se | stellar atmospheres | atomic | mixed |

**Notes**

Purpose: primary purpose for which the database was constructed.

Contents: major contents of the database.

Type: main means by which the data are generated. In practice most databases at least partly use a mixed approach.

adequate. Reliable line profiles for key species such as Na and K are, however, not always readily available especially for hot or high pressure environments (Burrows et al. 2001). More work is required on this problem.

HITRAN (Gordon 2017) is the best known and most-widely used database. The 2016 release comprises line lists for 49 molecules (122 isotopologues). It provides accurate spectroscopic data mostly for telluric applications as well as temperate planets. The data (both line positions and intensities) are mostly produced using semi-empirical, effective Hamiltonian or dipole techniques, allowing for extremely accurate description of the underlying experimental data. These techniques however do not extrapolate well to higher excitations (temperatures) limiting most of the line lists to the room temperature. The GEISA database (Jacquinet-Husson et al. 2016) is similar in spirit and coverage HITRAN.

HITRAN's high temperature spin-off is HITEMP (Rothman et al. 2010). The 2010 release contains less accurate but more complete line lists for only five molecules; better data sets for each of these species are now available. However, work on a new HITEMP release is now underway (Hargreaves et al. 2019, 2020); this release will provide a major update.

ExoMol is a spectroscopic database dedicated for studies of exoplanets and other hot atmospheres (Tennyson & Yurchenko 2012; Tennyson et al. 2020). It provides line lists for 80 molecules (190 isotopologues) with the special emphasis on completeness at high temperatures. The database contains line lists from two major sources: (i) line lists produced by the ExoMol group using variational nuclear-motion calculations (Tennyson & Yurchenko 2017b), (ii) hot experimental line lists from the MolLIST database produced by Peter Bernath's group (Bernath 2020), (iii) data on hot diatomics (CO, HF, HCl) taken from the same sources as HITRAN. The ExoMol line positions are based on accurate potential energy surfaces empirically tuned to the best experimental data available in the literature, while its intensities (Einstein $A$ coefficients) are based on high level *ab initio* dipole moment surfaces. It uses a package of accurate variational nuclear-motion codes Duo (Yurchenko et al. 2016), DVR3D (Tennyson et al. 2004), TROVE (Yurchenko et al. 2007), and WAVR4 (Kozin et al. 2004).

MolLIST (Bernath 2020) itself provides accurate line lists for 24 diatomic and two polyatomic molecules ($CH_4$ and $NH_3$) as well as experimental cross sections for eight polyatomic molecules. The line positions are either pure experimental or generated using an effective Hamiltonian approach. Most of the spectroscopic data are based on experiments performed at and applicable to high temperatures. The diatomic transition intensities are computed using *ab initio* dipole moment curves in combination with Le Roy's LEVEL program (Le Roy 2017).

TheoReTS (Rey et al. 2016) provides comprehensive empirical line lists for nine molecules and their isotopologues of atmospheric (terrestrial and exoplanetary) importance. TheoReTS follows a similar approach to ExoMol.

VAMDC (http://www.vamdc.org) is a data portal that federates data from a number of sources (Dubernet et al. 2016). VAMDC provides access to several individual databases containing empirical (effective Hamiltonian) line lists from Vincent Boudon's group for the molecules $CH_4$, $C_2H_4$, $CF_4$, $SF_6$, and $RuO_4$.

| | | | | | | | | | | | |
|---|---|---|---|---|---|---|---|---|---|---|---|
| AlH | AlCl | AlO | BeH | $C_2$ | CaO | CaF | CaH | CH | CN | CO | CP |
| CrH | CS | FeH | FeO | $H_2$ | $H_3^+$ | $H_3O^+$ | HCl | HF | LiCl | LiF | LiH |
| $LiH^+$ | KCl | KF | MgF | MgH | MgO | $N_2$ | NaCl | NaF | NaH | NaF | NiH |
| NH | NS | NO | OH | $O_2$ | PH | PN | PO | PS | ScH | SiC | SiH |
| SiS | SH | SiO | SiS | SO | TiO | TiH | TiF | VO | VN | YO | ZnS |
| $C_3$ | CaOH | $CO_2$ | $CH_2$ | $H_3^+$ | $H_2D^+$ | HCN | HNC | $H_2O$ | HDO | $H_2S$ | KOH |
| NaOH | $O_3$ | OCS | $PO_2$ | $SiH_2$ | $SiO_2$ | $SO_2$ | $AsH_3$ | $CH_3$ | $C_2H_2$ | HOOH | $H_2CO$ |
| $NH_3$ | $SiH_3$ | $PH_3$ | $P_2H_2$ | $PF_3$ | $SO_3$ | $CH_4$ | $CH_3D$ | $CH_3Cl$ | $CH_3F$ | $SiH_4$ | $HNO_3$ |
| $C_2H_4$ | $C_2H_6$ | $C_3H_8$ | | | | | | | | | |

**Figure 21.1.** Available molecular line lists: red indicates line lists in progress, blue corresponds to the line lists suggested for molecules specific for hot rocky exoplanets, and green indicates line lists that contain data applicable for high resolution.

The spectroscopic database SPECTRA (Mikhailenko & Babikov 2005) contains data from HITRAN, GEISA, as well as data produced in Tomsk, including $H_2S$, $CO_2$ (part of the CDSD database), $NO_2$, $CH_4$, $N_2O$, and $C_2H_2$ (ASD-1000).

Finally, Kurucz (2011) presents extensive atomic line data for stellar models; the Kurucz database also contains older and less accurate data for dozens of diatomics as well as $H_3^+$ and $H_2O$.

Figure 21.1 shows known molecular line lists of importance for exoplanets, available from the different databases discussed above. Indicated are the line lists in progress (red), line lists which are suitable for high resolution spectroscopic studies (green) and molecules, specifically for rocky super-Earths (blue).

These databases provide line positions, line intensities at some reference temperature (or Einstein *A* coefficients, line, or oscillator strengths), energies of the lower states, statistical weights (required for the temperature dependencies), and state assignments. Some of the databases also provide line profile parameters for some major broadeners; however, there is far from complete coverage of pressure effects (Fortney et al. 2019).

## 21.3 Important Questions, Goals, Opportunities and Challenges

A key role of laboratory astrophysics is to provide the data necessary to interpret and model astronomical observations. The rapidly expanding field of exoplanet characterization continues, and will continue for sometime, to throw up issues that demand new laboratory data. This can mean consideration of new species, improved accuracy, extended temperature ranges or, as not discussed above, extension of data to shorter wavelengths. It is clear that the challenge of meeting these demands will continue to drive the development of laboratory techniques, both theoretical (Tennyson & Yurchenko 2017b) and experimental (Bernath 2014; Fortney et al. 2019).

## 21.4 Future Developments

### 21.4.1 Spectroscopy of Lava-planets

Hundreds of super-Earths have been discovered in the recent years, many of them with very short orbital period. We know little about their nature, but it is expected that the rocky super-Earths are evaporating with water as a major constituent. The massive amounts of water in their atmospheres can melt rocks and put their constituents into the atmosphere. These objects will necessarily have different spectroscopic signatures than cooler planets or hot Jupiters, and this will influence interpretation of the atmospheric observations. The gases released from the rock soluble in steam, as it heats up and melts, should include silica and other rock-forming elements. Apart from the usual molecules such $H_2O$, $CO_2$, $O_2$, HF, $SO_2$, HCl, OH, CO, chemical models suggest less common constituents of steam-atmospheres such as ZnS, CaOH, MgOH, NaOH, KOH, SiO, $SiO_2$, $PO_2$ (see Figure 21.1 and also Schaefer et al. 2012). At present interpretation is severely impacted by the lack of the spectroscopic data specific to hot rocky super-Earths (Tennyson & Yurchenko 2017a). The massive number of absorbers in the atmosphere also affects the cooling and therefore the evolution of the young hot objects; comprehensive molecular data is also crucial to model these processes. It is only with good predictions of likely atmospheric composition allied to a comprehensive database of spectral signatures that the observed spectra can be deciphered.

### 21.4.2 High Resolution Data

The next generation of ground-based telescopes, which complement the space missions, are aimed at high resolution (HR) ($R > 100,000$) studies. Ground-based high resolution Doppler spectroscopy (HRS), pioneered by Snellen (2014), has been used to characterize the atmospheres of a growing number of exoplanets in recent years (see e.g., review by Birkby 2018). HRS yields secure detections of molecules but also potentially a wealth of other information for example on non-thermodynamic equilibrium effects, magnetic fields (through Zeeman splittings), isotopic composition, and wind speeds. HRS is in the process of revolutionizing the characterization of exoplanetary atmospheres and has the potential to shape exoplanetary science for the next decade. In the last few years there has been significant progress made with observational facilities and atmospheric modeling and inference methods for the observations. Planet surveys such as the current Transiting Exoplanet Survey Satellite (TESS) Mission will provide the most suitable nearby targets for atmospheric follow-up with such facilities. Hence we are on the cusp of atmospheric characterization of exoplanets with temperatures 500–1500 K with HR observations at a resolving power of $R > 50,000$.

None of this is possible without the necessary laboratory data. While spectroscopic and other data needs for studies of Earth-clone exoplanets are well met by databases constructed for studying our own atmosphere, such as HITRAN, this is not the case for those exoplanets that will be the first exoplanets to be characterized. These planets are hot and most are subject to strong bombardment by starlight (and presumably stellar winds). Such studies require data on hot molecules that are not generally available and that are in many cases not easily amenable to laboratory experiments. In particular, current developments in exoplanetary observations are fueling the requirement for HR data. This creates new challenges for laboratory spectroscopy projects like ExoMol, which have been mainly concerned with completeness (important for low resolution-based retrievals and opacity models) rather than HR studies. Indeed, not all existing hot line list are capable of providing the HR quality (see Figure 21.1). For example, most ExoMol line lists, though successful for modeling atmospheric opacity, have been shown not to be accurate enough for HRS of exoplanets (Brogi & Line 2019). The field of the HRS of exoplanets is growing extremely fast and urgently demanding high precision molecular data. Failures to detect molecules in atmospheres of exoplanets can often be attributed to the lack of the underlying quality of the line positions (Evans et al. 2016; Hoeijmakers et al. 2015). Most of the currently existing line lists were intended primarily for low resolution transit-like studies and therefore are not suitable for HR. It is important that these problem are addressed in the nearest future in order underpin the next great advance in our understanding of exoplanets.

The modern computational methods still cannot provide HR quality for all the lines. Therefore a new generation of line lists will require a new structure allowing for clear indicators showing which lines have the appropriate HR quality so they can be used for HR observations. This structure can be based on the associated uncertainties of the line positions or energy level. Uncertainties of the theoretical data (which constitutes the majority of the hot molecular data) are practically non-existent in modern line list databases such as ExoMol and TheoReTS. Their inclusion will allow astronomers looking for high accuracy transitions to be used in HR studies to identify which transitions can be used reliably. The associated spectral simulation or radiative transfer programs will need to be updated to allow HR lines to be easily identified.

A systematic provision of spectral data with experimental, HR quality has been a subject of the recent development of the MARVEL procedure (Furtenbacher et al. 2007; Császár et al. 2007; Furtenbacher & Császár 2012; Tóbiás et al. 2019). MARVEL (measured active vibration rotation energy levels) is a well-established algorithm and an online-app http://kkrk.chem.elte.hu/marvelonline/. It involves assembling all measured HR, assigned transition frequencies (wavenumbers), and inverting to produce a validated set of empirical energy levels of the HR quality. The MARVEL energies can be then used to replace the computed energies or line positions. Doing this significantly improves the associated line list as noted recently by the NASA Ames group (Huang et al. 2019).

Alternatively, the provision of HR spectroscopic data be achieved also via effective rotational Hamiltonians, e.g., PGOPHER (Western 2017) or SPFIT and SPCAT (Pickett 1991).

### 21.4.3 Pressure Broadening

Accurate and sufficiently complete sets of collisional (or pressure) broadening parameters are crucial for exoplanet atmosphere characterization and radiative transfer modeling.

While air-broadening parameters are known for many molecules as documented in HITRAN (Gordon 2017). The provision of line-shape parameters for other important broadeners, especially $H_2$ and He over an extended temperature range remains a major future challenge. Incorrect broadening data can significantly affect the atmospheric retrieval from exoplanet spectra (Gharib-Nezhad & Line 2019). Substantially more data is still needed to provide complete and reliable sets; steps in this direction are being taken (Tan et al. 2019).

### 21.4.4 Missing Data

Some of the existing line lists (especially from HITRAN) have limited temperature or wavelength coverage and will have to be extended, e.g., $N_2O$, $NO_2$, $C_2H_6$, $C_3H_6$, $O_3$, $SiH_3$, TiH, TiF, VN, YO (see Figure 21.1).

## References

Bernath, P. F. 2014, RSPTA, 372, 20130087
Bernath, P. F. 2020, JQSRT, 240, 106687
Birkby, J. L. 2018, in Handbook of Exoplanets, ed. H. Deeg, & J. Belmonte (Cham: Springer), 16
Brogi, M., & Line, M. R. 2019, ApJ, 157, 114
Burrows, A., Hubbard, W. B., Lunine, J. I., & Liebert, J. 2001, RvMP, 73, 719
Császár, A. G., Czakó, G., Furtenbacher, T., & Mátyus, E. 2007, ARCC, 3, 155
Dubernet, M.-L., Antony, B. K., Ba, Y. A., et al. 2016, JPhB, 49, 074003
Endres, C. P., Schlemmer, S., Schilke, P., Stutzki, J., & Müller, H. S. P. 2016, JMoSp, 327, 95
Evans, T. M., Sing, D. K., Wakeford, H. R., et al. 2016, ApJ, 822, L4
Fortney, J. J., Robinson, T. D., Domagal-Goldman, S., et al. 2019, arXiv:1905.07064
Furtenbacher, T., & Császár, A. G. 2012, JMoSt, 1009, 123
Furtenbacher, T., Császár, A. G., & Tennyson, J. 2007, JMoSp, 245, 115
Gharib-Nezhad, E., & Line, M. R. 2019, ApJ, 872, 27
Gordon, I. E., Rothman, L. S., Hill, C., et al. 2017, JQSRT, 203, 3
Hargreaves, R. J., Gordon, I. E., Rothman, L. S., et al. 2019, JQSRT, 232, 35
Hargreaves, R. J., Gordon, I. E., Rey, M., et al. 2020, ApJS, 247, 55
Hoeijmakers, H. J., de Kok, R. J., Snellen, I. A. G., et al. 2015, A&A, 575, A20
Huang, X., Schwenke, D. W., & Lee, T. J. 2019, JQSRT, 225, 327
Jacquinet-Husson, N., Armante, R., Scott, N. A., et al. 2016, JMoSp, 327, 31
Kozin, I. N., Law, M. M., Tennyson, J., & Hutson, J. M. 2004, CoPhC, 163, 117
Kurucz, R. L. 2011, CaJPh, 89, 417
Le Roy, R. J. 2017, JQSRT, 186, 167
Mikhailenko, S. N., Babikov, Y. L., & Golovko, V. F. 2005, AtOO, 18, 685
Pickett, H. M. 1991, JMoSp, 148, 371
Pickett, H. M., Poynter, R. L., Cohen, E. A., et al. 1998, JQSRT, 60, 883
Ralchenko, Y., & Kramida, A. 2020, Atoms, 8, 56
Rey, M., Nikitin, A. V., Babikov, Y. L., & Tyuterev, V. G. 2016, JMoSp, 327, 138
Rothman, L. S., Gordon, I. E., Barber, R. J., et al. 2010, JQSRT, 111, 2139
Ryabchikova, T., Piskunov, N., Kurucz, R. L., et al. 2015, PhyS, 90, 054005
Schaefer, L., Lodders, K., & Fegley, B. Jr. 2012, ApJ, 755, 41
Snellen, I. 2014, RSPTA, 372, 20130075

Tan, Y., Kochanov, R., Rothman, V. L. S., & Gordon, I. E. 2019, JGRD, 124, 11580
Tennyson, J., Kostin, M. A., Barletta, P., et al. 2004, CoPhC, 163, 85
Tennyson, J., & Yurchenko, S. N. 2012, MNRAS, 425, 21
Tennyson, J., & Yurchenko, S. N. 2017a, MolAs, 8, 1
Tennyson, J., & Yurchenko, S. N. 2017b, IJQC, 117, 92
Tennyson, J., & Yurchenko, S. N. 2018, Atoms, 6, 26
Tennyson, J., Yurchenko, S. N., Al-Refaie, A. F., et al. 2020, JQSRT, 255, 107228
Tóbiás, R., Furtenbacher, T., Tennyson, J., & Császár, A. G. 2019, PCCP, 21, 3473
Western, C. M. 2017, JQSRT, 186, 221
Yurchenko, S. N., Lodi, L., Tennyson, J., & Stolyarov, A. V. 2016, CoPhC, 202, 262
Yurchenko, S. N., Thiel, W., & Jensen, P. 2007, JMoSp, 245, 126

# Part IV

## Habitability and Biosignatures

AAS | IOP Astronomy

ExoFrontiers
Big questions in exoplanetary science
Nikku Madhusudhan

# Chapter 22

# Exoplanetary Habitability

**James F. Kasting**

To be habitable by conventional terrestrial definitions, a planet must possess a solid or liquid surface, a source of free energy, sufficient supplies of carbon and other bioessential elements, and liquid water. For life to be remotely detected on such a planet, that water must be at the surface. This leads to the definition of the habitable zone as the region around a star where a planet can maintain liquid water on its surface. Other factors may play an important role in planetary habitability, however. These include: the presence of a magnetic field, the operation of plate tectonics, the existence of continents, a stable obliquity, and the type of star around which a planet orbits. M-star planets, in particular, face a daunting array of challenges to supporting life. Progress in understanding planetary habitability depends heavily on improved observational capabilities. NASA's James Webb Space Telescope, which is scheduled for launch in October 2021, may allow us to observe and characterize a few transiting, potentially Earth-like planets. Space-based direct imaging telescopes are needed, though, to find such planets around the nearest Sun-like stars.

## 22.1 Introduction

Exoplanets are already of great interest to astronomers now that they can find them and begin to observe them. Many researchers in other disciplines, e.g., geosciences and biology, are most interested in a particular class of exoplanets that is harder to observe, namely, planets on which life could conceivably originate and evolve. This has renewed an old debate about what is necessary for a planet to be habitable and, more fundamentally, what is life itself? I will not attempt in this brief essay to delve into the latter question. Some astrobiologists (Seager et al. 2012; Walker et al. 2018) have suggested that there may be "universal" biomarkers that would allow us to search for alien life that is quite different from life here on Earth. All life on Earth is carbon-based and requires liquid water during at least part of its life cycle. That leads to specific requirements for planetary habitability. Life that did not share those

doi:10.1088/2514-3433/abfa8fch22  

requirements would necessarily have quite different (and poorly defined) requirements for habitability. So, here, I will assume that life everywhere is carbon-based and needs liquid water. Life on Earth shares many other common features, of course. For example, it relies on specific biomolecules such as DNA, RNA, ATP, ADP, along with various lipids and proteins. We do not need to be that anthropocentric here.

A further, arguably universal, requirement for habitability is that a planet must have a solid or liquid surface to provide a stable pressure–temperature regime within which life could evolve. Gas and ice giant planets have semi-infinite, convective atmospheres that provide no such stability. In his book (Sagan 1995) and television series, *Cosmos*, Carl Sagan speculated about giant "floaters" in Jupiter's atmosphere that could maintain their altitude by adjusting their internal pressure, much as some fish with air bladders maintain their depth in Earth's oceans. These are all highly evolved creatures. Regardless of what forms it may eventually take on, life must start from single-celled organisms which would be incapable of altitude regulation in a gaseous atmosphere. The search for life should therefore focus on rocky planets.

The big questions, then, are the following:

- Do habitable planets exist around other stars and can we observe and characterize them?
- What specific factors contribute to a planet's habitability or, alternatively, make it a less suitable abode for life?

The search for life on such planets is also obviously of interest, but that discussion is reserved for the next chapter of this book.

## 22.2 State of the Art

Based on the requirements outlined above, we can identify several factors that might determine whether a planet is, or is not, habitable. We start with liquid water. Liquid water is abundant on Earth and is thought to exist within the subsurfaces of other solar system planets and moons, including Mars, Europa, Ganymede, Enceladus, and Titan. The search for such subsurface life drives much of NASA's Solar System Exploration program. But subsurface life is not, or at least should not be, of great interest to astronomers because it is difficult or impossible to identify using a telescope. To be detectable, life must be present at a planet's surface so that it can modify the planet's atmosphere in a way that can be observed spectroscopically.

This leads directly to the concept of the habitable zone (HZ), defined as the region around a star where liquid water can exist on a planet's surface. This is not a new idea. The origin of the term goes back further than most researchers, including me, had realized. As recently shown by Ralph Lorenz (2020), the term "habitable zone" was coined by E. W. Maunder in a 1913 popular book (Maunder 1913). This is the same Maunder whose name is associated with the historical drop in the number of observed sunspots between 1645 and 1715 A.D. The famous astronomer Harlow Shapley wrote about the HZ concept in his 1953 book (Shapley 1953), although he called it the "liquid water belt." Michael Hart (1978, 1979) developed a simple model of the HZ. He also defined the term "continuously habitable zone," or CHZ, as the region around a star that remains habitable for some specified time. Modern concepts of the HZ (see Figure 22.1) are based on more elaborate, but still 1-D, climate models (Kasting et al. 1993; Kopparapu et al. 2013). These models incorporate the important assumption that atmospheric $CO_2$ builds up in a planet's atmosphere as its climate cools. On Earth, $CO_2$ is injected into the atmosphere by volcanism and is removed (mostly) by weathering of silicate minerals, followed by deposition of carbonate sediments in the ocean. ($CO_2$ is also lost by reduction to organic carbon, some of which is then buried. But this organic carbon cycle regulates, and is regulated by, atmospheric $O_2$ and hence cannot simultaneously regulate $CO_2$.) The silicate weathering rate is temperature-dependent, creating a stabilizing negative feedback loop that may have helped solve the "faint young Sun" problem on Earth (Walker et al. 1981; Kasting 1993). This same feedback causes the HZ to be relatively wide because $CO_2$ is assumed to build up near the outer edge, creating a stronger greenhouse effect and preventing the planet's surface from freezing.

Not all the relevant climate modeling is 1-D. The inner edge of the conservative HZ in Figure 22.1 incorporates corrections from a 3-D climate model calculation by Leconte et al. (2013). The 3-D model shows that the undersaturated upper branches

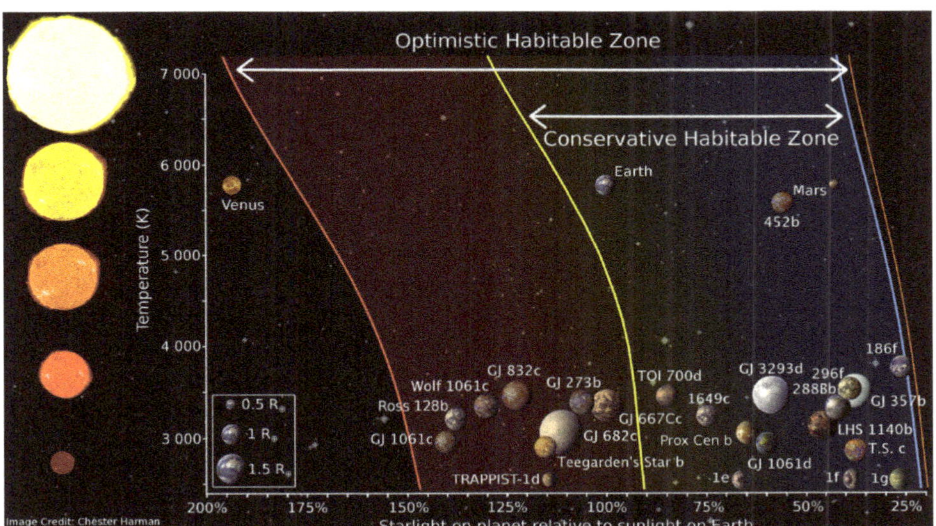


**Figure 22.1.** Diagram showing the habitable zone (HZ) around different types of stars. The *y*-axis is stellar effective temperature. The *x*-axis is shown in units of stellar flux relative to that incident on modern Earth. The boundaries are curved rather than straight because Rayleigh scattering, which is stronger at shorter wavelengths, increases the albedo of planets around hotter stars. The "conservative" HZ is based on climate modeling, as discussed in the text. The "optimistic" HZ uses the empirical "early Mars" and "recent Venus" limits from Kasting et al. (1993). Objects populating the diagram are planets of <1.6 Earth radii from the Kepler catalog and from other ground-based radial velocity and transit surveys. (Modified from Kasting et al. 2014 by C. Harman.)

of the tropical Hadley cells on a warm, Earth-like planet act like "radiator fins" to help the planet shed excess heat. This shifts the inner edge of the HZ around a Sun-like star from ~1.0 au back to 0.95 au, where it had been thought to lie previously (Hart 1978; Kasting et al. 1993). 3-D models have also been used to argue that the inner edge of the HZ could be much closer to the star for slowly or synchronously rotating planets because of increased cloud cover over the Sunlit hemisphere (Yang et al. 2013; Kopparapu et al. 2016; Way & Del Genio 2020). 3-D climate models are expensive to run and contain many adjustable parameters, but they are required to simulate phenomena such as clouds and polar ice caps.

The question has also been raised whether one should consider greenhouse gases other than $CO_2$ and $H_2O$. These two gases are the most important greenhouse gases on modern Earth. Methane, $CH_4$, is a poor third. On a molecule-per-molecule basis, its contribution to modern greenhouse warming is significant because it absorbs radiation within the 8–12 μm "window region" where absorption by $H_2O$ and $CO_2$ is weak. But adding modest amounts of methane to a $CO_2$-rich atmosphere causes little additional warming (Ramirez et al. 2014), and adding too much methane can cause cooling by triggering the formation of hydrocarbon aerosols (Haqq-Misra et al. 2016). By contrast, $H_2$ turns out to be an excellent greenhouse gas despite its lack of a permanent electric dipole moment. Collisions with other molecules create transient dipoles that allow $H_2$ to absorb radiation across the thermal-IR spectrum (Wordsworth & Pierrehumbert 2013). A captured 30 bar $H_2$ atmosphere around a 3 $M_E$ super-Earth could maintain liquid water on its surface out to ~10 au, near the orbit of Saturn (Pierrehumbert & Gaidos 2011; Seager 2013). Indeed, planets with hundreds of bars of $H_2$ could be kept warm by geothermal heat even if they were not attached to a star (Stevenson 1999). Do such planets exist, however, and would they be detectable? It seems prudent to focus initially on planets more like Earth that are already known to exist.

Just because a planet is within the habitable zone does not mean it must be habitable. Mars is a good example. Mars lies within the HZ today, according to Figure 22.1, and yet it is essentially uninhabitable, at least at its surface. (Paradoxically, early Mars may have been habitable despite lying outside the conventional HZ. Appreciable concentrations of atmospheric $H_2$ may help explain this observation; Ramirez et al. 2014.) It is not difficult to discern what Mars' problem(s) may be. Mars is a small planet, roughly 1/10th of an Earth mass. It lacks an intrinsic magnetic field, and this allows the solar wind to strip off its atmosphere (Jakosky et al. 2018). Equally importantly, because it was small Mars cooled down quickly, within its first 1–2 billion years, causing it to become volcanically inactive. Without volcanism, it had no way to replenish the $CO_2$ needed to keep its surface warm. We can conclude that there is a minimum mass limit for a habitable planet, and Mars is below it. Exactly what that lower mass limit is depends on how long one wants the planet to remain habitable; however, half an Earth mass seems like a reasonable value if one wishes to allow sufficient time to evolve complex life.

Following this same line of reasoning, we can identify multiple additional questions that must be answered about a HZ planet to determine whether it is actually habitable. The list that follows is by no means complete, but it provides an overview of issues that various authors have deemed important. Brief, also incomplete, answers to these questions are provided, as well.

1. Will HZ planets around other stars receive Earth-like endowments of carbon and water?

   Models of planetary accretion, e.g., Raymond et al. (2004, 2006), suggest that volatile delivery is stochastic, and hence HZ planets could receive widely differing complements of volatile materials.

2. Are planetary climates stabilized by feedback in the carbonate–silicate cycle that allows $CO_2$ to build up in a planet's atmosphere when the incident stellar flux is lower? Are continents required for this feedback to operate effectively?

   Up-to-date models of the carbonate–silicate cycle (Maher & Chamberlain 2014; Caves et al. 2016; MacDonald et al. 2019) suggest that the negative feedback between atmospheric $CO_2$ and climate does indeed operate but that $CO_2$ levels are also modulated by a variety of other factors. For example, the observed cooling during the last 40 m.yr may have been caused partly by the uplift of the Himalayas and the associated heavy rainfall from the Indian monsoon (Raymo & Ruddiman 1992) and partly by collisions between oceanic plates that created the islands of Indonesia, exposing highly weatherable oceanic basalts to the torrential rainfall of the intertropical convergence zone (MacDonald et al. 2019).

   Whether continents are required for the silicate weathering feedback to operate is a matter of debate. Continents are thought to form as a byproduct of plate tectonics, so this question is closely related to question 3. The answer depends on the efficiency and temperature dependence of silicate weathering on the seafloor. Most seafloor weathering occurs in the vicinity of hydrothermal vents, where warm seawater percolates through the highly fractured young seafloor. This process is thought to play a minor role (<10%) in $CO_2$ removal today, and it was initially assumed to be independent of ocean temperature (because it depended on the temperature of vent fluids). If that were true, no climate feedback would be provided. However, most of the seafloor weathering occurs within the flanks of the ridges, so the weathering rate *does* therefore depend on the deep ocean temperature (Coogan & Gillis 2013). These authors suggest that seafloor weathering might have accounted for as much as 50 percent of $CO_2$ removal during the Mesozoic Era when deep ocean temperatures were much warmer. Models of the carbonate–silicate cycle that include this process (Krissansen-Totton & Catling 2017) suggest that seafloor weathering contributes to climate feedback, but it does so in a different way than continental weathering because the temperature of the deep ocean is linked to the surface temperature in the polar regions where most deep water forms.

3. Is plate tectonics necessary to maintain climate stability?

   The role of plate tectonics in fostering climate stability has been studied using coupled mantle convection/carbon cycle models. Foley (2015) concluded that planets with plate tectonics were more likely to maintain stable climates because they recycled $CO_2$ more effectively. But Foley and

Smye (2018) argued that stagnant lid planets might also be habitable if their internal radiogenic heat production was relatively high. In both papers, the authors conclude that the feedback of climate on plate tectonics is relatively weak. By contrast, Lenardic et al. (2016) and Lenardic & Seales (2019) have argued that the coupling between plate tectonics and climate is quite strong. In their view the divergent evolutionary paths of Earth and Venus may have been caused as much by changes in their internal thermal/tectonic evolution as it was by the differing fluxes of solar radiation that each planet received. If so, then the pattern seen in our solar system, with habitable planets existing within the HZ and desiccated planets lying inside of it, may not be repeated elsewhere. This argument, like most of the questions listed here, would be best resolved by observations.

4. Is a planetary magnetic field required for a planet to retain its atmosphere?

   As pointed out earlier, Mars has lost much of its atmosphere, partly due to its small size and partly due to its lack of an intrinsic magnetic field. These factors are related because the absence of a magnetic field presumably resulted from the premature shutdown of the planet's magnetic dynamo. Evidence from crustal magnetic fields observed from Mars Global Surveyor and from MAVEN shows that a magnetic dynamo did exist early in the planet's history (Mittelholz et al. 2020). Other evidence shows that Mars has an iron core and that the core is at least partially molten (Yoder et al. 2003). But we do not yet know if a solid inner core exists, as on Earth (Helffrich 2017), and thus we do not fully understand why the dynamo shutdown. Seismic measurements from the current Insight probe may provide the answer if it can succeed in burying itself deeply enough to do its job effectively. The probe, or "mole," hit rock-hard duricrust on its way down and has been struggling for more than a year to get to its intended depth of 3–5 m.

   Venus also lacks an intrinsic magnetic field and is exposed to a stronger solar wind than either Mars or Earth, yet it has managed to hold onto an atmosphere that is almost 100 times denser than Earth's. It is better protected from atmospheric loss than non-magnetic Mars because its mass and surface gravity are much larger. So, one might argue that an Earth-mass planet can retain its atmosphere even without an intrinsic magnetic field. That argument has caveats, however. Venus' atmosphere, like that of Mars, is dominated by $CO_2$. $CO_2$ is a good absorber of thermal-IR radiation, making it a good greenhouse gas, but that means that it is also a good radiator. The upper atmosphere of Venus is therefore cold and compact (Prinn & Fegley 1987), making it less susceptible to solar wind stripping. Earth's oxygen-rich upper atmosphere, by contrast, is hot and extended (Banks & Kockarts 1973). So, if Earth lacked an intrinsic magnetic field, it could conceivably lose its atmosphere just as Mars has done. The magnetic field issue is important, particularly for $O_2$-rich planets that might harbor complex life.

5. Are waterworlds habitable?

   Planetary accretion models suggest that some HZ planets may receive enormous endowments of volatiles compared to Earth, leading to oceans that are tens, or even hundreds, of kilometers deep. Such hypothetical planets are termed *waterworlds*. On such a planet, continents would not be subaerially exposed and silicate weathering could only occur on the seafloor. Kite and Ford (2018) suggested that seafloor weathering is unlikely to occur on such worlds because of the extremely high pressures at the bottom of the oceans. Nonetheless, atmospheric $pCO_2$ in their models is coincidentally in the range 0.2–20 bar for which the habitable zone is as wide as possible. If this reasoning is correct, then such planets could conceivably remain habitable for billions of years even without a stabilizing negative feedback between $CO_2$ and surface temperature.
6. Can planets in the HZs of M stars be habitable? Would tidal locking be a serious problem for them? Would such planets lose too many of their volatiles during their parent star's pre-main-sequence evolution?

   The question of whether M-star planets can be habitable is of great interest because M stars are abundant and because their HZs lie close to the parent star, so the probability that a HZ planet will transit is relatively high. The TRAPPIST-1 planets shown in Figure 22.1, which orbit an M8V star, were found by this method and will likely be prime targets for observation by the James Webb Space Telescope (JWST) when it eventually launches. These planets are expected to be tidally locked, and perhaps (but not necessarily) synchronously rotating. (Mercury, for example, is within the Sun's tidal locking radius, but it is trapped in a 3:2 spin–orbit resonance.) This has led various authors, beginning with Dole (1964), to question whether this might render them uninhabitable by causing their atmospheres to freeze out on the planets' nightside. 3-D climate modeling has shown, however, that transport of heat by atmospheric winds or ocean currents should keep the nightside warm, provided that the planet's atmosphere is initially thick enough to get this process started (Joshi 2003; Joshi et al. 1997).

   Other, potentially more serious problems for M-star planets have since been identified. Their slow rotation rates may prevent the creation of a magnetic dynamo necessary to create a planetary magnetic field. M stars are generally more active than the Sun, so they have strong stellar winds and lots of flare activity. A planet within an M-star HZ also orbits close to the star, so solar wind stripping of its atmosphere is an acknowledged threat (Khodachenko et al. 2007; Lammer et al. 2007). An equally serious problem is that M stars go through a long gravitational contraction phase during which they can be as much as 10–20 times brighter than their eventual main sequence luminosity (Luger & Barnes 2015). Thus, a planet in an M-star HZ could experience an early runaway greenhouse that would strip it of its water. Both these problems might be overcome if the planet starts out with a much larger volatile abundance than did Earth. The best candidates for habitability could thus be the evaporated cores of mini-Neptunes that might still retain abundant volatiles despite losing much of their initial endowment (Luger et al. 2015).

7. Will otherwise Earth-like planets have stable obliquities even if they lack a large moon?

   Finally, much has been made of the observation, or calculation, that the presence of a large moon stabilizes Earth's obliquity (Laskar & Robutel 1993; Laskar et al. 1993). The moon pulls on Earth's equatorial bulge, causing Earth's spin axis to precess about three times faster than it would otherwise. This allows Earth to avoid a thicket of secular resonances with motions of the other planets (the precession of their orbital perihelia and line of nodes). Earth's obliquity is known to vary periodically from 22° to 24.5° on a 41,000 yr time scale. This is part of the Milankovitch forcing that has driven the timing of the Pleistocene ice ages (Imbrie & Imbrie 1986). If one were to magically remove the Moon from the present system, Earth's obliquity would vary chaotically from 0° to 85° on time scales of tens of millions of years (Laskar et al. 1993). Marine life might not be greatly affected because the temperature of the ocean is buffered by its large heat capacity. But life on land, especially animal life, would be severely impacted by such changes in obliquity. For this and other reasons, some authors have argued that complex life, meaning animal life, could be rare in the universe even if life itself is relatively common (Ward & Brownlee 2000).

   The question of whether the obliquity of a moon-less Earth should be chaotic or periodic is complicated because the answer depends on the planet's initial spin rate, as well as the orbital dynamics of the other planets (Laskar et al. 1993; Kasting 2001, 2010). If the initial daylength was <12 h, Earth's obliquity would vary regularly, even in the absence of the Moon. That said, large chaotic obliquity variations could occur on HZ planets orbiting other stars. So, this is yet another factor that must be considered in evaluating exoplanet habitability.

## 22.3 Important Questions/Goals

Most of the questions listed above would be best resolved by observations. These observations are difficult, though, as discussed in the next section. And some of the questions, such as whether a planet has plate tectonics, may be difficult to answer even *with* good observations. That is why the concept of the habitable zone has been emphasized here. Unlike most of the other parameters mentioned above, the boundaries of the HZ can be estimated simply by measuring the luminosity and effective temperature of the star. The answer depends on one's assumptions about greenhouse gas concentrations, as discussed earlier. But one can still use this concept to get an idea of where to look for potentially habitable planets. Elsewhere (Kasting et al. 2014), we have argued that the conservative HZ from Figure 22.1 should be used for that purpose. If we design our space telescope to be able to find rocky planets within this region, then we can be confident of obtaining a positive result. This does *not* imply that we should ignore planets that lie outside the conservative

HZ, as some of these may prove interesting. But we would not want to design the telescope to focus on these less promising regions of exoplanet parameter space.

That said, there are still many things we can do to advance our knowledge of exoplanetary habitability *before* we have new observations. A few of these are listed below.

- Continue to improve our 3-D climate models. Several good exoplanet climate models already exist but no such model is ever truly complete; rather, they are all works in progress. A comparison to the modern Earth climate modeling effort is illuminating. When the IPCC (Intergovernmental Panel on Climate Change) publishes their estimates of global warming, they compare results from 25–30 different 3-D models from around the globe. Each model is the product of many person-years of effort. Sustained climate modeling efforts at government laboratories like the Laboratoire Monde du Globe (LMD) in Paris and the Goddard Institute for Space Studies (GISS) in New York are examples of the kind of multi-decadal support needed to make progress in this area.
- Continue to study the long-term evolution of Earth. Earth is our only current example of a habitable planet, and we can study its evolution in considerable detail by examining its geological/geochemical/fossil record. We can test the theory of the weathering-climate feedback, we can attempt to ascertain the time history of atmospheric $O_2$, and we can try to determine when and why plate tectonics got started on Earth. Somewhat surprisingly, this is still a major unsolved problem in geophysics. Much of my own research over the years has been focused on understanding the evolution of Earth's atmosphere and climate. Someday, I hope this will help us to understand what we observe on exoplanets.
- Continue to develop and support the field of *astrobiology*. Astrobiology is sometimes defined as the search for life off the Earth. But it is more than just that. Astrobiologists also seek to understand the history of life on Earth. How did it originate, how does it evolve, and how widespread is it in the universe? From a practical standpoint, the key challenge for astrobiologists is to learn about subjects outside your home discipline. Astrobiology courses are now offered at many universities, including Penn State where I teach. Taking such courses and learning where your own research fits into the broader scheme of things is a key part of becoming an astrobiologist. Oftentimes, rewarding new insights come from applying skills acquired in one field to questions being asked in another.

## 22.4 Opportunities

As suggested above, the real opportunities in this field depend on future exoplanet observations. Ground-based radial velocity measurements and the space-based Kepler transit measurements have already revolutionized our knowledge of exoplanets. To learn more about exoplanetary habitability we need *spectroscopic* measurements of exoplanet atmospheres. Low-resolution spectroscopy of transiting exoplanets has already been performed using the Hubble and Spitzer space telescopes. These measurements have allowed us to study hot Jupiters and warm Neptunes in some detail. But to answer questions related to habitability, we need spectroscopic measurements of rocky planets more like Earth. In the near future, such measurements could be provided by JWST. JWST, like Hubble and Spitzer, will be limited to observing transiting exoplanets. This constrains both the types of stars (mostly M stars) that can be observed and the achievable spectroscopic resolution. But, with luck, we may learn interesting things about exoplanet habitability.

One very special, non-transiting, HZ exoplanet, Proxima Centauri b, may be studied spectroscopically from the ground using large facilities like the Extremely Large Telescope (ELT) that is currently under construction in Chile. The observational technique, called *direct imaging*, involves using a high-contrast coronagraph to block the light from the star, allowing one to see the reflected light from the planet. Proxima Centauri is close enough that the angular separation of the planet from the star can be resolved with a ~30 m aperture, even though the HZ is very close in. The large number of photons available from such a large telescope enables high-resolution spectroscopy, possibly allowing researchers to search for biosignatures like $O_2$ by using the planet's Doppler shift to separate the absorption lines created by the exoplanet from those of $O_2$ in Earth's atmosphere (Snellen et al. 2015).

The best opportunity to study exoplanet habitability, though, comes from future proposed space telescopes such as NASA's HabEx (the Habitable Planet Explorer) and LUVOIR (the Large UltraViolet–Optical–InfraRed space telescope). Both would be large, space-based, direct imaging telescopes, with a 4 m diameter aperture for HabEx and 8–15 m for LUVOIR. Both telescopes would use high-contrast coronagraphy to distinguish starlight reflected by planets from that emitted by the parent star. The smaller HabEx telescope could be assisted by an external starshade. Both telescopes could potentially find and characterize dozens of non-transiting, rocky planets in the HZs of their (nearby) parent stars. Meanwhile, ESA has recently begun to study a different type of direct-imaging mission called LIFE (Large Interferometer for Exoplanets). This would be a thermal-infrared interferometer building off the heritage of ESA's Darwin mission and NASA's TPF-I. This mission would be a natural follow-on for HabEx or LUVOIR, as it would provide complementary wavelength coverage of many of the same exoplanetary systems.

## 22.5 Challenges

The challenges in this field are the same as the opportunities. We need space-based direct imaging missions to find and observe potentially habitable planets around nearby stars. These missions are all expensive, with cost estimates ranging from ~\$7 billion for HabEx to \$13–\$16 billion for the 15 m LUVOIR-A option. Their chance of being selected and built depends critically on the success of JWST which, at the time of this writing, is still sitting on the ground. A successful deployment of JWST, coupled with promising observations of several M-star HZ planets, might provide the impetus to get a direct imaging mission funded. Alternatively, spectra of

Proxima Centauri b obtained by the ELT might prod NASA and ESA into action. Public support for the search for habitable/inhabited exoplanets is essential. As my colleague Sara Seager at MIT is fond of saying, we are positioned on the verge of the 2nd Copernican revolution. It is up to our generation to ensure that it actually happens.

## References


Banks, P. M., & Kockarts, G. 1973, Aeronomy (New York: Academic)

Caves, J. K., Jost, A. B., Lau, K. V., & Maher, K. 2016, E&PSL, 450, 152

Coogan, L. A., & Gillis, K. M. 2013, GGG, 14, 1771

Dole, S. H. 1964, in Habitable Planets for Man (New York: Blaisdell Publishing), 158

Foley, B. J. 2015, ApJ, 812, 36

Foley, B. J., & Smye, A. J. 2018, AsBio, 18, 873

Haqq-Misra, J., Kopparapu, R. K., Batalha, N. E., Harman, C. E., & Kasting, J. F. 2016, ApJ, 827, 120

Hart, M. H. 1978, Icar, 33, 23

Hart, M. H. 1979, Icar, 37, 351

Helffrich, G. 2017, PEPS, 4, 24

Imbrie, J., & Imbrie, K. P. 1986, Ice Ages: Solving the Mystery (Cambridge, MA: Harvard Univ. Press)

Jakosky, B. M., Brain, D., Chaffin, M., et al. 2018, Icar, 315, 146

Joshi, M. 2003, AsBio, 3, 415

Joshi, M. M., Haberle, R. M., & Reynolds, R. T. 1997, Icar, 129, 450

Kasting, J. F. 1993, Sci, 259, 920

Kasting, J. F. 2001, Perspectives in Biology and Medicine, 44, 117

Kasting, J. F. 2010, in How to Find a Habitable Planet (Princeton, NJ: Princeton Univ. Press), 326

Kasting, J. F., Kopparapu, R., Ramirez, R. M., & Harman, C. E. 2014, PNAS, 111, 12641

Kasting, J. F., Whitmire, D. P., & Reynolds, R. T. 1993, Icar, 101, 108

Khodachenko, M. L., Ribas, I., Lammer, H., et al. 2007, AsBio, 7, 167

Kite, E. S., & Ford, E. B. 2018, ApJ, 864, 75

Kopparapu, R. K., Ramirez, R., Kasting, J. F., et al. 2013, ApJ, 770, 82

Kopparapu, R. K., Wolf, E. T., Haqq-Misra, J., et al. 2016, ApJ, 819, 84

Krissansen-Totton, J., & Catling, D. C. 2017, NatCo, 8, 15423

Lammer, H., Lichtenegger, H. I. M., Kulikov, Y. N., et al. 2007, AsBio, 7, 185

Laskar, J., Joutel, F., & Robutel, P. 1993, Natur, 361, 615

Laskar, J., & Robutel, P. 1993, Natur, 361, 608

Leconte, J., Forget, F., Charnay, B., Wordsworth, R., & Pottier, A. 2013, Natur, 504, 268

Lenardic, A., Crowley, J. W., Jellinek, A. M., & Weller, M. 2016, AsBio, 16, 551

Lenardic, A., & Seales, J. 2019, AsBio, 19, 1398

Lorenz, R. D. 2020, RNAAS, 4, 79

Luger, R., Barnes, R., Lopez, E., et al. 2015, AsBio, 15, 57

Luger, R., & Barnes, R. 2015, AsBio, 15, 119

MacDonald, F. A., Swanson-Hysell, N. L., Park, Y., Lisiecki, L., & Jagoutz, O. 2019, Sci, 364, 181

Maher, K., & Chamberlain, C. P. 2014, Sci, 343, 1502

Maunder, E. W. 1913, Are the Planets Inhabited? (London: Harper and Brothers)

Mittelholz, A., Johnson, C. L., Feinberg, J. M., Langlais, B., & Phillips, R. J. 2020, SciA, 6, eaba0513
Pierrehumbert, R., & Gaidos, E. 2011, ApJL, 734, L13
Prinn, R. G., & Fegley, B. 1987, AREPS, 15, 171
Ramirez, R. M., Kopparapu, R., Zugger, M. E., et al. 2014, NatGe, 7, 59
Raymo, M. E., & Ruddiman, W. F. 1992, Natur, 359, 117
Raymond, S. N., Mandell, A. M., & Sigurdsson, S. 2006, Sci, 313, 1413
Raymond, S. N., Quinn, T., & Lunine, J. I. 2004, Icar, 168, 1
Sagan, C. 1995, in Cosmos (Avenel, NJ: Wings Books), 365
Seager, S. 2013, Sci, 340, 577
Seager, S., Schrenk, M., & Bains, W. 2012, AsBio, 12, 61
Shapley, H. 1953, Climatic Change: Evidence, Causes, and Effects (Cambridge, MA: Harvard Univ. Press)
Snellen, I., de Kok, R, Birkby, J. L., et al. 2015, A&A, 576, A59
Stevenson, D. J. 1999, Natur, 400, 32
Walker, J. C. G., Hays, P. B., & Kasting, J. F. 1981, JGR, 86, 9776
Walker, S. I., Bains, W., Cronin, L., et al. 2018, AsBio, 18, 779
Ward, P., & Brownlee, D. 2000, Rare Earth: Why Complex Life is Uncommon in the Universe (New York: Copernicus)
Way, M. J., & Del Genio, A. D. 2020, JGRE, 125, e2019JE006276
Wordsworth, R., & Pierrehumbert, R. 2013, Sci, 339, 64
Yang, J., Cowan, N. B., & Abbot, D. S. 2013, ApJL, 771, L45
Yoder, C. F., Konopliv, A. S., Yuan, D. N., Standish, E. M., & Folkner, W. M. 2003, Sci, 300, 299

AAS | IOP Astronomy

ExoFrontiers
Big questions in exoplanetary science
Nikku Madhusudhan

# Chapter 23

## Exoplanetary Astrobiology

**Victoria Meadows**

### 23.1 Introduction

Are we alone in the universe? This age-old question drives one of humanity's most profound scientific goals: the search for life beyond the Earth. While solar system worlds such as Mars, Europa, and Enceladus may harbor life in their subsurfaces, terrestrial exoplanets provide an opportunity to discover more Earth-like worlds—with surface oceans and atmospheres that interact strongly with their biospheres. Exoplanet science plays a critical role in our search for life beyond the solar system by developing the modeling, instrumentation, and observations needed to discover and explore the atmospheric and surface environments of terrestrial exoplanets. The field of astrobiology—which encompasses studies of the origin, evolution and distribution of life in the universe—provides the interdisciplinary scientific foundation needed to select the most promising exoplanetary targets, and to interpret any potential signs of habitability or life observed in planetary environments.

Much of exoplanet astrobiology is encapsulated in a single question: How do we recognize whether an exoplanet can or does support life? Examining life's fundamental physics and chemistry (Hoehler et al. 2020) reveals the requirements and limits of life, and the planetary environments that are more likely to support it. Understanding and comparing the formation and evolutionary processes that allow a planet to acquire, maintain and lose habitability (Meadows & Barnes 2018; Kopparapu et al. 2020) helps us to identify those stellar, planetary, and planetary system characteristics that are most likely to lead to a habitable planet. This in turn informs the selection of the most promising planets to search for life. The coevolution of life with its environment is now known to impact the nature and evolution of Earth's atmosphere, surface and interior, including its habitability (Stueeken et al. 2020). By understanding how life might globally modify a planetary environment we can identify signs of life, biosignatures, that could be visible in a planet's atmosphere or on its surface, even over interstellar distances (Robinson & Reinhard 2020). The work of scientists in multiple disciplines has culminated in one

doi:10.1088/2514-3433/abfa8fch23  

of astrobiology's clearest lessons: that the nature of life and life's environmental impact cannot be divorced from the planet on which it originated and evolved. Consequently the search for life in the universe must be informed by planetary, stellar, and planetary system context.

Several key questions provide stepping stones in the multi-decade quest for habitable environments and life on extrasolar planets:

- How common are terrestrial planets in the HZ of nearby stars? Answering this question will help us understand the frequency of potentially habitable planets throughout the Galaxy, and also identify nearby targets for habitability and biosignature searches.
- What is the nature of terrestrial exoplanet atmospheres and surfaces? New observations and theory would open up a new realm of comparative planetology of terrestrial planets.
- What factors affect the evolution and maintenance of planetary habitability? How planets interact as systems, and how they interact with their star and planetary system to acquire and maintain habitability helps us identify when planetary habitability may be more likely.
- How do we recognize if a planet is more likely to be habitable? The identification and search for observable signs of liquid surface water help prioritize targets for biosignature searches.
- How do we identify new signs of life to search for? Expanding the range of biosignatures to target helps us broaden our search.
- How do we recognize signs of life in the context of the planetary environment? A probabilistic framework for biosignature assessment is needed that takes into account the likelihood that a planetary process, other than biology, is responsible for a possible biosignature.
- How do we as a community assess and increase the credibility of potentially detected biosignatures? A community protocol for the verification and reporting of a putative biosignature detection is needed to enhance the credibility, and manage the societal impact, of the potential detection of alien life on exoplanets.

## 23.2 State of the Art

The origins of astrobiology-relevant exoplanet science came 60 years ago, well before the discovery of exoplanets, when researchers developed Earth systems science and the concept of a circumstellar habitable zone, and realized that life's impact on its environment could be global and potentially visible from space. Early researchers introduced the concept of a habitable zone (Huang 1959; Dole 1964), a region around a star where surface liquid water was possible based on planetary and stellar properties. This framework was later developed to become the habitable zone concept that is used today (Kasting et al. 1993; Kopparapu et al. 2013). Around the same time, researchers explored life's impact on the global environment, considering Earth's abundant oxygen, and the existence of the atmospheric $O_2/CH_4$ disequilibrium pair (Hitchcock & Lovelock 1967) driven by the combined presence of photosynthetic and methanogenic metabolisms interacting with the atmosphere. The vegetation red edge was also measured, and posited as a globally-visible biosignature (Gates et al. 1965).

Since these initial explorations, exoplanet astrobiology has grown into a field with planetary targets that are potentially observable in the next few years, and theoretical models with the increased complexity and rigor needed to plan for and interpret upcoming observations. Exoplanet demographics obtained by the Kepler mission provided the first clues that terrestrial-sized exoplanets in the habitable zones of their parent stars were likely common in our Galaxy (Dressing & Charbonneau 2015). Observational and theoretical work that combined information on planetary sizes and masses helped to constrain planetary characteristics that were more likely to be indicative of a rocky exoplanet (Rogers 2015; Weiss & Marcy 2014), and preliminary observations of these candidate terrestrial exoplanets suggested that they did not have $H_2$-dominated atmospheres (Delrez et al. 2018; de Wit et al. 2018; Moran et al. 2018). Ground-based telescopic searches and the K2 and TESS missions have continued to make discoveries of planets orbiting M and late K dwarf stars that have planetary sizes, densities, and star–planet separations that make them promising targets (Gillon et al. 2017; Grimm et al. 2018; Ducrot et al. 2020; Suissa et al. 2020) for habitability assessment and biosignature searches with upcoming missions.

To support these future observations, exoplanet astrobiology has harnessed the power of interdisciplinary research to start to identify many of the planetary, stellar, and planetary system characteristics and processes that determine whether a planet can acquire and maintain surface liquid water (Shields et al. 2016; Meadows & Barnes 2018). In particular, work on star–planet interactions, both radiative and gravitational, and their impacts on climate (Kopparapu et al. 2016; Turbet et al. 2018), photochemistry (Lincowski et al. 2018; Chen et al. 2019) and planetary atmospheric evolution (Luger & Barnes 2015; Bolmont et al. 2017; Owen & Mohanty 2016) have been explored using a suite of planetary atmospheric models. This work in particular emphasizes the importance of evolutionary processes on habitability, including ocean and atmospheric loss (Luger & Barnes 2015), interior-atmosphere exchange (Stueeken et al. 2020), and stellar evolution and activity (Segura et al. 2010; Tilley et al. 2019). This has moved the field of habitability assessment beyond the more static habitable zone concept into a much richer multi-dimensional view of habitability and its evolution.

The theme of increasing rigor and the development of a more systems-level view has also characterized recent biosignature research (Schwieterman et al. 2018). While the impacts of oxygenic photosynthesis on the planetary environment remain attractive biosignatures due to the sheer energy advantage accorded to a metabolism that can harness the energy from its star (Meadows 2017), recent interdisciplinary research has sought new biosignatures for novel metabolisms in new environmental contexts. These have included possible biosignatures for Earth's early metabolisms (Pilcher 2003; Domagal-Goldman et al. 2011; Arney et al. 2018) including new disequilibria centered around methanogenesis (Krissansen-Totton et al. 2018a, 2018b). Possible biosignatures for terrestrial atmospheres with a high $H_2$ fraction, including $NH_3$ (Seager et al. 2013) and $PH_3$ (Bains et al. 2019; Sousa-Silva et al. 2020) are also being explored, as are broader searches for volatile molecules that may serve as biosignatures (Seager et al. 2016). In an exciting cross-disciplinary effort, work is also underway to understand the viability of exoplanetary searches for "agnostic biosignatures", anomalous patterns in a planetary environment, such as atmospheric chemical networks, for example, that are of sufficiently high complexity that they are unlikely to be due to planetary processes (Walker et al. 2020).

In addition to efforts to identify novel biosignatures to search for, new frameworks for a more robust interpretation of observed biosignatures are being developed. These frameworks use both potential biosignatures and their environmental context to arrive at a probabilistic assessment of whether or not life is present (Catling et al. 2018; Walker et al. 2018, 2020). Although still under development, key pieces of this approach, including the identification of "false positives"—abiotic planetary processes that could generate a targeted biosignature gas—have been extensively studied for $O_2$ (see Meadows 2017, Meadows et al. 2018 for reviews of this topic). The history of atmospheric $O_2$ on our planet also illuminates so called "false negatives", environmental processes that suppress the detectability of a biosignature by sequestering or destroying it (Reinhard 2017; Glaser et al. 2020). These promising new approaches are a first step toward a robust, consensus assessment of habitability and biosignatures.

## 23.3 Important Questions

In the upcoming decades, exoplanet astrobiology can advance its broader goal of understanding whether nearby stars host habitable or inhabited planets through both observational and theoretical studies. In the near term, observations will focus on terrestrial exoplanets orbiting nearby M dwarf stars, as these targets can be studied with near-term facilities such as JWST and ground-based telescopes. The potential availability of large aperture, space-based telescopes could greatly expand the study of habitable exoplanets to those orbiting more Sun-like stars (Space Studies Board et al. 2019), which may face fewer challenges to habitability.

**Are terrestrial planets orbiting M dwarfs able to support life?** Key research includes exploring how the star's evolution impacts the planet's atmosphere and ocean, and whether M dwarf HZ planets can lose primordial atmospheres, and produce and keep secondary atmospheres. Habitability could be severely compromised if the planet's primary and/or secondary atmosphere is stripped completely during the super-luminous pre-main-sequence phase, or under an extended period of early high activity (Garcia-Sage et al. 2017; Dong et al. 2017) but habitability is also compromised if the planet's dense primordial $H_2$ envelope cannot be removed (Owen & Mohanty 2016). Comprehensive, interdisciplinary modeling is needed that takes into account the impact of stellar evolution, planetary photochemistry and ion chemistry, magnetic fields, tidal heating, and the replenishment of secondary atmospheres via outgassing. Similarly, the nature of the delivery, cycling, early loss, and possible replenishment of water for M dwarf terrestrial planets is a key question for both modeling and observation that bridges and synthesizes studies of protoplanetary disks, volatile delivery, atmospheric escape, and stellar and planetary evolution.

**Are there habitable planets orbiting nearby Sun-like stars, and how do they differ from Earth and M dwarfs HZ planets?** The evolutionary path of M dwarf HZ exoplanets is likely different to those of planets orbiting more Sun-like stars, as M dwarf planets are subjected to an extended (up to 1 Gy) super-luminous pre-main sequence phase from their host star (Luger & Barnes 2015). M dwarf HZ planets may therefore provide fascinating glimpses into potentially diverse outcomes and challenges to habitability. However, to place our Earth in context and increase our chances of finding more habitable planets, studying planets orbiting more Sun-like stars is an important key step, but one that remains a significant technological challenge that is unlikely to be met by currently-planned facilities such as JWST and the ground-based extremely large telescopes.

**How do we identify planets that are most likely to be habitable and how do we best determine that they are habitable?** Significant work is needed to understand what characteristics and processes most impact planetary habitability, and which of these can be observationally identified using a variety of observing techniques (e.g., transmission, emission, and direct imaging reflected light; Figure 23.1). Terrestrial planetary characteristics that increase the likelihood of habitability include the presence of an atmosphere, and atmospheric greenhouse and biosignature gases. In addition, indicators of the presence or absence of a surface ocean could be sought via mapping and ocean glint in direct imaging observations (Williams & Gaidos 2008; Robinson et al. 2010; Lustig-Yaeger et al. 2018) or indirectly (e.g., in transmission) by searching for gases or fractionation indicative of past ocean loss or the lack of a current ocean (Lincowski et al. 2018, 2019; Lustig-Yaeger et al. 2019; Loftus et al. 2019), or via other, yet to be realized means.

**How do we improve our ability to detect and interpret signs of life?** First attempts to characterize M dwarf terrestrial exoplanet atmospheres will also provide opportunities to search for possible biosignatures on a handful of targets, including the $CO_2/CH_4$ disequilibrium pair in JWST transmission observations (Krissansen-Totton et al. 2018a), and $O_2$ bands in high-resolution ELT transmission or reflection observations (Lovis et al. 2017; López-Morales et al. 2019; Serindag & Snellen 2019; Leung et al. 2020). However, considerable work is still needed to identify a larger range of potential biosignatures, including those known to be produced by different Earth metabolisms over the Earth's history, as well as less Earth-centric "agnostic" biosignatures—which can focus on patterns and complexity in a planets' environment, rather than specific products. Complementing efforts to broaden known potential biosignatures, in depth study of promising biosignatures increases our confidence in the interpretation of the known candidates. These in depth studies have been pioneered on photosynthetic $O_2$, where modeling of stellar interactions and planetary processes have identified those that suppress the detectability of a biosignature (false negatives) or abiotically generate biosignatures (false positives). These studies will form the basis of the development of a more probabilistic framework for biosignature assessment, and the

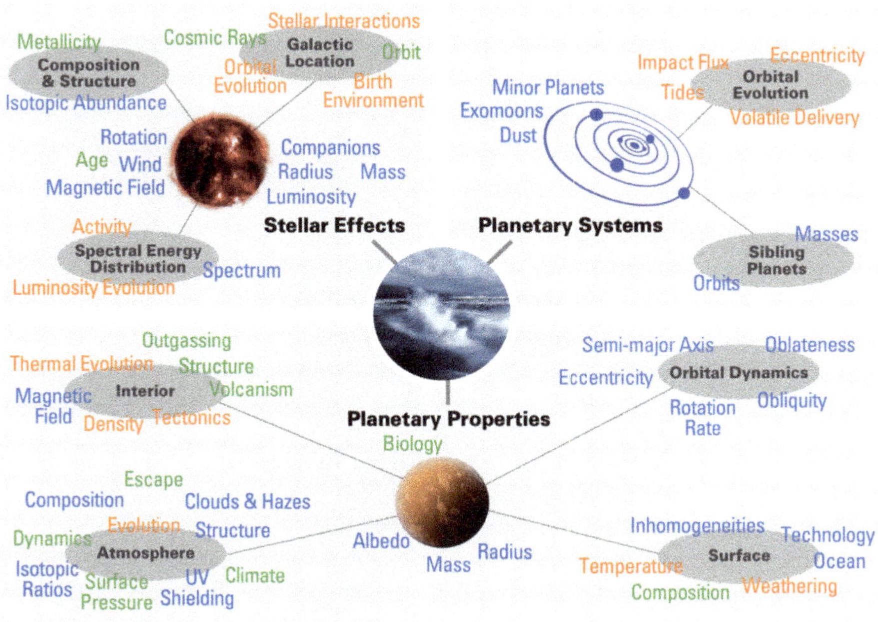


**Figure 23.1.** How do we identify the planets that are most likely to have surface liquid water? This diagram provides a summary of the known planetary, stellar, and planetary system properties that may impact planetary habitability. We can advance our assessment of habitability for a given habitable zone planet by constraining as many of these factors as possible. Font color denotes characteristics that could be observed directly with sufficiently powerful telescopes (blue), those that require modeling interpretation, possibly constrained by observations (green), and the properties or processes that are accessible primarily through theoretical modeling (orange). Reprinted by permission from Springer Nature: Meadows & Barnes (2018) © 2018.

corresponding development of detailed observing protocols for biosignature detection and interpretation based on characteristics of the planet, its star, and its planetary system.

## 23.4 Opportunities

There are several key near-term opportunities to advance exoplanet astrobiology including promising nearby M dwarf targets for study, increasing interest in enabling interdisciplinary studies both for astrobiology and other exoplanet research areas, and mutual efforts from the exoplanet, solar system, and heliophysics communities to break down disciplinary barriers to enrich these fields and address outstanding issues in exoplanet science.

The most exciting observational opportunity for exoplanet astrobiology is the potential for spectroscopic reconnaissance of the TRAPPIST-1 system with JWST (Figure 23.2). The discovery of the TRAPPIST-1 system of seven Earth-sized planets (three in the HZ) orbiting a very late-type M dwarf has advanced what was previously considered the decades-distant goal to study the atmospheres of HZ

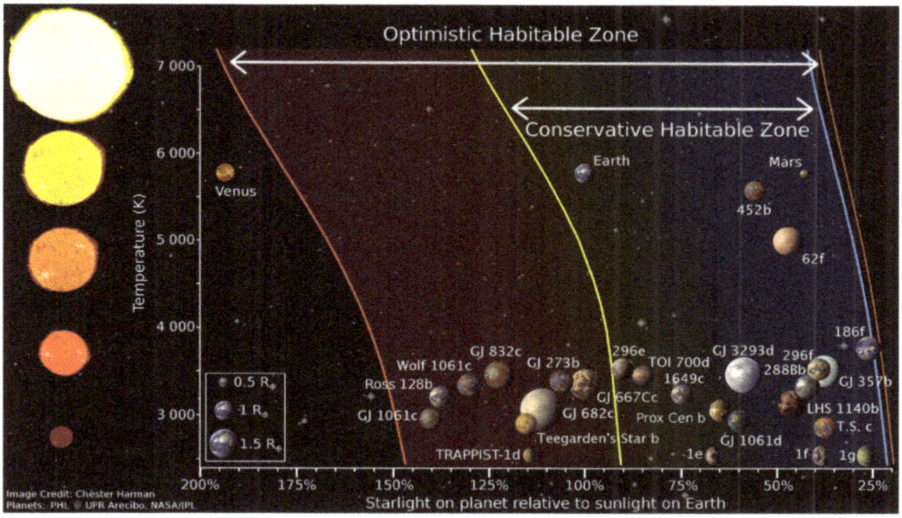


**Figure 23.2.** Potentially terrestrial habitable zone targets. This diagram shows known planets of less than 1.5 Earth radii in the range of incident stellar radiation consistent with the conservative (yellow and blue lines) and the optimistic (red and orange lines) habitable zone. The TRAPPIST-1 system is a high priority target for exoplanet astrobiology as the small size of the host star enhances the potential detectability of atmospheric features seen in transmission, and three planets in this seven planet system are within the optimistic habitable zone. Image credit: Chester (Sonny) Harman.

terrestrial exoplanets, to one that may be achieved in the next five years. Simulations from multiple teams suggest that transmission spectroscopy with JWST may be able to detect the presence of a terrestrial-type atmosphere, and confirm the presence of key atmospheric molecules in two to tens of transits (Morley et al. 2017; Lincowski et al. 2018; Lustig-Yaeger et al. 2019; Wunderlich et al. 2019; Krissansen-Totton et al. 2018a). If detected, these molecules could potentially reveal the outcomes of M dwarf planetary evolution, including the action of atmospheric and ocean-loss processes (Lincowski et al. 2018; Lustig-Yaeger et al. 2019). Nearby early-type M dwarf HZ planets discovered by TESS (Suissa et al. 2020), and non-transiting targets found by ground-based surveys (Anglada-Escudé et al. 2016) may also be accessible to atmospheric study using large ground-based telescopes in the next 5–10 years (Lovis et al. 2017; López-Morales et al. 2019; Serindag & Snellen 2019; Leung et al. 2020).

To realize a system- and process-based approach to planetary habitability and biosignatures, research at the intersection of exoplanet and solar system research is needed, and there is also growing interest in breaking down barriers to communication and collaboration between these disciplines (Meadows et al. 2020; Arney et al. 2007). Solar system science provides the detailed study of potentially-universal planetary processes, and the analog observations needed to validate models of exoplanet atmospheres and surfaces. In turn, exoplanet science provides the statistics—and the diverse environments and evolutionary outcomes—needed to put our solar system in a broader context, and deepen our understanding of the solar system's evolution and processes.

Stronger links with the stellar, heliophysics, and Earth science communities would also enhance our understanding of the "space weather" experienced by terrestrial planets as a function of stellar type, age, and activity levels. Recent conferences have been organized that specifically address areas of mutual scientific interest and mechanisms to improve the interaction of the solar system/exoplanet/heliophysics communities (e.g., https://http://www.hou.usra.edu/meetings/exoplanets2020/EioB-Findings.pdf). NASA's development of interdisciplinary research coordination networks like the Nexus for Exoplanet System Science (NExSS) have also helped to catalyze interdisciplinary science for exoplanets. Efforts like these need to be supported and expanded to capitalize on the momentum and current opportunity to forge a more holistic understanding of both solar system planets and exoplanets.

## 23.5 Challenges

Perhaps the biggest challenge facing exoplanet astrobiology at this time is the need for the technologically-challenging observations that can constrain what is currently largely a theory-based field—and ultimately provide the data needed to explore nearby planets for signs of habitability and life. To start to address this, we can take advantage of relevant observations that can be made in the solar system to help advance models of terrestrial planet evolution, and improve techniques for observing signs of habitability and life on distant exoplanets (Robinson et al. 2014; Mayorga et al. 2016). The field of terrestrial exoplanet observations around M dwarf stars is also just taking off, with recent observations that constrain planetary masses and densities (Grimm et al. 2018), and the molecular weights of terrestrial exoplanet atmospheres taken with ground-based telescopes, HST, Spitzer (de Wit et al. 2018; Delrez et al. 2018; Moran et al. 2018), and soon, hopefully, JWST. However, the extremely large telescopes needed on the ground and in space that can increase the opportunities to identify habitable planets and detect life by surveying larger samples of planet orbiting more Sun-like stars, are still one to two decades away. To meet these technological challenges, advances in launch vehicles, mirror capabilities, detector technology, and a host of other capabilities will be needed (The LUVOIR Team 2019; Gaudi et al. 2020; Meixner et al. 2019; Snellen et al. 2019) to ultimately lead to the successful launch, deployment, and operation of a large space-based telescope to survey HZ exoplanets.

Even with an influx of new data, this field still faces significant modeling and community challenges due to the massive interdisciplinarity required to make progress on planning, analyzing and interpreting these types of observations. A systems-level approach to planetary habitability that also recognizes that life itself is a planetary process requires continued work across disciplines to understand the formation and evolution of terrestrial planets. Robust life detection needs multiple lines of evidence to strengthen the conclusion that a potential biosignature indeed arises from life, including taking environmental context into account. Given the impact life detection will have on science and society we will also need to develop rigorous, uniformly adopted community standards and processes for biosignature detection, assessment, and verification (Meadows et al. 2020). Despite its advances over the last two decades, there is still considerable room for astrobiology to grow and combine needed research from multiple disciplines to enable and inform the search for life beyond our solar system.

## References


Anglada-Escudé, G., Amado, P. J., Barnes, J., et al. 2016, Natur, 536, 437

Arney, G., Domagal-Goldman, S. D., & Meadows, V. S. 2018, AsBio, 18, 311

Arney, G. N., Izenberg, N. R., Kane, S. R., et al. 2007, arXiv:2007.09231

Bains, W., Petkowski, J. J., Sousa-Silva, C., & Seager, S. 2019, AsBio, 19, 885

Bolmont, E., Selsis, F., Owen, J. E., et al. 2017, MNRAS, 464, 3728

Catling, D. C., Krissansen-Totton, J., Kiang, N. Y., et al. 2018, AsBio, 18, 709

Chen, H., Wolf, E. T., Zhan, Z., & Horton, D. E. 2019, ApJ, 886, 16

de Wit, J., Wakeford, H. R., Lewis, N. K., et al. 2018, NatAs, 2, 214

Delrez, L., Gillon, M., Triaud, A. H. M. J., et al. 2018, MNRAS, 475, 3577

Dole, S. H. 1964, Habitable Planets for Man (New York: Blaisdell Publishing Co.)

Domagal-Goldman, S. D., Meadows, V. S., Claire, M. W., & Kasting, J. F. 2011, AsBio, 11, 419

Dong, C., Lingam, M., Ma, Y., & Cohen, O. 2017, ApJL, 837, L26

Dressing, C. D., & Charbonneau, D. 2015, ApJ, 807, 45

Ducrot, E., Gillon, M., Delrez, L., et al. 2020, A&A, 640, A112

Garcia-Sage, K., Glocer, A., Drake, J. J., Gronoff, G., & Cohen, O. 2017, ApJ, 844, L13

Gates, D. M., Keegan, H. J., Schleter, J. C., & Weidner, V. R. 1965, ApOpt, 4, 11

Gaudi, B. S., Seager, S., Mennesson, B., et al. 2020, arXiv:2001.06683

Gillon, M., Triaud, A. H., Demory, B.-O., et al. 2017, Natur, 542, 456

Glaser, D. M., Hartnett, H. E., Desch, S. J., et al. 2020, ApJ, 893, 163

Grimm, S. L., Demory, B.-O., Gillon, M., et al. 2018, A&A, 613, A68

Hitchcock, D. R., & Lovelock, J. E. 1967, Icar, 7, 149

Hoehler, T. M., Bains, W., Davila, A., Parenteau, M. N., & Pohorille, A. 2020, in Planetary Astrobiology, ed. V. Meadows, G. Arney, B. Schmidt, & D. Des Marais (Tucson, AZ: Univ. Arizona Press), 37

Huang, S. S. 1959, AmSci, 47, 397

Kasting, J. F., Whitmire, D. P., & Reynolds, R. T. 1993, Icar, 101, 108

Kopparapu, R. K., Ramirez, R., Kasting, J. F., et al. 2013, ApJ, 765, 131

Kopparapu, R. K., Wolf, E. T., & Meadows, V. 2020, in Planetary Astrobiology, ed. V. Meadows, G. Arney, B. Schmidt, & D. Des Marais (Tucson, AZ: Univ. Arizona Press), 449

Krissansen-Totton, J., Garland, R., Irwin, P., & Catling, D. C. 2018a, AJ, 156, 114

Krissansen-Totton, J., Olson, S., & Catling, D. C. 2018b, SciA, 4, eaao5747

Kopparapu, R. K., Wolf, E. T., & Haqq-Misra, J. 2016, ApJ, 819, 84

Leung, M., Meadows, V. S., & Lustig-Yaeger, J. 2020, AJ, 160, 11

Lincowski, A. P., Lustig-Yaeger, J., & Meadows, V. S. 2019, AJ, 158, 26

Lincowski, A. P., Meadows, V. S., Crisp, D., et al. 2018, ApJ, 867, 76

Loftus, K., Wordsworth, R. D., & Morley, C. V. 2019, ApJ, 887, 231

López-Morales, M., Ben-Ami, S., Gonzalez-Abad, G., et al. 2019, AJ, 158, 24

Lovis, C., Snellen, I., Mouillet, D., et al. 2017, A&A, 599, A16

Luger, R., & Barnes, R. 2015, AsBio, 15, 119

Lustig-Yaeger, J., Meadows, V. S., & Lincowski, A. P. 2019, AJ, 158, 27
Lustig-Yaeger, J., Meadows, V. S., Mendoza, G. T., et al. 2018, AJ, 156, 301
Mayorga, L. C., Jackiewicz, J., Rages, K., et al. 2016, AJ, 152, 209
Meadows, V. S. 2017, AsBio, 17, 1022
Meadows, V. S., & Barnes, R. K. 2018, in Handbook of Exoplanets, ed. H.J. Deeg, & J.A. Belmonte (Berlin: Springer), 57
Meadows, V. S., Reinhard, C. T., Arney, G. N., et al. 2018, AsBio, 18, 630
Meadows, V. S., Arney, G. N., Des Marais, D. J., & Schmidt, B. E. 2020, in Planetary Astrobiology, ed. V. Meadows (Tucson, AZ: Univ. Arizona Press), 505
Meixner, M., Cooray, A., Leisawitz, D., et al. 2019, arXiv:1912.06213
Moran, S. E., Hörst, S. M., Batalha, N. E., Lewis, N. K., & Wakeford, H. R. 2018, AJ, 156, 252
Morley, C. V., Kreidberg, L., Rustamkulov, Z., Robinson, T., & Fortney, J. J. 2017, ApJ, 850, 121
Owen, J. E., & Mohanty, S. 2016, MNRAS, 459, 4088
Pilcher, C. B. 2003, AsBio, 3, 471
Reinhard, C. T., Olson, S. L., Schwieterman, E. W., & Lyons, T. W. 2017, AsBio, 17, 287
Robinson, T. D., & Reinhard, C. T. 2020, in Planetary Astrobiology, ed. V. Meadows, et al. (Tucson, AZ: Univ. Arizona Press), 379
Robinson, T. D., Maltagliati, L., Marley, M. S., & Fortney, J. J. 2014, PNAS, 111, 9042
Robinson, T. D., Meadows, V. S., & Crisp, D. 2010, ApJ, 721, L67
Rogers, L. A. 2015, ApJ, 801, 41
Schwieterman, E. W., Kiang, N. Y., Parenteau, M. N., et al. 2018, AsBio, 18, 663
Seager, S., Bains, W., & Hu, R. 2013, ApJ, 777, 95
Seager, S., Bains, W., & Petkowski, J. J. 2016, AsBio, 16, 465
Segura, A., Walkowicz, L. M., Meadows, V., Kasting, J., & Hawley, S. 2010, AsBio, 10, 751
Serindag, D. B., & Snellen, A. G. 2019, ApJ, 871, L7
Shields, A. L., Ballard, S., & Johnson, J. A. 2016, PhR, 663, 1
Snellen, I., Albrecht, S., Anglada-Escudé, G., et al. 2019, arXiv:1908.01803
Sousa-Silva, C., Seager, S., Ranjan, S., et al. 2020, AsBio, 20, 235
Space Studies Board, National Academies of Sciences, Engineering and Medicine, et al. 2019, Exoplanet Science Strategy (Washington, DC: National Academies Press)
Stüeken, E., Fournier, G., & Eyster, A. 2020, in Planetary Astrobiology, ed. V. Meadows, et al. (Tucson, AZ: Univ. Arizona Press), 93
Suissa, G., Mandell, A. M., & Wolf, E. T. 2020, ApJ, 891, 58
The LUVOIR Team, 2019, arXiv:1912.06219
Tilley, M. A., Segura, A., Meadows, V., Hawley, S., & Davenport, J. 2019, AsBio, 19, 64
Turbet, M., Bolmont, E., Leconte, J., et al. 2018, A&A, 612, A86
Walker, S. I., Bains, W., Cronin, L., et al. 2018, AsBio, 18, 779
Walker, S. I., Cronin, L., Drew, A., et al. 2020, in Planetary Astrobiology, ed. V. Meadows, et al. (Tucson, AZ: Univ. Arizona Press), 477
Weiss, L. M., & Marcy, G. W. 2014, ApJL, 783, L6
Williams, D. M., & Gaidos, E. 2008, Icar, 195, 927
Wunderlich, F., Godolt, M., Grenfell, J. L., et al. 2019, A&A, 624, A49

# Part V

## New Horizons

AAS | IOP Astronomy

ExoFrontiers
Big questions in exoplanetary science
Nikku Madhusudhan

# Chapter 24

# The Search for Spectroscopic Signs of Life

**Sara Seager**

## 24.1 Introduction

Humankind has pondered questions about life beyond Earth for decades, if not millennia. Thousands of years ago the Greek philosophers posed the ancient questions "Do other Earths exist?," "Are they common?," "Do any have signs of life?". A century ago, astronomical observation tools matured enough to advance concepts and even initiate the search for life by way of remote sensing observations. In 1912 Arcichovsky proposed looking for vegetation signatures in earthshine (on the Moon) as a reference case for vegetation or chlorophyll searches on other planets (Arcichovsky 1912). In 1930 Jeans described the concept of oxygen as an atmospheric biosignature gas.[1] Mars has always been of huge interest for the search for life, and early remote sensing included not just Lowell's sensational reports of canals on Mars but also atmospheric spectroscopic analysis (e.g., Sinton 1957; Rea et al. 1965). Today, *in situ* exploration for direct discovery of life or study of life's building blocks is a main thrust of mission concept studies to Europa,[2] Enceladus,[3] and Titan,[4] and the search for past life on Mars is one of the primary goals of NASA's Perseverance.[5]

The search for signs of life on exoplanets is a culmination of decades of speculation, science, and engineering. The goal is to search for "biosignature gases," gases produced by life that can accumulate in exoplanet atmospheres to remotely

[1] "It seems at first somewhat surprising that oxygen figures so largely in the Earth's atmosphere, in view of its readiness to enter into chemical combination with other substances. We know, however, that vegetation is continually discharging oxygen into the atmosphere, and it has often been suggested that the oxygen of the Earth's atmosphere may be mainly or entirely of vegetable origin. If so, the presence or absence of oxygen in the atmosphere of other planets should show whether vegetation similar to that we have on earth exists on those planets or not."

[2] https://europa.nasa.gov/resources/58/europa-lander-study-2016-report/

[3] https://trs.jpl.nasa.gov/bitstream/handle/2014/45905/15-4603_A1b.pdf

[4] https://www.nasa.gov/dragonfly/dragonfly-overview/index.html

[5] https://Mars.nasa.gov/mars2020/

doi:10.1088/2514-3433/abfa8fch24

detectable levels. The study of exoplanet atmospheres relies on remote sensing, blending decades of foundational work on solar system planets with new techniques for exoplanetary atmospheres observed at limited spectral and spatial resolution. The birth and evolution of the scientific field of exoplanets is moving science fiction to science fact—as the community has lofty plans to find and characterize habitable worlds.

## 24.2 State of the Art

State of the art for the study of exoplanet atmospheres falls into three categories: telescopes and instrumentation; observation techniques and data reduction tools; and atmosphere modeling for data interpretation. Recent advancements and future outlook on all these aspects are discussed in other chapters in this volume. For example, the Hubble Space Telescope instruments STIS and WFC3 are currently the workhorses for reliable high-quality transmission spectra of exoplanets. For ground-based direct imaging, the Gemini Planet Imager (GPI; Macintosh et al. 2014) and VLT SPHERE (Vigan et al. 2016) are enabling studies of self-luminous giant planets far from their host stars. Observational techniques for transmission spectra presently reach down to ~25 ppm precision levels and are making it feasible to study super-Earths and mini-Neptunes (e.g., Kreidberg 2018; Benneke et al. 2019). Modeling exoplanet atmospheres and atmospheric retrieval is now a fairly mature field, at least for one dimensional atmosphere models (e.g., Madhusudhan 2018, 2019; Barstow & Heng 2020). Publicly available computer codes have lowered the entry barrier to the field (e.g., Batalha et al. 2017; Malik et al. 2017; Mollière et al. 2019).

No current facilities or models have successfully observed and interpreted terrestrial-like exoplanet atmospheres—an area that remains in the future.

## 24.3 Important Questions/Goals

The key questions are those that drive our observational searches for detection of biosignature gases.

The first key question is, **"What kind of planets can host life?"** The answer is such a wide variety. We may qualitatively consider the range of planets that might support life by looking through the lens of the requirements of life. Theory, data, and experiments suggest the requirements for life, in decreasing order of certainty, are (National Research Council 2007):

1. A thermodynamic disequilibrium;
2. An environment capable of maintaining covalent bonds, especially between carbon, hydrogen, and other atoms;
3. A liquid environment; and
4. A molecular system that can support Darwinian evolution.

The search for habitable worlds focuses on planets with the right surface temperature to support liquid water (Item 3). Liquid water is not only needed for all life as we know it but is also expected to be the most cosmically abundant liquid (Bains 2004). The temperature for liquid water also satisfies the condition for the right temperatures and pressures for planet atmospheres and surfaces to maintain covalent bonds (Item 2). Suitable temperatures for liquid water further support an environment for Darwinian evolution (moderate temperatures enable molecular interactions while too cold temperatures limit molecular interactions). For exoplanets, a thermodynamic disequilibrium (Item 1) is maintained by the stellar energy incident on the planet. A rocky planet may also have geothermal and potential chemical energy gradients as Earth does. We may on the whole exclude giant exoplanets from the list of life-supporting planet types as beneath their atmospheres is no solid surface but compressed phases of hydrogen and helium at temperatures far too high to support covalent bonds.

In principle many different planet types will satisfy the above criteria for supporting life. More quantitative research, however, needs to be done to determine how wide or narrow the range of life-hosting planet types actually are. For example, rocky planets with thin $H_2$-dominated atmospheres (if they exist) can support life (Seager et al. 2020). In contrast, water worlds, with deep water oceans and no continents may be too dilute for Darwinian conditions (especially with respect to concentrated ingredients needed for the origin of life). Water worlds furthermore will have high-pressure ice as their "crust and mantle" layers, which prevents life-needed nutrients that on Earth come from continental weathering (nutrients such as silicates and metals).

The second key question in the search for life on exoplanets is, **"How can we identify signs of life?"** We make an underlying assumption that life elsewhere uses chemistry as life on Earth does, to extract energy from the environment, store energy, and use the energy in a way that might generate a byproduct chemical. In this way we assume that byproduct gases are the biosignature gases that further might accumulate in an exoplanet atmosphere to levels detectable remotely by telescopes.

The general consensus is to look for gases that are present in too high amounts, that are also out-of-context with their atmosphere environment. Researchers are working hard to come up with many scenarios and false positives (see the comprehensive review in Schwieterman et al. 2018). A key challenge is that we may not be able to detect all gases of interest. A touted example is that the simultaneous presence of $CH_4$ and $O_2$ is a clear indicator of life, because a reduced gas should not exist in a highly oxidized environment; however both were never simultaneously at high enough levels on Earth for this idea to be validated (Reinhard et al. 2017; Olson et al. 2016).

Beyond $O_2$ and $CH_4$ (e.g., Léger et al. 1996; Des Marais et al. 2002; Kaltenegger et al. 2007), a growing list of gases are being considered as biosignatures for simulated predictions of exoplanet spectra. The gases include: dimethyl sulfide (DMS or $(CH_3)_2S$; Domagal-Goldman et al. 2011; Seager et al. 2012; Arney et al. 2018); nitrous oxide ($N_2O$; Des Marais et al. 2002); methyl chloride ($CH_3Cl$; Segura et al. 2005; Seager et al. 2013b); phosphine ($PH_3$; Sousa-Silva et al. 2020); isoprene ($C_5H_8$; Zhan et al. 2021); and ammonia $NH_3$ (Seager et al. 2013a) and other amines (Huang et al. 2021). A unique approach by Seager et al. (2016) aims to exhaustively sort through all classes of potential biosignature gases.

## 24.4 Opportunities

The key opportunities lie in the upcoming ground- and space-based telescopes. Opportunities are tremendous and only limited by our imaginations.

### 24.4.1 Traditional Astronomy

We are on the brink of operational next-generation telescopes, each of which will open brand new opportunities for rocky exoplanet discovery and characterization. We may divide the observational approaches into three categories: transits; direct imaging from ground-based telescopes; and direct imaging from space-based telescopes (Figure 24.1). Each category already has known planets or promising planet candidates available for observations.

**Transiting planets:** The NASA/ESA/CSA James Webb Space Telescope (Webb) is scheduled for launch in 2021. Webb holds the promise for a generation of hundreds of exoplanet astronomers. A main driver is the study of rocky planet atmospheres including the search for life by way of atmospheric biosignature gases.[6] Webb has near-infrared capability at moderate spectral resolution. There is, however, a growing concern that the Webb noise floor (adopted as 10 ppm) is a limiting factor to obtaining transmission spectra for low molecular weight atmosphere rocky planets. That said, some of the ultracool star Trappist 1 planets (Gillon et al. 2017) are argued as the most observationally favorable system of potentially habitable planets, assuming they have maintained planet atmospheres (Turbet et al. 2020).

**Direct imaging from the ground:** Habitable planets orbiting (and not necessarily transiting) M dwarf stars are accessible with the right coronagraph instrumentation and extreme adaptive optics on the large ground-based telescopes now under construction.

**Figure 24.1.** Schematic outlining the three dominant categories of observatories capable of the search for life by way of biosignature gases on exoplanets.

[6] In addition, ESA's Atmospheric Remote-sensing Infrared Exoplanet Large-survey (ARIEL; Tinetti et al. 2018) may reach down to super-Earths around the very smallest M dwarf stars.

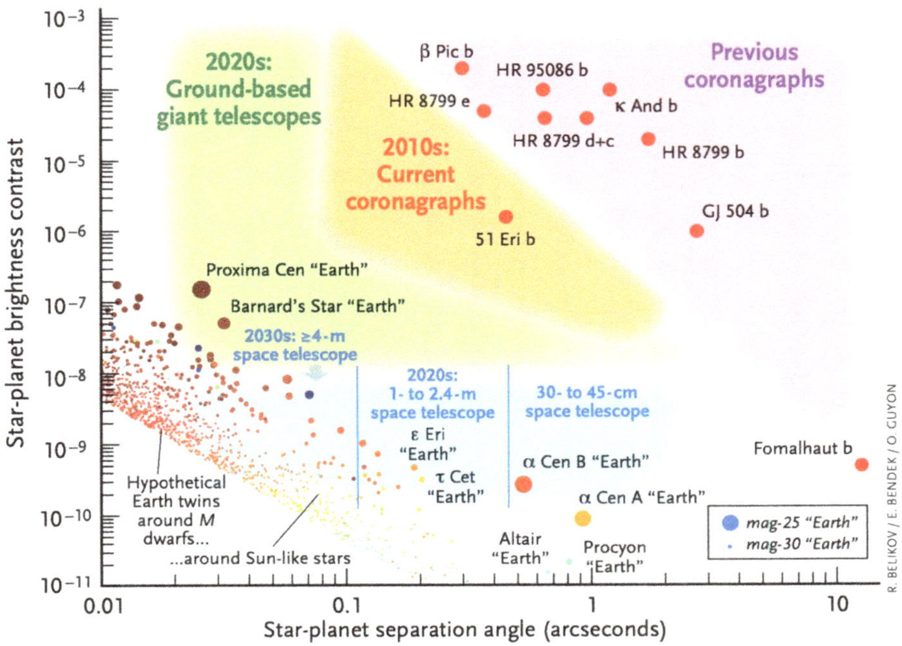


**Figure 24.2.** Overview of the field of exoplanet direct imaging in visible and infrared light. Capabilities of past, current, and future direct imaging instruments are shown. Red dots in the upper right are known directly imaged giant exoplanets (from Belikov and Bendek 2015). Figure and caption credit: R. Belikov, E. Bendek, O. Guyon, and NASA. https://www.nasa.gov/content/core-capability-3-exoplanet-characterization-enabling-nasa-s-search-for-life.

Near infrared imaging and spectroscopy of reflected light from rocky planets in their host star's habitable zone is anticipated to be possible for up to 100 nearby low-mass stars (primarily mid M dwarf stars; Guyon 2018; Guyon et al. 2019; Artigau et al. 2018). The challenge is the high planet-star contrast at $10^7$ to $10^8$ levels; see Figure 24.2. Direct imaging in the thermal infrared for about five Sun-like stars may also be possible (Guyon et al. 2019). The telescopes include: the Extremely Large Telescope (EELT, 39 m aperture diameter; Gilmozzi & Spyromilio 2007; Tamai & Spyromilio 2014); the Thirty Meter Telescope (TMT, 30 m aperture diameter; Gary 2013); and the Giant Magellan Telescope (GMT, 20 m aperture diameter; Johns et al. 2012; Bernstein et al. 2014). Other than direct imaging, the large ground-based telescopes might be capable of a combination of high-dispersion, high-spectral resolution (R~100,000) spectroscopy with moderate high-contrast imaging to observe spectra of a few rocky planets orbiting Sun-like stars (Snellen et al. 2015). Proxima Centauri b, an Earth-mass planet orbiting the nearest star (an M5.5 dwarf; Anglada-Escudé et al. 2016) is a prime candidate for high-contrast, near-infrared direct imaging reflected light study.

**Direct imaging from space:** To truly open up the Earth analog[7] discovery space, we must make space-based high-contrast direct imaging a reality. Only

[7] A true Earth analog is an Earth-size planet in an Earth-like orbit about a Sun-like star.

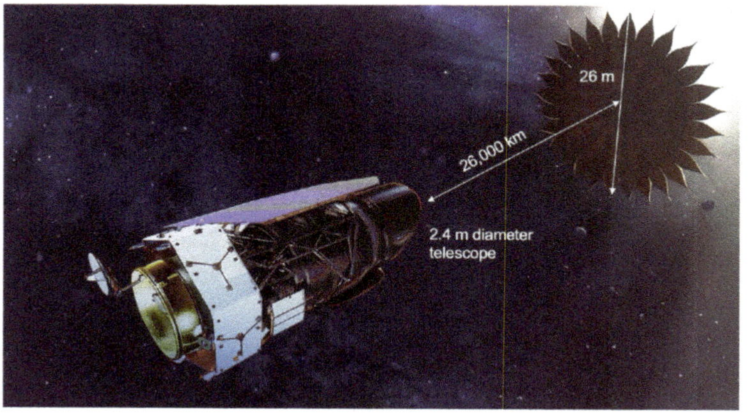


**Figure 24.3.** The Starshade Rendezvous Probe Mission consists of the Roman Telescope and a formation-flying starshade. The mission will enable direct imaging of the habitable zones of nearby Sun-like stars with sufficient sensitivity to detect and identify Earth-like exoplanets. Image not to scale. Credit: NASA.

space-based high-contrast direct imaging enabled by starlight suppression can give us access to a large enough list of Sun-like stars for detecting Earth analogs in reflected light with their extreme planet-star contrast of $\sim 10^{-10}$; see Figure 24.2. In space, the blurring effects of Earth atmosphere can be avoided and more importantly the search for biosignature gases will not be contaminated by the same gases in Earth's atmosphere itself.

There are several developed mission concepts, starting with the Starshade Rendezvous Probe Mission (Seager et al. 2019; Figure 24.3). Starshade is a giant specially shaped screen that would formation fly thousands of km from a space-based telescope. Starshade would block out the starlight so that only planet light enters the telescope. Starshade Rendezvous would be built and launched separately from the Roman Telescope[8] a 2.4 m aperture diameter telescope scheduled to launch in the mid-2020s. Starshade Rendezvous Probe could access about 8–12 bright Sun-like stars. The Habitable Exoplanet Observatory (HabEx; Gaudi et al. 2020) is a mission concept for a 4 m aperture diameter, monolithic mirror, off-axis telescope, with both an internal coronagraph and an accompanying Starshade for starlight suppression. The internal coronagraph would enable the telescope to efficiently discover terrestrial-size exoplanets, whereas the Starshade would enable efficient spectroscopy. LUVOIR is a more advanced concept for an 8 or 15 m aperture diameter segmented mirror telescope, pushing the envelope for what can fit in existing rocket fairings (The LUVOIR Team 2019). HabEx could survey about 50 Sun-like stars and LUVOIR could reach a few hundred stars.

The bright, nearby Sun-like star $\tau$ Ceti is reported to host a multiple planet system (Tuomi et al. 2013; Feng et al. 2017), including two habitable planets suitable for space-based high-contrast direct imaging.

[7] A true Earth analog is an Earth-size planet in an Earth-like orbit about a Sun-like star.

[8] https://www.nasa.gov/content/goddard/nancy-grace-roman-space-telescope

### 24.4.2 Futuristic Concepts

Studying exoplanet atmospheres for biosignature gases, no matter how presently frontiers, will only get us so far in establishing the existence of life on another world (see Section 24.5). Ultimately we will need a detailed look at any exoplanet that shows some evidence of hosting life. There are two interesting concepts under development. Notably, both ideas require a new operating timescale paradigm, 25 years from launch until scientific results.

**Breakthrough Starshot:**[9] is a Breakthrough Initiatives project to develop small spacecraft dubbed "starchips" to be sent to a nearby star, namely Alpha Centauri B. The goal is to launch thousands of starchips, each of which will deploy a ~4 m diameter solar sail. A bank of coherent ground-based lasers with a combined power output of Gigawatts will accelerate the starchips to 0.2 c on their journey to Alpha Centauri. After about 20 years' travel to cover the 4.4 light years, some of surviving starchips will rapidly flyby Alpha Centauri B and any planets in the system. The starchips will take images and relay those images back to Earth, but even at the speed of light the data transmission time will be 4.4 years. The initiative is funding groups to solve nineteen technology challenges to make the mission possible within the next generation.

**The Solar Gravitational Lens Telescope:** is a concept to use the Sun as a powerful gravitational lens (Eshleman 1979; Landis 2017; Turyshev et al. 2020). The Sun would be so highly magnifying it could image exoplanet surfaces at a resolution of 10 km. The telescope has to observe an exoplanet from the Sun's focal point, about 550 au, a distance to which no spacecraft has yet traveled. To reach 550 au in a reasonable time (~25 years) the spacecraft would have to travel an order of magnitude faster than the Voyager spacecraft. On reaching the Sun's gravitational lens focal point, the spacecraft must follow the motion of the Sun-exoplanet focal line with a tailored trajectory in order to monitor the exoplanet. The planet image would be huge—on order of kilometers—which is far larger than any conceivable telescope focal plane. For these and other reasons a computer deconvolution would be needed to reconstruct the planet image. Because each planet has its own focal point with the Sun's gravitational lens, the telescope would be single-purpose, only able to study the chosen exoplanet along the line of sight. Multiple telescopes could be sent in different directions, each for a pre-chosen target (Turyshev et al. 2020). Despite the list of challenges, the revolutionary possibility for direct imaging and spectroscopy of an Earth-like exoplanet up to 30 pc away at 10 km resolution makes this concept worth pursuing.

## 24.5 Challenges

Without question the greatest challenge in the search for life by way of atmospheric biosignature gases is that of robustly determining whether or not a gas is produced by life. Indeed this is a challenge we likely cannot overcome with remote sensing alone, leaving us with "inhabited planet candidates." As scientists, we are used to living with uncertainty, but when it comes to the search for life is this enough for us

[9] https://breakthroughinitiatives.org/concept/3

or the public? If not, we will need to implement brand new types of futuristic telescopes (Section 24.4.2) to follow up on any planets that show signs of life.

In the meantime, our generation of exoplanet astronomers are pinning our hopes in the search for signs of life for what will be a very small number of habitable planets–those with host stars accessible to next-generation telescopes (Seager 2018). The challenge is, if we don't find exoplanets with signs of life, how can we scale up our efforts for a further generations of telescopes? To reach larger apertures than have been considered (e.g., LUVOIR), aperture sizes needed to reach even more distant stars, the next generations of astronomers will have to make space-based manufacturing and space-based assembly a reality.

The discovery and characterization of exoplanets has come a long way in the millennia since humans have pondered the mysteries of the multitude of stars. We are lucky to be the first generation who will not just hope, but can truly explore the nearest stars for habitable and inhabited worlds.

## References

Anglada-Escudé, G., Amado, P. J., Barnes, J., et al. 2016, Natur, 536, 437
Arcichovsky, V. M. 1912, Ann. Inst. Polytech. Don Tsar. Alexis (Novotcherkassk), 1, 195
Arney, G., Domagal-Goldman, S. D., & Meadows, V. S. 2018, AsBio, 18, 311
Artigau, E., Bernstein, R. A., Brandt, T., et al. 2018, arXiv:1808.09632
Bains, W. 2004, AsBio, 4, 137
Barstow, J. K., & Heng, K. 2020, SSRv, 216, 82
Belikov, R., & Bendek, E. 2015, Sky & Telescope, October 2015, 17
Batalha, N. E., Mandell, A., Pontoppidan, K., et al. 2017, PASP, 129, 064501
Benneke, B., Wong, I., Piaulet, C., et al. 2019, ApJL, 887, L14
Bernstein, R. A., McCarthy, P. J., Raybould, K., et al. 2014, Proc. SPIE, 9145, 91451C
Des Marais, D. J., Harwit, M. O., Jucks, K. W., et al. 2002, AsBio, 2, 153
Domagal-Goldman, S. D., Meadows, V. S., Claire, M. W., & Kasting, J. F. 2011, AsBio, 11, 419
Eshleman, V. R. 1979, Sci, 205, 1133
Feng, F., Tuomi, M., Jones, H. R. A., et al. 2017, AJ, 154, 135
Gaudi, B. S., Seager, S., Mennesson, B., et al. 2020, arXiv:2001.06683
Gillon, M., Triaud, A. H. M. J., Demory, B.-O., et al. 2017, Natur, 542, 456
Gilmozzi, R., & Spyromilio, J. 2007, Msngr, 127, 11
Guyon, O. 2018, ARA&A, 56, 315
Guyon, O., Bottom, M., Chun, M., et al. 2019, BAAS, 51, 203
Huang, J., Seager, S., Petkowski, J. J., Ranjan, S., & Zhan, Z. 2021, arXiv:2107.12424
Johns, M., McCarthy, P., Raybould, K., et al. 2012, Proc. SPIE, 8444, 84441H
Kaltenegger, L., Traub, W. A., & Jucks, K. W. 2007, ApJ, 658, 598
Kreidberg, L. 2018, in Handbook of Exoplanets, ed. H.J. Deeg, & J.A. Belmonte (Berlin: Springer), 100
Landis, G. A. 2017, in AIAA Science and Technology Forum and Exposition 2017, Vol. 2017, 1679, arXiv:1604.06351
Léger, A., Mariotti, J. M., Mennesson, B., et al. 1996, Icar, 123, 249
Macintosh, B., Graham, J. R., Ingraham, P., et al. 2014, PNAS, 111, 12661
Madhusudhan, N. 2018, in Handbook of Exoplanets, ed. H. J. Deeg, & J.A. Belmonte (Berlin: Springer), 104

Madhusudhan, N. 2019, ARA&A, 57, 617
Malik, M., Grosheintz, L., Mendonça, J., et al. 2017, AJ, 153, 56
Mollière, P., Wardenier, J. P., van Boekel, R., et al. 2019, A&A, 627, A67
National Research Council, 2007, The Limits of Organic Life in Planetary Systems (Washington, DC: National Academies Press)
Olson, S. L., Reinhard, C. T., & Lyons, T. W. 2016, PNAS, 113, 11447
Rea, D. G., O'Leary, B. T., & Sinton, W. M. 1965, Sci, 147, 1286
Reinhard, C. T., Olson, S. L., Schwieterman, E. W., & Lyons, T. W. 2017, AsBio, 17, 287
Sanders, G. H. 2013, JApA, 34, 81
Schwieterman, E. W., Kiang, N. Y., Parenteau, M. N., et al. 2018, AsBio, 18, 663
Seager, S., Schrenk, M., & Bains, W. 2012, AsBio, 12, 61
Seager, S., Bains, W., & Hu, R. 2013a, ApJ, 775, 104
Seager, S., Bains, W., & Hu, R. 2013b, ApJ, 777, 95
Seager, S., Bains, W., & Petkowski, J. J. 2016, AsBio, 16, 465
Seager, S., Huang, J., Petkowski, J. J., & Pajusalu, M. 2020, NatAs, 4, 802
Seager, S. 2018, IJAsB, 17, 294
Seager, S., Kasdin, N. J., Booth, J., et al. 2019, BAAS, 51, 106
Segura, A., Kasting, J. F., Meadows, V., et al. 2005, AsBio, 5, 706
Sinton, W. M. 1957, ApJ, 126, 231
Snellen, I., de Kok, R., Birkby, J. L., et al. 2015, A&A, 576, A59
Sousa-Silva, C., Seager, S., Ranjan, S., et al. 2020, AsBio, 20, 235
Tamai, R., & Spyromilio, J. 2014, Proc. SPIE, 9145, 91451E
The LUVOIR Team, 2019, arXiv:1912.06219
Tinetti, G., Drossart, P., Eccleston, P., et al. 2018, ExA, 46, 135
Tuomi, M., Jones, H. R. A., Jenkins, J. S., et al. 2013, A&A, 551, A79
Turbet, M., Bolmont, E., Bourrier, V., et al. 2020, SSRv, 216, 100
Turyshev, S. G., Shao, M., Toth, V. T., et al. 2020, arXiv:2002.11871
Vigan, A., Bonnefoy, M., Ginski, C., et al. 2016, A&A, 587, A55
Zhan, Z., Seager, S., Petkowski, J. J., et al. 2021, AsBio, 21, 765

www.ingramcontent.com/pod-product-compliance
Ingram Content Group UK Ltd.
Pitfield, Milton Keynes, MK11 3LW, UK
UKHW061858100926
8556IPUK00008B/916

9 780750 314701